D0960750

SAMSUNG
RISING

SAMSUNG
RISING

The Inside Story of the South Korean Giant

That Set Out to Beat Apple

and Conquer Tech

GEOFFREY CAIN

CURRENCY
New York

Published in the United States by Currency, an imprint of
Random House, a division of Penguin Random House LLC, New York.

CURRENCY and its colophon are trademarks of
Penguin Random House LLC.

Hardback ISBN 978-1-101-90725-2
International edition ISBN 978-0-593-23670-3
Ebook ISBN 978-1-101-90726-9

Printed in the United States of America on acid-free paper

randomhousebooks.com

9 8 7 6 5 4 3 2 1

First Edition

Book design by Debbie Glasserman
Photos courtesy of the author unless otherwise noted.

FOR DON AND SHIRLEE

Contents

CONTENTS

Acknowledgments

TO MOM, DAD, AMY, and Natalie, for their patience as I traveled across oceans and disappeared.

To Roger Scholl, my editor, for elevating my manuscript into something worthy of a book.

To Alan Rinzler, my mentor and consulting editor, for teaching me what it takes to be an author.

To David Halpern, my agent, for believing in me.

To Max Soeun Kim, my right-hand man, for making this book possible with skilled research and translations.

To Jieun Choi, Hyejin Shin, Junyoub Lee, Jihye Lee, Jeong-min Park, Lucinda "Ping" Cowing, and Simon Denny for research, fact-checking, and translations.

To Haeryun Kang, Calvin Godfrey, and Daniel Tudor, for reading and critiquing my manuscript.

To Hal Husrevoglu, for giving me a workspace in his cigar bar, Burn (where my lungs have since aged).

To David Swanson, Krista Mahr, Simon Long, Hugo Restall, Colum

Murphy, David Case, and Emily Lodish, for teaching me to be a foreign correspondent.

To Don Kirk, for giving me a roof.

To the people of Samsung, for telling me their stories with little benefit to themselves.

To the people of Korea, for hosting me.

Note on Research, Translations, and Korean Names

I INTERVIEWED MORE THAN four hundred current and former Samsung employees, executives, politicians, businesspeople, board members, journalists, activists, and analysts, as well as a member of Samsung's founding Lee family. Samsung did not cooperate with this book. Most of my interviews were through unofficial channels. Samsung, however, cooperated with my earlier magazine reporting for *Time* and *Fast Company*. Those official interviews are included.

The interviews took place in South Korea, Japan, China, New York, New Jersey, Texas, and California. Many interviewees did not want to be named or go on the record. I have respected their wishes. Many graciously offered to help me, knowing their careers were at great risk, and with little benefit to themselves.

Interviews were conducted in Korean, Japanese, and English, depending on the needs of the interviewee. Many Korean figures of speech don't translate fluidly into English. I have double-checked these statements with the subjects to ensure that the original intent was not lost, with the help of my researcher, Max Soeun Kim, talented young *Guardian* stringer and native Korean speaker.

This book follows no single convention for Korean names. It follows the personal preferences of the people I interviewed and wrote about. Most Koreans prefer the standard usage: family name first, given name second. On second mention, some write their given names as initials. Lee Byung-chul, for instance, becomes B.C. Lee. Sometimes Koreans prefer to hyphenate their given names (Lee Kun-hee). Sometimes they leave out the hyphen (Park Chul Wan). The Revised Romanization system is used throughout.

Researching this book, I used a mixture of investigative journalism and ethnography methods, drawing on my academic background in anthropology. For archival research, I consulted the collections at my graduate-school alma mater, the School of Oriental and African Studies at the University of London, as well as the Presidential Archives and National Assembly Library of Korea.

Cast of Characters

I. THE KOREANS[1]

Chairman Lee Byung-chul (B.C. Lee, also known as Chairman Lee I) (1910–1987). Founder and first chairman of the Samsung Group, from 1938 until his death in October 1987.

Chairman Lee Kun-hee (K.H. Lee, also referred to as Chairman Lee II) (b. 1942). Son of B.C. Lee and second chairman of the Samsung Group, from 1987 to the present, who transformed Samsung into a global nameplate brand. He suffered a heart attack in May 2014 and has not been seen in public since.

Vice Chairman Lee Jae-yong (Jay Y. Lee, tapped to become Chairman Lee III) (b. 1968). Son of K.H. Lee and heir apparent to the Samsung empire, he was sentenced to five years in prison for bribery and embezzlement. Released early from prison on appeal, a panel of judges upheld but lessened the extent of his bribery conviction. He awaits retrial after Korea's Supreme Court voided his second ruling, and may be sent back to prison with possibly expanded charges.

Lee Maeng-hee (1931–2015). Oldest brother among B.C. Lee's three sons.

---

[1] I've listed the family name first and the given name second.

He was the favored heir to the Samsung Group before he resigned in 1969, accused by his father of mismanagement.

Lee Mie-kyung (Miky Lee). Daughter of Lee Maeng-hee, niece of Samsung chairman Lee Kun-hee, and Samsung's emissary to the U.S. entertainment and design industries before leaving to start her own production house, CJ Entertainment, in a partnership with DreamWorks.

Hong Jin-ki (1917–1986). Samsung founder B.C. Lee's political ally and former head of Samsung's *JoongAng* newspaper. His family forged an alliance with Samsung's founding Lee family. His daughter Hong Ra-hee married Samsung chairman K.H. Lee.

President Park Chung-hee (1917–1979). Dictator of South Korea starting with his military coup d'état in 1961 and ending with his assassination in 1979. He laid the economic and political foundation for the Korean economic miracle, bolstering companies like Samsung and Hyundai.

President Park Geun-hye. Daughter of dictator Park Chung-hee. As the democratically elected president of South Korea from 2013 to 2017, she was implicated in organizing almost $38 million in bribes from Samsung for her political ally. President Park was removed from office and jailed in March 2017.

Lee Ki-tae (K.T. Lee). No relation to the ruling Lee family. K.T. Lee was the CEO of Samsung Electronics' mobile phone unit from 2000 to 2007. He started out as a factory floor manager; as CEO of Samsung's mobile phone unit, he elevated the quality and durability of Samsung mobile phones and brought them to the U.S. market.

Hwang Chang-gyu. President of Samsung's semiconductors unit from 2004 to 2008, and chief technology officer of Samsung Electronics from 2008 to 2010. Hwang made a key deal with Apple's Steve Jobs in 2005 to supply chips for the iPod and later the iPhone, spurring Samsung's explosive growth.

Choi Gee-sung (G.S. Choi). Formerly the powerful lieutenant to Samsung's ruling Lee family and head of the Future Strategy Office, Samsung's highest body that houses many of the elite executives, known as "the Tower." Now serving a five-year prison sentence for bribery and embezzlement.

Shin Jong-kyun (J.K. Shin). CEO of Samsung's mobile unit from March 2013 to December 2015. He oversaw the kick-starting of the Galaxy smartphone line and helped to initiate the smartphone wars against Apple.

Koh Dong-jin (D.J. Koh). Successor to J.K. Shin and CEO of Samsung's mobile unit from December 2015 to the present. He oversaw the recall and cancellation of the Galaxy Note 7 after the product began catching fire.

Sohn Dae-il ("Dale"). CEO of Samsung Telecommunications America from 2006 to August 2013. He helped lead the smartphone wars against Apple.

II. THE AMERICANS

Peter Arnell. Hired by Chairman Lee II to bring cutting-edge, fashion-based advertising to Samsung products in the mid-1990s, when the company was known as a manufacturer. Former head of the Arnell Group in New York.

Gordon Bruce. Design professor at ArtCenter in Pasadena, California. Co-founder of the Innovative Design Lab of Samsung (IDS) from 1995 to 1998.

Pete Skarzynski. Vice president for sales and marketing at Samsung's American unit from 1997 to 2007. With K.T. Lee, he brought Samsung mobile phones to the American market.

Eric Kim. Executive vice president and then chief marketing officer at Samsung Electronics from 1999 to 2004. Through his branding and advertising campaigns, Samsung overtook Sony in brand value and sales by the mid-2000s.

Todd Pendleton. Chief marketing officer at Samsung's American mobile unit from 2011 to 2015. Led advertising efforts against Apple during the Samsung-versus-Apple smartphone wars.

Daren Tsui and Ed Ho. Two Silicon Valley entrepreneurs who got their start working with Elon Musk and later sold their music software, mSpot, to Samsung in May 2012. After the sale, they joined Samsung as vice presidents for content and services; they ran the grand experiment Milk Music until its closure in September 2016.

Paul Elliott Singer. Founder and CEO of Elliott Management, an ultrasecretive hedge fund in New York. Known to many in the industry as "the Vulture." He challenged Samsung in court and in shareholder votes as the company set out to install its heir, Jay Lee, as the next chairman.

III. THE JAPANESE[2]

Tameo Fukuda. Design adviser to Chairman Lee II and author of the June 1993 "Fukuda Report" that forced Samsung to improve its product quality.

---

[2] Given name first, family name second.

SAMSUNG
RISING

1

Galaxy Death Star

"LEAVE YOUR BAGS. GET off the plane immediately!" shouted the flight attendant.

Brian Green felt he'd entered a nightmare. The morning of October 5, 2016, Green had taken his seat on Southwest flight 994 at Kentucky's Louisville International Airport, where he was preparing for a business trip in Baltimore. Ten minutes before takeoff, during the safety demonstration, he powered down his new Samsung Galaxy phone and put it in his pocket.

"I heard some popping that sounded like a ziplock popping open," Brian later told a television crew in his southern accent, "and looked around to see what that was. There was smoke just billowing, pouring out of my pocket."

He yanked his smartphone from his pants pocket and threw it on the carpeted floor.

"I didn't want it to explode in my hand," he said. Angry, thick, green-gray smoke was billowing out of it. The smoke spread several rows in front of him and behind him before dissipating throughout the rest of the cabin. The Southwest Airlines crew decided it was time to get everyone out.

By 9:20 A.M., the flight attendants had evacuated all seventy-five passengers and crew members. Emergency crews arrived to retrieve the smoldering device and to check the passengers for injuries. Fortunately, no one lost "a finger or a hand over it," as Green put it—he had tossed the phone away just in time. The smoking piece of metal, plastic, and circuitry was so hot that it had burned right through the carpet. When airline mechanics pulled the layer of carpet away to reveal the subfloor underneath, it was seared and blackened.

Investigators from the arson unit at the Louisville Fire Department showed up on the tarmac, seized the device, and questioned Green. But Green, they quickly realized, had done nothing wrong. The problem lay in the phone that he had bought, loved, and admired.

THE GALAXY NOTE 7 had been a source of trouble for two months in South Korea, in the United States, and around the world. But everyone had assumed the problem had been resolved. Since late August, Samsung had documented ninety-two instances of its new, much-heralded Galaxy Note 7 device overheating in the hands and homes and cars of its customers. A number of them caught fire thanks to what Samsung claimed were faulty batteries.

After three weeks of stumbling and stammering around the faulty device, Samsung had begun to recall the Galaxy Note 7s in the United States. As the company had advised, Brian Green had exchanged his new Note 7 at the AT&T store two weeks before his flight.

Green had studied the replacement phone and its packaging carefully. All indications on Samsung's packaging were that the device was safe to use. The box was marked with a black square, indicating a replacement device rather than an original Note 7. When he punched the new phone's IMEI—a unique fifteen-digit number on every device—into Samsung's recall eligibility website, he got this recorded response: "Great News! Your device is NOT in the list of affected devices."

After the evacuation, Brian called Samsung's customer service line.

"I did everything I was supposed to," he explained to the Samsung rep. "This was a recalled phone."

The rep patched his message into a ticketing system. Green wondered when he'd hear back from Samsung. The company was slow to treat the incident as a public-safety issue. Instead, when journalists followed up on the incident, the company sounded skeptical that the phone was at fault.

"Until we are able to retrieve the device, we cannot confirm that this incident involves the new Note 7," company representatives wrote to journalists repeatedly.

Investigators from the Consumer Product Safety Commission (CPSC), the federal agency whose job it is to test faulty and dangerous products, saw things differently. They initiated decisive and unusually strong legal measures. Citing "exigent circumstances," CPSC investigators obtained a court subpoena and seized Green's phone from the Louisville Fire Department the day after the fire. The team drove it to a laboratory in Bethesda, Maryland, where they got to work on an urgent succession of tests. The gravity of the situation was becoming clearer. It's one thing for a company to issue a recall. It's another for that same company to reissue replacement products—labeled safe—that continue to pose a severe danger to the public.

While Samsung remained unmoved by the public inquiry, thirteen-year-old Abby Zuis was picking up her siblings two days later at North Trail Elementary School in Farmington, Minnesota. Playing with her replacement Galaxy Note 7, she suddenly felt a strange burning sensation on her hand, "like pins and needles," she recalled, "except a lot more intense."

Her immediate reflex was to throw the phone on the floor—thankfully with no more than a burn mark on her thumb. The school principal raced forward and kicked the smoldering device out of the building.

"I'm glad it was in my hand and not my pocket," Zuis told the media later.

"We thought we were safe with the new phone," her father said.

MICHAEL KLERING AND HIS wife woke up at 4:00 A.M. in his Kentucky home to a hissing noise. "The whole room just covered in smoke, smells awful," Klering told a local radio station. "I look over and my phone is on fire." Later that day, Klering started vomiting black fluid; he checked in to the emergency room, where he was diagnosed with acute bronchitis. Doctors determined that he suffered from smoke inhalation.

A Samsung representative contacted Klering and asked him to return the Note 7. Klering refused. Then he received a text message sent accidentally to him by a Samsung employee.

"I can try and slow him down if we think it will matter," the text read, "or we just let him do what he keeps threatening to do and see if he does it."

Klering was aghast. What the heck was going on?

"The most disturbing part of this is that Klering's phone caught fire on Tuesday"—one day before the Southwest flight—"*and Samsung knew about it and didn't say anything*," wrote *The Verge*'s Jordan Golson. Warned *Gizmodo*'s Rhett Jones: "The evidence suggests that Samsung . . . now appears to be suppressing the information that replacements are dangerous."

The reports continued to stream in with no decisive statement or action from Samsung.

A woman in Taiwan was walking her dog when her Galaxy Note 7 caught fire in her back pocket. In Virginia, another replacement Note 7 ignited on Shawn Minter's nightstand at five forty-five in the morning.

"It filled my bedroom with . . . smoke. I woke up in complete panic."

After the incident, Mintner visited the local Sprint store, where a salesperson offered him *another* Samsung Galaxy Note 7. Um, thanks but no thanks?

Then, hours later, another Galaxy Note 7 owned by an eight-year-old girl in Texas caught fire at a lunch table. Regulators, journalists, and the public were looking for answers from Samsung. Puzzled by the inaction of the global powerhouse, Samsung's carrier partners started abandoning the company's products. AT&T announced on October 9, four days after the evacuation of the Southwest Airlines flight, that it was discontinuing all sales and exchanges of the Galaxy Note 7. Other carriers followed suit.

The Samsung Galaxy Note 7.

Samsung announced that day it was "temporarily pausing" shipments of the Galaxy Note 7 to an Australian carrier. But the messages from the company were still hazy and unclear. Tens or possibly hundreds of thousands of people were still tapping away on these potentially explosive devices in their purses and pockets. But evidently the honor of the Samsung corporation came first.

"Samsung is confident in the replacement Note 7 and says they have no reason to believe it's not safe," the Australian carrier that paused shipments, Telstra, said in an internal memo.

"In other words," *CNNMoney* correspondent Samuel Burke responded, "the phone wasn't good enough for them [Samsung] to keep making it for now, but was okay for consumers to keep on using."

As the crisis grew, customers seeking an exchange for the Galaxy Note 7 opened their email in-boxes to garbled, incomprehensible emails from Samsung's customer service department—at times with the wrong order number attached, as well as other errors.

"As this is going to another company, when these exchanges are submitted, we cannot check the status of them for you until they submit you an order number for the new phone or tracking information," wrote a customer service rep to a Note 7 customer who had been requesting a refund for almost a month with no response. "We have limited information on a lot of the process at this time. I hope this information is helpful and resolves the issues soon." (Translation: We're clueless about what we're doing or what you should do.)

The brand's reputation was in free fall, yet Samsung was failing to act.

"Does anyone here have a Samsung Galaxy Note 7?" Stephen Colbert asked on *The Late Show with Stephen Colbert.* "If so, please calmly remove yourself from the theater. Hazmat teams are waiting for you in the lobby.

"Didn't Samsung recall those phones that were catching fire? Yes, they did," he continued. "It's like the old saying: Fool me once, shame on you. Fool me twice, *oh my god! My crotch is on fire!*"

Even Sprint, a longtime Samsung ally, took a swipe at its partner.

"Hey @sprint, what if I don't trust @SamsungMobile devices anymore?" a frustrated customer tweeted.

Sprint's response?

"Greetings," wrote the carrier from its Twitter account. "You still can trust Apple, HTC, LG or Alcatel."

Every time Samsung announced a new television or washing machine or phone on its Facebook and Twitter pages, or customers talked about Samsung products on Web forums, an endless deluge of sarcastic zingers appeared in the comments sections.

"I'm sitting in front of a Samsung LCD monitor. I hope it doesn't explode!"

Airlines around the world banned the Note 7 from flights. Pranksters posted YouTube videos of *Grand Theft Auto V* showing characters buying Note 7s at the local guns-and-ammo shop and hurling them at people and cars like grenades on the streets of Los Santos, a fictional Southern California city.

Samsung issued a copyright takedown notice to YouTube—even though it did not own the copyright to the videos. The attempt at censorship only backfired—more of these recordings proliferated.

"Samsung doesn't want you to see video of this *GTA V* exploding phone mod," proclaimed *Ars Technica,* posting a video for all to see.

"It appears Samsung took the easy path to removing the content it did not like by making a copyright claim where none existed," wrote the Electronic Frontier Foundation, the group that defends free speech on the Internet, on October 26.

It was, in other words, one of the greatest brand disasters in recent history: a recall, followed by the release of replacement products that were equally dangerous, followed by increasingly confused and desperate attempts to save face.

FOR ME, KOREA WAS a journey of discovery.

I first settled in the country in September 2009. I lived on and off in South Korea until the fall of 2016, between reporting stints in Vietnam and Cambodia.

I was immediately fascinated with my new, adopted home. It was a nation divided between the authoritarian North and the democratic South, polar opposites in terms of their societies, economies, and politics.

Yet they were essentially one nation and one people, divided by an artificial border.

Thirty minutes north of my home in Seoul was the demilitarized zone, one of the world's most heavily mined borders, an area I would visit dozens of times. Beyond that was North Korea, where Marshal Kim Jong Il was threatening war against the United States and had set up a network of brutal prison camps.

Yet my South Korean friends practically nodded off when I brought up North Korea.

"We really don't care about North Korea," one friend said, repeating what I had heard from so many others. "We've dealt with North Korea for decades. Nothing's changed. It's not like we wake up thinking about it. I have my own life to worry about. My parents want me to get a job at Samsung.

"You get a job with Samsung, Hyundai, LG, SK. Then your parents, and society, say you're a success and brag about you to their friends. If you don't get into Samsung, well, then you're a failure and you don't exist."

I was captivated by comments like these. Two generations earlier, South Korea had been a dictatorship, and poorer than North Korea. Today it is among the richest and most tech savvy democracies, a success story that few would have predicted a few decades ago. Locals referred to South Korea as the "Republic of Samsung."

How could South Korea go from an agrarian nation to one of the world's most successful economies in two generations? What were the effects on its people, its businesses, and its culture? I enrolled at Seoul's Yonsei University for an intensive period of Korean study, poring over the most highly regarded academic books on the country. I befriended locals in an effort to understand the Korean people better.

"Everyone on this street corner has a Samsung product somewhere on them, or in their homes," a former Samsung vice president told me proudly. "In their pocket, in their living room, at their workplace, the parts inside their Apple smartphones. Everyone."

I discovered he was right.

Disembarking at airports in Mexico City, London, and Budapest, I was hit with Samsung signage and advertising everywhere. I saw people in Cambodia, Cuba, Russia, and Venezuela using Samsung phones. Outside the airport in Vientiane in the sleepy Southeast Asian nation of Laos, I saw a giant ad for the Note 7 months after the device had been pulled from the market. On a trip to Pyongyang, North Korea, my regime minder showed me a photo on his smartphone of his living room—where I spied a flat-screen television with a Samsung logo. Samsung was the pride of South Korea; to most, Korea's success rested on Samsung's success.

IN SEOUL, NEWS ABOUT the exploding Galaxy Note 7 hit the news shows and websites on a minute-by-minute basis. I was inundated with calls from CNN, NPR, BBC, and Bloomberg TV—all desperate to know what was going on, absent any clear statement from Samsung. Its public relations people, one reporter complained, were refusing to hold a candid on-record chat.

"Samsung won't tell us anything," a news anchor at one of the big media outlets said.

I'd seen this all before. I had spent the past six years reporting on the company, both through official sources within Samsung and through my own backdoor sources. Samsung, I knew, was a strange labyrinth of a company, the product of a vastly different business culture from what Americans were accustomed to.

Samsung wanted people to believe the exploding Galaxy Note 7 was the result of a problem with the batteries the company used. But my years of research into how Samsung operated had revealed that the underlying problem wasn't just a snafu with the power source. It was a problem that began with the corporate culture, which the company had long been trying to reform.

In more than four hundred interviews, I documented an organization with a unique, almost military-style management system that conducted business as if it were issuing battle orders. Samsung was no Apple, with its personalized relationships among engineers, designers, marketers, and the vast millions of users who had fallen in love with its elegant cellphones, iPads, and computers and integrated them into every aspect of their lives.

Samsung, on the other hand, was highly regimented. But the Samsung Way, as employees called it, was becoming a liability in the face of the Note 7 debacle. The company's defensive approach to its public relations disaster was only hurting it further. Engineers and designers were discouraged from speaking up about potential problems. The Note 7 was a casualty of that culture.

Over decades of struggle, growth, and gradual success, Samsung had instilled a sense of respect, loyalty, and fear in its employees, as well as a reluctance to challenge Samsung management, either internally or publicly. Samsung had connected its fate to the fate of its homeland.

"[Samsung's South Korean headquarters] didn't want anyone touching anything or anyone saying anything," a senior marketing consultant told me, exasperated by how the company was handling the exploding Galaxy Note 7. "So you've got these long delays in being able to actually talk to the world about what was happening. In addition to honor and ego and all of the other things that [come into play] when you kind of screw up."

SOON I DISCOVERED THAT Samsung had taken offense at what I was saying about it and had launched a defensive attack on me personally. After offering my candid take to CNN and NPR and other media about the Note 7, I opened my email in-box to a message that had been forwarded to me written by David Steel, an old dinner friend who served as Samsung's executive vice president and chief of global communications. A genteel and well-spoken Englishman who had a PhD in physics from MIT, he was complaining about what I'd been quoted as saying on NPR. I had repeated on prime-time news what Samsung's employees were telling me.

"These one-sided, sensational views," Steel wrote about my commentary, "do not fit with NPR's reputation for fairness and balance." He called me one of two "self-declared Samsung critics" and decried my commentary as "particularly inappropriate."

I was becoming persona non grata over the Note 7 fiasco. It's a risk confronted by any journalist who goes against the official line of the company they cover. But I knew this was the only way to write honestly about the company.

ON THE MORNING OF October 11, 2016, a day after Samsung halted production of the Note 7, I opened my smartphone to a message from a Samsung marketing manager.

"Can you please check what's going on with our Apple iPhones?" she wrote. "It seems like they have various defects and the media keeps silent." She was looking to shift the heat to a competitor.

I realized that Samsung wasn't ready to admit to failure. For years I'd heard in-house conspiracy theories from friends and contacts at the company that Samsung saw itself as misunderstood and mistreated—that the press and the shareholders were out to get it. When a young business journalist dropped into Seoul to write about the Note 7 fires, a public relations executive told her the negative media reports on his company came from the fact that "everyone loves a good Apple story."

The same morning, I was told, Samsung had gathered executives at its campus at Suwon, an hour's drive south of Seoul, for a meeting on the devices that were catching fire. They were to hear a briefing from a Samsung executive who had returned from Bethesda, Maryland, where his delegation had met with the CPSC in an attempt to smooth things over. The American officials were concerned over consumer complaints about how Samsung was handling its recall and exchange program.

But the real purpose of the meeting, I was told, was to attempt to keep morale up.

"Stay calm and confident," my source told me was the message of the meeting, after being briefed by her boss.

Her boss told her that the police had not confirmed that the fires or explosions were due to the Note 7. "We did lots of tests here," she was told, "but nothing more than smoke comes from the device. So the media made it up. The media keeps silent about Apple products and converges on Samsung only."

On the defensive, Samsung had been failing in its efforts to repair its damaged relationship with its customers; instead, based on what I was hearing from my sources, it resorted internally to a counterattack on Apple in an effort to divert attention. To me it demonstrated the company's arrogance—as well as its underlying shame, insecurity, and desperation. The company knew it needed to turn the debacle of the Note 7 around fast—or see its brand, and even the prestige of South Korea, sullied by its new smartphone.

"I AM ASHAMED TO be Korean," a shop owner named Mr. Park told me as he bagged my goods at a store around the corner from my apartment.

Mr. Park's convenience store—up the hill from the nearby U.S. military base—was the regular start of my hour-long morning walk through Seoul, a helter-skelter neon-lit megacity of brick homes, hills, and mountains. I would pass the military base and stroll through the rowdy drinking district of Itaewon, walking by the transgender prostitutes finishing assignations from the previous night, approaching the modern glass headquarters of Samsung's in-house advertising agency, Cheil.

Around the corner and to the left of the ad agency headquarters was Samsung's Leeum Museum of Art, founded by the wife of Samsung's chairman, Lee Kun-hee. A prestigious landmark, the museum was home to paintings by Rothko and art pieces by Damien Hirst and a magnificent collection of medieval Korean pottery. After grabbing a cup of coffee nearby, I would continue up through the posh hillside neighborhood of Hannam, stacked with luxury homes that resembled military bunkers—some the homes of Samsung's powerful ruling family.

One of the first houses was that of Jay Y. Lee, son of Chairman Lee and heir to the Samsung dynasty. His home was down the hill from the home of his father, pursuant to strict Confucian hierarchy. The chairman's home was located on a scenic overlook according to the Korean art of geomancy; it was a structure of stunning traditional beauty. The chairman, known to be a recluse, had a breathtaking view of the city below.

After my morning exercise, I would head down and grab a bus to the government's Foreign Press Center, with its view of a public square, Gwanghwamun. From my offices on the tenth floor, I usually took notice of the Samsung and LG Electronics ads hanging over the neighborhood. In the square below, there were never-ending political demonstrations for or against Samsung, or about this or that political issue. Those on the left typically hung out around the statue of King Sejong, a beloved scholar-king who invented South Korea's alphabet, in the center of the square, while the right-wing groups picketed across the street at the edges of the square.

On any given day, there could be dozens to thousands of these demonstrators, depending on the political mood. Most of the protesters were legitimate, though the South Korean media exposed some of the probusiness conservatives for taking payments for their protests from the Federation of Korean Industries, a major business lobbying organization. I would go back and forth between the rival camps during coffee breaks.

"Samsung built this nation! Fed our people! Clothed us when we had nothing! Gave us jobs! Brought us global prestige! Only leftist zombies dare oppose Samsung!" shouted one elderly protester, a patriotic military veteran wearing an insignia cap.

I wandered across the street to the tents of those on the left—some of them had camped out for weeks—for their response.

"Did Marie Antoinette say, 'Let them eat cake'? Samsung bought our nation! We say, 'Off with their heads!'"

From the right-wing group, I heard: "The Samsung unionists are communists. Send them back to North Korea!"

In the wake of the botched recall of the Galaxy Note 7, I approached one of the pro-Samsung demonstrators, who appeared dejected and downtrodden.

"The world is watching and the Note 7 has failed," he said. "I am ashamed for our country. I am ashamed for Samsung."

On the basis of the failure of a single line of smartphones, the nation's political leaders were seriously contemplating the possibility of a South Korean economic decline.

"At a recent meeting with her aides, President Park showed a great deal of concern after being briefed on the Galaxy Note 7," a government official told reporters. She called on everyone to "focus all our efforts on minimizing the damage" to smaller firms.

Opposition leader Moon Jae-in, who would soon be elected president, proclaimed: "This is not just Samsung's trouble. It's trouble for the entire economy. Because people take pride in Samsung as a brand representing South Korea, it is their trouble, too."

THE BACKSTORY OF THE failure of the Galaxy Note 7 actually began five years earlier, in July 2011. I was on a morning coffee break in Gwanghwamun Square, on a sweltering summer day. I opened my (Samsung) phone to an email from Samsung's public relations team. It was a response to a request that would define my work for the next decade.

"The interview is a go," wrote Nam Ki-young, a staffer for Samsung global communications. A few days later, I took a forty-five-minute subway ride south, getting out at the nouveau riche district of Gangnam, home to plastic surgery clinics and designer handbag shops. I emerged through a revolving door straight into the basement of Samsung's sleek, towering compound—a quadrant of four glass buildings designed to look like interlocking jigsaw pieces, inspired by the Korean craft of woodworking. Samsung had built this center as a symbol of its mission: to solve the world's puzzles, a feat to be tackled by its greatest minds.

I was covering South and North Korea for *Time,* and Samsung had invited me to come in so it could make an urgent case in its defense against Apple.

Three months earlier, in April 2011, Steve Jobs had initiated a slew of lawsuits accusing Samsung of "slavishly" copying the iPhone and iPad, demanding $2.5 billion in damages. His action ignited a spectacular legal war between the two industry giants, involving more than fifty lawsuits around the world.

Samsung Electronics building in Gangnam, Seoul, one of four buildings designed like fitting puzzle pieces. November 2010.

Nam, as he called himself, pulled me aside at a coffee shop to explain what was going on.

"I just want you to know that the CEO approved this interview," he said, emphasizing the gravity of the invitation. Samsung, he said, typically does not give interviews on sensitive topics like lawsuits against it.

After riding in a company car through the grubby industrial town of Suwon and getting cleared past Samsung's security checkpoint— where guards checked my old Samsung phone to be sure it had not been tampered with—we passed within the campus walls, where I saw more seemingly obligatory coffee shops. After taking a tour of the company museum, we arrived at the closely guarded premises of the Samsung Electronics mobile phone unit.

A portly Samsung executive named D.J. Koh, vice president for mobile research and development, waited for me at the conference table. D.J. had an imposing presence, a booming voice, and a way of lightening the mood with his cheesy jokes.

"I feel like the movie star George Clooney. But I will admit my look and feel are totally different," he said in the spotlight at a product launch.

You laugh at his jokes not because they're funny but because they're hilariously silly. But D.J. was deadly serious as he faced me at the table and made a case for Samsung and its strategy against its archrival, Apple.

"Instead of presenting a product and saying, 'This is it, follow us,' and leading the way, we want to satisfy the needs of different preferences of all the different people, to support their lifestyles."

Though he never mentioned Apple by name, D.J. was tacitly revealing Samsung's strategy in the recently launched smartphone war against Jobs and Apple. At the time, Apple was the control-freak company with a far narrower catalog of product lines. The company had little diversity. You had the choice between the most recent version of the iPhone and the much bigger iPad. And that was it. They made those who used them cool; if you didn't own them, you weren't cool. As Jobs was famous for saying: "People don't know what they want until you show it to them."

Samsung's market researchers in the United States thought Apple's

"Think Different" motto was pompous and presumptuous. Their research suggested that Android users—who used Samsung products—
considered themselves smart and independent in their choices. In
contrast, they saw fans of Apple products as consumers who fancied
themselves creative but were, in reality, sheep—followers. Samsung
had decided to embrace the opposite strategy to Apple's: tailoring a
version of each product for everyone who wanted something different.

Want a small screen? Samsung had it (the Galaxy S). Do you prefer
a bigger screen? Samsung was preparing to release that too (the Galaxy
Note). Are you looking for a massive screen? Samsung had that as well
(the Galaxy Tab). And many other products at every price point in between.

"We strongly believe we have our own competitiveness," D.J. said,
pointing to the fact that Samsung had registered far more patents in
the United States than Apple. In fact, D.J. said, Samsung was the inven-
tor of a multitude of hardware technologies. Steve Jobs had elected to
use these technologies in the guts of the iPhone; Samsung's chips, in
fact, made the iPhone possible.

In August 2012, a California court ruled that Samsung had copied
the patents and the "look and feel" of Apple products. Samsung, however, won legal victories against Apple in the UK, Japan, and South
Korea.

Samsung executives saw the lawsuits as the subplot of a bigger
story, a debate over where to draw the lines between inspiration, imitation, and outright copying. Samsung was known in the industry as a
fast executor and an incremental innovator, different from Apple's
style of creating the one, big product.

Samsung watched how disruptive products like the iPhone fared on
the market and then, when the path to success was clearer, released its
own smartphones. Seeking an edge, its mission was to improve the
smartphone's hardware features in small steps: a bigger screen, a
longer-lasting battery, a water-resistant exterior. *Gaeseon* is what Koreans call this process of "incremental innovation." To the Japanese, it is
kaizen.

Apple did not invent the smartphone, a product category dominated by BlackBerry. It was inspired by other companies to disrupt the

industry. Steve Jobs was an admirer of Sony and its corporate culture. Apple designers borrowed Sony designs that changed the direction of the iPhone. Pulling together a thread of ideas and technologies that already existed, Jobs created the iPhone.

Samsung executives felt Apple was trying to create a monopoly with generic patents like the iPad's black rounded rectangle shape, a patent so silly that a court threw it out. "We are going to patent it all," Jobs once said.

He also blatantly mocked Samsung and other competitors, calling their larger phones "Hummers." "No one's going to buy that," he said at a press conference in July 2010.

Samsung's management team didn't take Jobs's attacks lightly.

"I am talking to you on a phone right now that Apple just copied," Brian Wallace, Samsung's former vice president for strategic marketing, told me years later. "I've got a Note Edge. It's a giant fuckin' phone that Steve Jobs, to his dying breath, made fun of. Who was right? Fuck Steve. He's dead and we were right. Samsung was right."

The Samsung designers felt they were onto something. With the first Galaxy Note, released in October 2011, they'd created a phone that was halfway between tablet and smartphone, and it caught on. The larger screens pioneered by Samsung became a defining element in the evolution of the smartphone. Eventually Apple released iPhones with larger screens in each iteration. In this area, Samsung was leading and Apple was following.

But Samsung managers continued to be in a frenzy to beat Apple in the marketplace, to replace Apple as the world's most famous and successful electronic devices company. And this desperate desire would expose a critical hole in their corporate culture. Samsung was still benchmarking Apple products and was being labeled a "fast follower." The "copycat" accusation against Samsung was especially touchy for the company's executives. In February 2014, Samsung sued the British vacuum cleaner company Dyson for defamation after Dyson claimed, in its own lawsuit, that Samsung had ripped off its patents.

In planning the Note 7, Samsung managers were looking for openings in the market to exploit, listening for rumors, dreaming up new products the company could use to beat Apple. But the pressure to overtake the Cupertino company would lead to sloppiness, and it would ultimately bring the wildly successful brand of Samsung Electronics, and its prized Galaxy, to the brink of disaster.

In early 2016, D.J. Koh, the executive I had met with in 2011, was promoted to CEO—one of three CEOs at Samsung Electronics. For the past five years, D.J.'s research and development team had pushed the limits of smartphone technology through rapid-fire incremental improvements. Samsung released phones with bigger screens and stronger hardware, sold at greater volumes and with better marketing, overtaking Apple's market share. Now he and the other executives had heard that the next iPhone was slated to be a modest release. Its physical design wasn't changing much, and its look and feel were going to be fundamentally the same.

Though the rumors proved to be not quite accurate, it seemed like an opportunity to forge ahead of Apple with a new Samsung phone packed with features. The Galaxy Note 7, as it became called, was encased in an eye-catching glass and metal casing, giving it a luxurious rather than cheap plastic feel. Its large wrap-around screen with curved edges disguised an eye iris security scanner, something the iPhone didn't have. It also had a water-resistant casing that could survive drops and spills—one of the criticisms of the iPhone—and a faster-charging and longer-lasting battery.

Consumers had long desired more battery life, and D.J. wanted to use that feature in the Note 7 to draw new consumers to Samsung. So the Note was built with a 3,500-milliampere-per-hour battery, compared with the 3,000-milliampere the company had used in the previous model, giving the Note 16 percent more battery life. Apple, meanwhile, had only a 2,900-milliampere-per-hour battery in the iPhone 7 Plus.

Samsung Electronics settled on two suppliers. For its U.S. phones, it tapped fellow Samsung affiliate Samsung SDI, one of the world's largest and most successful battery manufacturers.

In a uniquely South Korean arrangement, SDI was its own company on paper but not its own company in reality. Samsung Electronics owned a fifth of SDI, and SDI reported to the same ruling family under a complex shareholding structure. SDI, in fact, was merely one corporation in a web of cross-shareholdings and family ownership interests, tied together under the massive Samsung Group. It consisted of more than fifty companies in shipbuilding, fashion, advertisements, food courts, and a hospital. "The Samsung Group" was often mistaken for a conglomerate. But "the Samsung Group" was merely a term to express the way these firms are tied together under a founding family.

For phones sold in China and elsewhere around the world, Samsung turned to an outside supplier, Hong Kong–based Amperex.

Beating the iPhone to the market was essential to the company, and Samsung's executives demanded an accelerated deadline.

"We were sensitive to the iPhone's release date," a mobile manager told me. "We wanted to beat them to [market]. Then we could get iPhone customers."

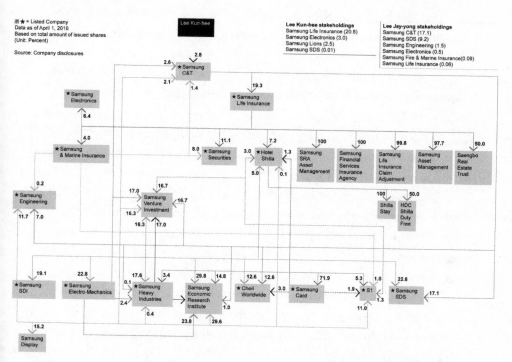

Samsung's ownership structure in 2016. The Samsung Group consists of more than fifty affiliates in chipsets, shipbuilding, a hospital, smartphones, a theme park, and fashion. But the "Samsung Group" isn't an actual business entity, nor is it a holding company. It describes the bewilderingly complex web of cross-shareholdings, above, that the Lee family uses to keep control of Samsung with a relatively small stake.

DIAGRAM DESIGNED BY GORDON BRUCE AND JON CRAINE. USED WITH PERMISSION.

Its unveiling date? August 3, 2016—ten days earlier than the previous year's Galaxy phone release date. Confident in a head-to-head battle against Apple, D.J. Koh's executives opted to skip over the number six for the new Galaxy Note—there would be no Galaxy Note 6—and go straight to the name "Galaxy Note 7." The company hoped that would place the Note 7 directly in comparison with the iPhone 7 to consumers.

As *Bloomberg Technology* put it, "Apple's taunts that Samsung was a copycat would be silenced for good."

In Samsung's meetings with suppliers, executives demanded breakneck speed with tight deadlines and aggressive specs.

"The pressure was tremendous," the exasperated Samsung mobile manager told me on the phone. "Tremendous! We were having shortages [of the display] in the Galaxy S6, and now there was huge pressure to get our supply [for the Note 7]. Curved displays, batteries, you name it."

Executives made things worse by repeatedly changing their minds on specifications and schedules. One supplier admitted to *Bloomberg* it was struggling to keep up.

"The pressure was huge," a Samsung Electronics mobile phone manager, who was directly involved in the orders from suppliers, told me. "It was a rush. It was chaos."

After months of whip cracking, Samsung's managers got their hands on an early version of the finished Galaxy Note 7 and gushed in amazement. The stylus pen, the screen, the battery—the phone was a work of art. They sent off models to carriers for testing, and the carriers found no flaws. After a brief delay for quality testing, the Galaxy Note 7 was declared good to go. This was Samsung's greatest strength: its ability to manufacture superior hardware, faster than any of its competitors, through its vast, strict, top-down management system and its superior supply chain.

"We faced skeptics who doubted us," D.J. proclaimed at the product unveiling, giving a victory speech in New York City on August 3, 2016. "We heard from critics who dismissed the large screen and our new S Pen. But we stayed true to our vision and we persevered."

Others weren't so smitten.

"You know the kind of reality-warping bubbles those guys live in," an exhausted Samsung marketer explained to me.

It was a dig at Samsung's military-like hierarchy, a characteristic shared by many South Korean companies. Untouchable "generals" charged headlong into each new project, and even when things looked iffy, the field troops were expected to praise them to the skies, convincing themselves of their company's and leaders' greatness. The marketing team member who spoke with me was a non-Korean. He had more leeway to speak freely, as he could always get a job elsewhere with a company back in the States. A Korean at Samsung, I knew, wouldn't have been so blunt, for fear of retribution. A handful of employers dominated the job market in South Korea, and the companies were notorious for holding grudges.

The Galaxy hit the shelves on August 19, 2016, to surging demand. It put so much pressure on the supply chain that Samsung had to push back launches in Russia and Malaysia. But there were no signs of trouble; the company's stock price was hitting record highs, and the product reviews were sparkling.

"The Galaxy Note 7 is a beautiful, capable Android phone that showcases Samsung's best in design, battery life, speed and features," wrote *CNET*'s Jessica Dolcourt. *TechRadar* raved that the Note 7 "took Samsung's best phone to date and added even more features, like a larger curved display and stylus pen." *Engadget* awarded the Note 7 one of its highest review scores ever, proclaiming, "The company wound up making its best phone yet in the process."

But it was a short-lived victory. Within days the YouTube videos and social media cacophony started trickling into Samsung's headquarters.

"How can you believe that? It's all nonsense," a senior manager told a mobile phone marketer who dared to ask about the reports. Younghee Lee, Samsung's executive vice president for global marketing, told *Bloomberg Businessweek* that she didn't believe the reports at first.

But eventually a gallows humor settled over employees.

"One mobile division executive described the Galaxy Note 7 as a 'radioactive' topic," reported *The Wall Street Journal*, "with staffers afraid of even discussing it in the company canteen."

Several floors up, near the top of Samsung's mobile-phone building in Suwon, C-suite executives were frantic with fear for their jobs. The company formed a task force under CEO D.J. Koh to contain the damage, meeting every day at 7:00 A.M. for the next four months to coordinate responses. The powerful enforcer for the ruling family, Vice Chairman Choi Gee-sung (G.S.), also stepped in.

As South Korea's *EBN* reported, "Now, speculation that DJ Koh . . . will be replaced seems to hold more water."

Although D.J. Koh was a CEO, Samsung treated its CEOs more like COOs. While Koh headed the investigation, he ran one company within the Samsung Group and still required supervision from an executive above him. Ascend through the labyrinth of the Samsung empire, and the officers become more secretive, their decisions more guarded and hidden. They rarely speak in public. They descend on trouble spots within the company with little forewarning and sometimes grave consequences to the executives in charge.

Vice Chairman Choi was a lifelong Samsung employee who had once been a semiconductors salesman and television marketer. He had a methodical and calculating demeanor. One former vice president called Choi a "terrorist" for his aggressive ways. Admired and feared, he was grand master of the ruling dynasty's palace court, an office called the Future Strategy Office (FSO) that acted as disciplinarian for the myriad business units.

"The Tower is watching, the Tower is watching," a former vice president once put it to me, using the internal nickname for Samsung's circle of tight-knit leaders, many from the Future Strategy Office.

Vice Chairman Choi and D.J. met in D.J.'s office in a building called R5 at the Suwon campus, where they examined a laboratory report containing X-ray and CT scans of Galaxy Note 7s, as they set out to determine the cause of the fires. The scans showed Samsung SDI's battery had a protruding bulge through the casing, as well as heat damage to the battery's internal structure. They could see no bulge in the phones powered by batteries from the other supplier, Amperex.

"It wasn't a definitive answer," reported *The Wall Street Journal*. But with customers complaining and carriers demanding answers, Koh thought they had enough information to act. They pinned the blame on Samsung's battery supplier for manufacturing a faulty component and instigated a sweeping recall.

On September 2, two weeks after the Galaxy Note 7's release, D.J. entered a conference room to flashing media cameras for a grim press conference. He bowed deeply, a Korean act of apology.

"By putting our top priority on customer safety," he said, "we've decided to halt sales of the Galaxy Note 7 and offer new replacement handsets to all customers, no matter when they bought it."

The Galaxy Note 7, the shining light of the Samsung empire, was now its greatest disaster. Of the 2.5 million Note 7s on the market, Samsung was set to recall an estimated 1 million phones that used the Samsung SDI battery. But as it turned out, not everyone was satisfied with Samsung's conclusion. Samsung had told *The Wall Street Journal* that the phones with Amperex batteries, used in China and elsewhere, were safe. Yet consumers in China were reporting fires in those phones, too. Samsung dismissed those reports as fake, citing its own laboratory tests.

The half measure left people scratching their heads. "It wasn't a recall at all, since the company didn't get the CPSC involved," wrote *Gizmodo*'s Matt Novak.

"Without guidance from the CPSC, other agencies, like the FAA [Federal Aviation Administration], were paralyzed," he wrote. "I contacted the FAA about whether the phones were still allowed on airplanes. Because there hadn't been a proper recall yet, the FAA couldn't flatly say that the explosive smartphones shouldn't be allowed on planes."

The federal government concurred.

"In my mind," warned CPSC chairman Elliot Kaye at a press conference, "anyone who thinks that the best way for a recall is for a company to go out on its own needs to have more than its phone checked."

Samsung took a full week to start working with the CPSC, announcing its cooperation with the agency on September 9. No official recall was yet in place, and there continued to be more reports of phones bursting into flames.

A Galaxy Note 7 that was being recharged ignited a Florida man's Jeep Grand Cherokee. The fire started on the dashboard and quickly spread to the highly combustible air bags and eventually the engine and gas tank, turning his car into a twisted hunk of metal that looked like a prop in a Hollywood movie.

The media coverage in South Korea was relatively docile. I read through a dozen laudatory news reports—one in Korea's largest newspaper, the *Chosun,* which marveled at "the power of communication that led to the achievement of the recall."

On December 5, 2016, I got a batch of six leaked documents from the Korean Agency for Technology and Standards (KATS), the nation's product safety regulator. An advisory panel of ten experts was gingerly going along with Samsung's official line, save a few recommendations for consumers.

"When checking the information held by Samsung," the panel declared, "there is no need for a separate test.

"It is judged that company A's batteries [Samsung SDI's], which were apprehended by Samsung, contain the faulty factors of batteries," but that company B's (Amperex's) batteries were "safe." At the panel's request, Samsung agreed to modify some of its recall plans in South Korea. But even the six outside advisers on the panel—some of them esteemed university professors—concurred that Samsung's decision "was sufficient to remove the harm in the product."

"They're buying time," speculated Park Chul Wan, an unassuming engineer and former government battery explosion investigator. Chul Wan had met me and my assistant, Max, in a noisy coffee shop in an obscure neighborhood of Seoul. "They don't know what's going on. They're not a testing agency. They're a decision-making body."

Chul Wan was drawing on his experience in a long career as a battery detective, with stretches as a government regulator and with the nation's battery industry association. He was one of the first Koreans to establish the country's industry for lithium ion cellphone batteries.

As a PhD candidate at the prestigious Seoul National University in the early 1990s, he'd photocopied and collected thousands of dissertations in Japanese and English—"so many that I forgot to learn English and Japanese for speaking, only English and Japanese for batteries," he joked. His lesson? Establishing causality between a battery, a fire, and the aftermath is difficult. It's like going on an arduous hike up a mountain only to fall on your leg and think you've broken it when you've actually fractured your skull.

"In 2007 I was looking into the death of a heavy-machinery operator," he recounted. His body was found one winter morning at a construction site with a charred flip phone in his shirt pocket and burns on his clothing and flesh. "The media played up the exploding-battery story, the first documented death from an exploding lithium ion battery in a cellphone."

The truth, however, came out later that day. Another machinery operator admitted to accidentally slamming a piece of heavy machinery into the man's chest. Unbeknownst to him, the metal had hit the injured operator right where his phone was placed in his shirt pocket, causing the battery to explode after the machine operator had fled— and after the injured man had died.

Chul Wan had spent his career in the obscure world of lithium ion battery engineering. He dealt not in political gossip but in materials science.

Overwhelmed with interview requests from his country's equivalent of CNN, and penning a popular column in *Chosun*'s technology magazine, *IT Chosun*, he quickly became his country's loudest rebel and critic, not always a sound position in the Republic of Samsung.

"A lot of Korean battery experts won't risk . . . going up against Samsung," he explained to Max and me. "They depend on Samsung for their careers."

But, he told me, "you can't just blame the batteries that easily!" It was an extraordinary claim that could have easily risked his credibility and career. Chul Wan had gotten hold of a small shopping bag's worth of Note 7s and was attempting to re-create the fires at home. He pored over the YouTube videos and studied the testimony of the victims, whose stories were full of clues as they recounted the immolations step by step. The public materials were still too scant to conclude exactly what the culprit was—smartphones are incredibly complex, and hundreds of thousands of phones would ideally need to be laboratory tested to establish the true cause—but he had reason to challenge what he thought was Samsung's hasty conclusion.

CEO Koh's error, he thought, was blaming a single component.

"Big mistake," Chul Wan told us. He believed that the *interactions* of hundreds of pieces of circuitry, parts, architecture, and even software code could have caused a malfunction somewhere and ignited the phones. He published his hypothesis in his column and expounded on it at our table.

"This is the most complex Samsung phone ever made. You can't assign these explosions to one component alone," he said. "In the chase to beat Apple, it looks like Samsung crammed the Note 7 with features until it became uncontrollable."

There were inconsistencies in Samsung's claims that stuck out to Chul Wan, and one was glaring.

"Usually, when there's a battery short circuit, it's a short blast, a very short, intense fire," Chul Wan said. "This was a much longer process"—with the phone heating up for about thirty seconds before igniting into a long, drawn-out burn, discoloring the screen.

"Journalists," he said, "called me and told me that Samsung asked them not to quote me." But behind the scenes, people in the Tower were starting to listen.

"We're also observing Professor Park Chul Wan. Some parts of his recent writing are very persuasive," a powerful staffer named Lee Soo-hyung at the Future Strategy Office wrote to another Samsung executive in a text message, forwarded to Chul Wan.

He showed us another message from an industry colleague. "According to an acquaintance I met at my home today, [Samsung SDI] is busy looking for a sacrificial lamb. . . . The quality control team even suggested seeking your advice, but apparently [the higher-ups] are ignoring it." Clearly a lot was at stake here, and no one wanted to take the blame.

IN THE UNITED STATES, CPSC chairman Elliot Kaye was finally able to put his foot down and intervene after a frustrating two weeks of inaction from Samsung, stating at a press conference on September 15, "Now it is the time to act."

The official replacement program, backed by the federal government, was good to go. Harried Samsung executives raced to get out safe replacement phones. But their haste caught up with them once again. They sent off "safe" Note 7s to a passenger on the now-infamous Southwest flight that was evacuated in Kentucky, to the student who came close to setting fire to her school, and to the homeowner who vomited from smoke inhalation and was hospitalized.

"I knew Samsung hadn't solved it," the battery pundit Park Chul Wan later told me. "Samsung's corporate culture is too rigid. This is the result of decades of fast following with poor fundamentals." The precise factors that had made Samsung a runaway success—its speed and agility—had now become a devastating liability.

It was nothing short of an existential problem for Vice Chairman Jay Y. Lee, the crown prince of Samsung, heir to the empire and one of his country's most powerful men.

Traversing the United States, Jay was getting live updates from Samsung's "Tower" and D.J.'s task force. But Jay spent much of his time outside the company, traveling and meeting CEOs overseas. As he spoke with American mobile carriers, he got the clear message that the carriers were fed up. Verizon CEO Lowell McAdam and others were urging Jay to kill the phone altogether.

"The executives told Mr. Lee the smartphone was becoming increasingly unsalable," *The Wall Street Journal* reported.

The decision to end the Galaxy Note 7 didn't come ultimately from Samsung, but from its customer-minded carriers. Jay called CEO Koh and ordered him to abort the Note 7.

D.J. sent out a memo to all employees later that day, calling this "one of the toughest challenges we have ever faced."

This wasn't a second recall or a chance for redemption. Samsung had exhausted its credibility with the Galaxy Note 7. If Samsung continued manufacturing the Note 7, few people would trust it. The damage to the name "Samsung" could become permanent.

"When the family spoke, you didn't challenge it," dozens of executives told me. "[Jay Y. Lee] was like a god."

Jay's intervention sealed the phone's fate. Once a symbol of corporate pride, the Galaxy Note 7 had been consumed quite literally in flames.

Little did they know that the Samsung empire itself was teetering.

Four months later, on February 16, 2017, viewers turned on their televisions to the alarming headlines that Jay Lee, the crown prince and heir apparent, was under arrest, accused of bribing Korea's president with tens of millions of dollars. He would await trial from a jail cell.

Whenever Samsung had a crisis, employees and executives sought guidance from Jay's father, Chairman Lee Kun-hee. They memorized his sayings and listened to his speeches. The chairman thrived in crisis. He inspired Samsung's employees to imagine the future possibilities, to embrace a fighting spirit, to overcome limitations and take on the world.

But this time the chairman was nowhere to be seen. In fact, no one had heard from him in public for more than two years. He'd suffered a heart attack in May 2014 and had disappeared into a hospital suite. Some feared the worst. This time, it appeared, the chairman was not going to save Samsung. The chairman, it was rumored, was dead.

2

Shadow of Empire

I FIRST VISITED DAEGU, a city in southeastern South Korea, in the autumn of 2013, wandering the city after its early sunsets and long, frigid evenings. It was a place noisy with cars and people and yet tinged with a forlorn emptiness. I was there to document the early history of Samsung. But my quest proved difficult.

"Where's the home of Lee Byung-chul?" I asked a taxi driver, pointing to the identifying dot on a tourist map.

Despite my hard-earned ability to speak passable Korean, he shrugged and drove off.

Daegu was a trading hub under the Japanese empire, where prosperous Korean traders and businessmen, taking advantage of its link to the nation's early railroad system, built their early fortunes on the back of their colonial master's might. I was surprised at the apathy toward the home of Samsung's founder.

I wandered for at least an hour through the biting wind of a ramshackle industrial neighborhood.

"I'm looking for Lee Byung-chul's home," I would ask random passersby, pointing to my map, which listed no address but only a rough location. The house was unmarked on my GPS; no one seemed to know what I was talking about.

"Lee Byung-chul? Samsung?"

I got more shrugs.

Then I got lucky.

"Ah, Lee Byung-chul, the founder of Samsung! But you are a foreigner. Why are you interested in Samsung?" a frumpy middle-aged woman with a perm, called an *ajumma* in Korean, asked me in English. She identified herself as a professor at a local university.

"It's a global company," I explained. "There's more interest these days."

"It's Korean at its heart," she insisted, "not global." With her droopy eyes, she pointed me toward a street. "It's that one," she said, ushering me to a discreet traditional home with a curved wooden roof, set amid a thicket of low-slung, dilapidated apartments. "My mom knew him."

The gate was locked. No one was home.

"You must understand that when we think of Samsung in Korea, we think of the future," she said. "We don't think about history. The past is shameful. This is why you will find history is buried here in Daegu."

She hailed me a taxi. "I'll tell him to take you to the founding site of the first Samsung company. Samsung Sanghoe [pronounced 'Sanghwaey'] we called it. They sold vegetables. Now they make smartphones."

After a short drive through a gray and forlorn clash of high-rises, I arrived at sunset at a Samsung holy site: a sized-down wooden replica of the first Samsung trading shop, subtly and artfully lit near a drab street corner. Samsung's original shack-like building was deemed a safety hazard and razed in 1997.

I read the museum plaques in this miniature shrine to the corporation.

"I think people are most happy when they know what gives their life purpose," read a quote from Chairman Lee Byung-chul (B.C. Lee) on a plaque. Lee had died in 1987. "I am unshakable in my faith that strengthening the nation through business is the path I must walk."

———

IN 1936 B.C. LEE had an epiphany.

"I had returned home late after gambling with dominoes. The bright moon was seeping into the room through the window. I was 26, by then a father of three. When I saw my children, drenched in moonlight and peacefully sleeping, I felt as though I had awoken from a nightmare," he wrote in his memoir. "I've idled away my time. It's time to set an aim in life."

B.C. Lee lived in a rugged, dusty Japanese colony. The Japanese military, in its bid to unify Asia under its "co-prosperity sphere," was throwing its weight across the continent. The world was on the cusp of an inferno, in the form of World War II, that would claim the lives of millions. B.C. had both financial and patriotic reasons for entering business—even if it meant working with his Japanese overlords.

"Koreans have no solidarity. Therefore, they cannot be expected to manage something like a joint venture," he wrote. "At the time, the Japanese looked down on us in this way. . . . My decision to embark on a joint venture had financial reasons, but there was also an unyielding desire to prove such scorn wrong."

Born in a rustic rural town called Uiryeong, B.C. had an upper-class upbringing; he studied economics at Tokyo's elite Waseda University, a pipeline for Korea's future leaders. He had to drop out of college after a year due to illness, returning home to a country in disarray. Koreans would soon be forced to worship at Japanese shrines in reverence to the Japanese emperor. They were required to speak Japanese in public. Korean laborers and prostitutes were being shipped off in the service of the Japanese military.

After B.C. was forced to shut down a failed rice-trading venture in 1937, he took time off to travel in China and Korea for a year and study the markets. During this time he noticed a gap in the fresh produce market. In March 1938 he opened a vegetable and dried fish shop at the site where I would stand on that chilly autumn day in October 2013. It had nineteen employees, bringing in fresh produce from the countryside and shipping it to China and Manchuria, the industrial base of the Japanese war effort, in what is today northeast China.

He named his vegetable shop "Samsung Sanghoe," the "Three Stars Shop."

"In placing my hopes in this new business venture, I chose the name

B.C. Lee in 1950.

myself. Looking back even now, it was a name that brimmed with drive," he wrote in his memoir. "The 'sam' ['three'] in 'Samsung' symbolizes the big, the many, and the strong, and is a number that our people like the most. 'Sung' ['star'] means to shine brightly, high above, its light pristine for all of eternity. Be big, strong, and eternal."

A year later he expanded to buy a Japanese-founded beer brewery, Chosun, which he later sold off for a small fortune.

B.C. was probably eyeing the success of Mitsubishi ("Three Diamonds"), the Japanese carmaker, given the company's remarkably similar name and early "three diamonds" logo. Longevity, strength, and size were some of the characteristics of Japanese *zaibatsu* ("wealth clan") groups that B.C. admired the most.

The *zaibatsu* were some of the largest and most powerful corporations in the world, as industrial and financial conglomerates run by family dynasties, that transformed Japan from a backwater into a world power in fifty years. Mitsui, a *zaibatsu*, was the world's largest private business during World War II, employing 1 million non-Japanese Asians. Admired for their immense wealth and national prestige, the *zaibatsu* were also reproached for their power. They put politicians into office, set directions for the nation, and profited immensely from World War II.

B.C.'S WORLD WOULD BE upended in August 1945, when American bomber planes dropped two atomic bombs on Hiroshima and Nagasaki, forcing the once-mighty Japanese empire—its industry and population already devastated by months of relentless Allied fire bombings—to surrender. General MacArthur arrived in Tokyo and became the virtual shogun, running the defeated kingdom and attempting to tear down the *zaibatsu*. MacArthur attempted to institute a new order of American-style capitalism and democracy.

"If this concentration of economic power is not torn down and redistributed peacefully . . . there is no slightest doubt that its cleansing will eventually occur through a bloodbath of revolutionary violence," he wrote to a congressman.

Six days after the Japanese surrender, American occupation troops landed in Korea to enforce the Japanese emperor's abandonment there. B.C., as a provincial tycoon, began establishing relationships with the new American military government. Using his father's inheritance and his small fortune from the grocery store and beer brewery, he bought up a local university and a newspaper. He invited Korean managers from the American military to his new business guild, called "Ulyuhoi," a patriotic name referring to Korea's independence from Japan.

"We met every week to seriously discuss business philosophies and the future of our country and society," he wrote. "It was an opportunity to deeply reflect about business and what it means to live as a human."

In 1947 he relocated to the Korean capital, Seoul, setting out to build a nationwide business. In February 1950 he visited downtrodden Tokyo, where he had an epiphany while getting a haircut.

"Japan should have been completely despondent from their defeat in the war, but [the barber] was calmly living on, following his lone generational path," he wrote. "I was impressed by his sharp work ethic."

B.C. admired the Japanese people's resilience. Already a new generation of Japanese entrepreneurs, trained in the military and toiling away in their bomb-damaged offices, were experimenting with new technologies that would bring them global success in the decades to come: Sony, Toyota, and Honda. These new companies were called the *keiretsu,* and they were run not always by families but by shareholding collectives between companies, centered around a private bank, marking a break in Japanese business tradition. They decided to embrace the *keiretsu* ("group") corporate structure because the U.S. military government banned holding companies, in its attempt to remove the *zaibatsu* families.

South Korea followed with its own ban on holding companies. But its business leaders were determined to keep the *zaibatsu* practice of top-down family rule. So they embraced the cross-shareholding practices similar to Japan's newer *keiretsu* companies, and found loopholes to pass those cross-shareholdings to their children, through charitable donations and mergers within their business empires.

Two months after his tour of more than fifty factory and business sites in Japan, B.C. came home to South Korea to shocking news.

On June 25, 1950, North Korean communist forces had crossed the thirty-eighth parallel and invaded the South.

"Past noon and until evening," B.C. wrote, "trucks filled with soldiers continuously raced through Seoul, headed north. Citizens were clapping for them, wishing them victory and good fortune. What is to become of the business, or beyond that, my life?"

After staying up all night with his family, B.C. heard a thunderous sound. He looked outside to see an unfamiliar tank displaying the North Korean flag on its turret. The enemy, he realized, had entered Seoul.

Four days later, party officials knocked on the door, demanding to inspect his assets and test his ideology. Capitalists were being accused of collaborating with the Americans and Japanese. They were brought before "People's Courts" in public squares, where they were executed by firing squad. North Korean soldiers and local opportunists broke into Samsung's warehouse and looted the company of its inventory.

Two weeks later B.C. saw a powerful communist chief riding around in a late-model American-made Chevrolet. "I recognized that it was my car," he wrote. "I felt an anger that could not be put into words."

IN SEPTEMBER 1950 UNITED Nations forces arrived and recaptured Seoul. Three months later, B.C. decided to sell off Samsung's remaining assets—whatever hadn't been looted—and use the proceeds to buy five trucks to evacuate Samsung employees and their families, as the communist forces advanced on Seoul for a second time.

"I loaded up company employees and their families, and we left Seoul."

They joined the deluge of refugees headed south. He was en route to his family home of Daegu.

The North Korean army, reinforced by China's million-man armed forces, retook Seoul. B.C. fled to the country's southern tip at the port town of Busan—the last remaining South Korean stronghold.

In rebuilding his company, B.C. learned the art of reading people as a means of survival. He established a human resources–centered approach modeled on Japanese corporate practices. Talented people were in short supply during the war. Managerial aptitude became a prized Korean asset.

"[The Japanese] steadfastly valued loyalty, and prioritized the cosmic self over the individual and the public over the private," he wrote. "The Japanese capacity for unity and diligent work comes from that patriotism, which values a greater public cause."

Since B.C. prized lifelong loyalty, he believed in a careful and cautious approach to each new hire. He personally sat in on almost every employee interview—reportedly about 100,000 in his tenure. He hired professional physiognomists, or face readers, to help him interpret the structure of a candidate's eyes, nose, lips, ears, and face to get a sense of who they were.

"Be prudent in hiring someone," he wrote. "But once you've hired them, be bold in entrusting them with tasks."

His philosophy was similar to that of the Japanese *zaibatsu*, where the ruling family sets the vision and the executives execute it without micromanaging employees.

While American companies were keen to snap up technical special-ists and give them short-term projects rewarded with short-term in-centives, Samsung (like many East Asian companies) became obsessed with cultivating lifetime generalists—creating what became known as "Samsung Men"—rather than short-term employees, future managers shuffled into new roles every few years by a powerful HR department. Company was family, and family was company.

A "Samsung Man" commanded respect in Korea as a genteel and hardworking husband.

"Samsung treats you the best. Thus, you are the best," a company motto declared, instilling a sense of corporate pride.

Taking a cue from the Japanese, B.C. formed a powerful human resources department at Samsung, giving it a privileged position among his Samsung Group affiliates.

"HR actually is the leading department, the leading function," a senior human resources executive at Samsung told me. "The role of HR is very different from, let's say, Western companies." The role of the department similar to HR in East Asia, even before industrial times and Samsung, "was really to help the emperors to select and evaluate and appoint government officials, as well as to train and inculcate Ko-rean ideals in its recruits."

"They said *nunchi* is important to succeed in Samsung," said former product manager Scott Seungkyu Yoon, who was trained in 2010. *Nun-chi* is the Korean art of gauging others' moods, even if the other per-son says nothing. Unlike America's talkative, more straightforward business culture, B.C. made Samsung a place of terse and sparse com-munications. You had to feel out your boss's emotions. To survive, you had to master *nunchi*.

THE KOREAN WAR CAME to a halt with a cease-fire in 1953. Though the fighting was finished, the war never formally ended, since neither side signed a peace treaty. More than 36,000 American servicemen and 5 million Koreans were killed, making it one of the deadliest civil wars of the twentieth century, on par with the Vietnam War.

At the time, the West did not view South Korea as a country with a great deal of promise. Its GDP was about the size of Sudan's. When the South Korean government attempted economic planning, the International Monetary Fund derided its thinking as absurd. It had virtually no natural resources. Its communist cousin, North Korea, was faring better, itching for the chance to conquer its vulnerable cousin.

Snobbish stereotypes toward Koreans were especially prevalent in Japan, where the country's now-liberated Korean slave laborers were denied citizenship.

"The Koreans . . . are the cruellest, most ruthless people in the world," James Bond's nemesis Goldfinger says in the Ian Fleming novel (but not the film adaptation), explaining why he hired Korean henchman Oddjob, whose famous top hat could slice through a marble statue.

B.C. built his first fortune using political savvy, and by playing into a sudden postwar boom. He established political links to South Korea's first president, Syngman Rhee, and was awarded a crucial but

An aged Korean woman pauses during her search for salvageable materials in Seoul, South Korea, 1950. For much of its history, South Korea was a poor nation with low prospects for development, devastated by the Korean War of 1950 to 1953.

difficult-to-get government license as a foreign currency recipient, allowing him to embrace an economic growth model called import substitution, importing raw materials like wool and transforming them into finished products like clothing.

In the 1950s, people began using the word *chaebol,* or "wealth clan," to describe family empires like Samsung and Hyundai. The Chinese characters for the Korean word are the same as those used in the Japanese word *zaibatsu,* the family conglomerates that ruled over Japanese industry in the race to World War II.

As his wealth grew, B.C. set up a sugar refinery in 1953 and a wool-spinning plant the next year with Japanese and West German help. He used the profits to acquire shares in a bank (in 1957), an insurance company, a department store, and a Confucian university, Sungkyunkwan, founded in 1398. By the end of the decade, B.C. was reputed to be the nation's richest man. But he was also a symbol of South Korean corruption.

People called him the man with the "golden touch," and "Mr. All-Wool" for his luxurious wool coats. But for all the unflattering criticism of the company, Samsung's business projects were symbols of the postwar reconstruction. They boosted national morale in times of dire need.

I met with a boisterous Jersey-ite named Tom Casey, who had lived in Korea since 1968, running a rambunctious bar called the Sportsmen's Club near the base for servicemen and local elites and celebrities. His notoriety once earned him the ire of U.S. senators upset over a gambling operation for American soldiers overseas. He put me in touch with Henry Cho.

Henry was a grandson of B.C. Lee, as well as honorary chairman at a chemical company called Hansol. Henry was one of the scions of the Samsung empire. In his younger years, he'd managed the chemical business, one of the five spin-off companies from the original Samsung empire that still exist today. It took months to arrange a meeting with Henry. At first, my request was flatly rejected. His people didn't think the time was right or that it was wise to speak with a journalist.

"The family, they protect each other," Tom told me. But I persevered and Henry finally agreed to meet.

I first met Henry over Chinese dim sum in a five-star hotel. We met a second time at Seoul's Grand Hyatt hotel overlooking the cityscape. They were the sort of places where *chaebol* princes dine. Henry brought two of his assistants to our first meeting, perhaps to monitor my questions.

In both our meetings, Henry made clear that his family's crest of honor, called a *kahoon,* was of the upmost importance to him.

"Do you know what that is?" he asked of his family's motto.

I didn't, I admitted.

"First, serving the nation through business. Second, people and talent come first. Third, the pursuit of the reasonable."

He repeated his family's three values during both of our chats. Then he explained what made Samsung's culture unique—a devotion to detail that came from his grandfather.

"One of the good habits I picked up from my grandfather is taking notes," Henry told me. "I take notes. Every detail."

"Do you tell your employees to take notes?" I asked.

"I don't tell. They're supposed to notice. They're supposed to notice what I do, what I think, how I think. They pick it up," he explained of his family's style of imparting their vision in their companies, while letting their "Samsung Men" manage the day-to-day business.

Detail, precision, and cautious record keeping were B.C.'s trademarks as well.

"At meals he would tear his napkin in half and save the other half for later," Henry told me. Like the heads of many Korean companies, he wanted to discourage waste, a quality that was of particular importance since Korea was relatively devoid of natural resources. Meetings and decisions at Samsung were documented and recorded with an obsessive flair. Employees were required to write reports on each meeting within incredibly specific parameters, from margins to word count to length.

B.C.'s face was often stone flat, Henry told me; he rarely smiled or showed emotion. And he chose his words carefully.

"He was so cool, you know?" he told me. "I went in there as a kid, and he never even blinked."

B.C. loved golf and Chinese calligraphy, activities that contributed to his almost meditative business philosophy.

"And his political connections?" I asked.

"As far as politics was concerned," he said of B.C.'s powerful allies in government, "keep them 'not too far, not too close.'"

"B.C. arrives at his downtown Seoul office at 9 A.M. sharp, ready to meet with his executives in exhaustive planning sessions," *Time* reported. "Twice a week he breaks the routine and plays golf. Lee returns to his palace, pottery and peacocks by 5 P.M. He usually dines alone, then plots new ways to increase his wealth. Preferring the glitter of Seoul, his wife [Park Du-eul], eight children and 20 grandchildren live apart from Korea's richest man."

B.C. lived in a palatial estate an hour south of Seoul in Yongin, where a muster of peacocks roamed, and spent a lot of time in Japan with his second wife, Kurata Michiko, a hairdresser, and the son and daughter he had with her.

But the tycoon had a darker side. When a government trading official named Daniel Lee recommended the government stop earmarking American aid money for Samsung's sugar business—he thought American assistance was better put to buying fertilizer and wheat, rather than sugar, a luxury for the wealthy—his partner was abruptly accused of small-scale corruption and jailed. "Critics," Daniel Lee wrote, "called us puppies who did not fear a tiger."

The Republic of Samsung, for both good and bad, was on the ascendant. Henry Cho recounted, "They treated me like a god."

3

Dynasty Ascendant

"I THINK THEY [SAMSUNG] are mimicking exactly the old ways," Jisoo Lee, a corporate governance lawyer who'd tussled with Samsung at shareholder meetings, told me over coffee after I'd first talked with Henry. He gave me a comparison that Koreans often make.

"If you look at the first king of [Korea's] Yi dynasty, Taejo, he had eight sons, and there was this battle between the fifth and the youngest son," he said. "The fifth son later became the third king of the Yi dynasty. But he had to fight against his father's decision to make the youngest son the crown prince."

He then made a comparison to North Korea. After the dictator Kim Jong Un rose to power when his father died in December 2011, he conducted purges. He ordered the execution of his prominent uncle. Later, his half-brother was assassinated with VX, a deadly nerve agent used in chemical warfare, as he was boarding an airplane in Malaysia.

A dynastic battle within Samsung started growing as Samsung's rising corporate kings turned to another aristocratic tool of alliance: marriage.

B.C. had five daughters and three sons, and almost all of them married into influential families in the 1950s and 1960s. His oldest son married the daughter of an insurance executive and provincial governor; his second son married the daughter of a Japanese businessman; his third son, the future Chairman Lee Kun-hee, married a woman named Hong Ra-hee, who was the daughter of a presidential cabinet member named Hong Jin-ki.

The marriage of Lee Kun-hee to Hong Jin-ki's daughter established a blood union between the families. They confided in each other for decades, and the relationship gave Samsung a line to the government's major power players.

Hong Jin-ki was a colorful lawyer with a love for poetry and words. In 1965, B.C. founded Samsung's newspaper, the *JoongAng Ilbo* ("The Central Times") and hired Hong to run it. *JoongAng* would become one of South Korea's three major news outlets, modeled on Japan's daily papers. The newspaper was meant to represent corporate voices in addition to reporting the news.

For Samsung, the newspaper was a way of defending itself against political attack. As B.C. told *Time* magazine, "Mass communications are the best way to prevent bad politics."

As head of the company newspaper, Hong was able to control B.C. "He was so good at holding my grandfather by the balls," Henry Cho said.

"How did he pull that off?" I asked.

"By being next to him," Henry said. "Giving [him] lots of interesting information about politics and what's going on in the world."

Hong, in other words, served as the court whisperer at Samsung.

IN THE MORNING DARKNESS of May 16, 1961, South Koreans in Seoul woke up once again to the sound of tanks rolling through the streets.

"We shall rise up against the government to save the country," announced General Park Chung-hee to his soldiers before the coup d'état. "We can accomplish our goals without bloodshed. Let us join in this Revolutionary Army to save the country!"

"That day . . . around 7:00 A.M.," wrote B.C. Lee, who was in Japan, "I set out to enjoy some early-morning golf, after a long time. As I climbed into the car at the hotel entrance, my Japanese driver, Ku-

General Park Chung-hee, center, on the day of the coup d'état, May 16, 1961.

wabara, said to me: 'Have you heard the news that there's been a military revolution in Korea?'"

As Samsung's founder and a reviled symbol of illicit wealth in South Korea, B.C. was on the interrogation list. Two Korean government agents showed up at B.C.'s Japanese hotel. They left a note telling him to return to South Korea—or else.

So B.C. flew back to Seoul, where "a fresh summer rain was pouring out of the pitch-black night sky." A young man picked him up in a Jeep.

After a night spent in his hotel suite—he felt lucky he wasn't joining his business colleagues in jail cells—he went before the "Supreme Council for National Reconstruction." B.C. did not know what awaited him.

"Passing the secretary's office, I entered a large room. . . . From the other side, several military personnel and a man wearing black glasses who gave off an air of integrity walked over. Immediately I recognized the man in the black glasses as Vice Chairman Park Chung-hee." He was the new dictator of South Korea and the leader of the coup d'état.

"The room was filled with a tense and heavy atmosphere," B.C. wrote.

Short, stern, and almost always wearing either a suit or fatigues and aviator sunglasses, General Park had a small frame. A lifetime soldier who'd flirted with communism as a youngster, he was not a man to question.

"You can say anything, so talk without reservations," Vice Chairman Park, who also held the title of General in the army, told B.C.

B.C. made his case and pleaded for leniency.

"The people will not accept this solution," President Park shot back.

The dictator had a grand bargain in mind. Samsung, he believed, was chaired by a white-collar criminal. Prison time, however, would be pointless, stripping Samsung of its use to the nation. Instead, B.C. would have to give up large portions of his three banks to the state and cough up $4,400,000 in unpaid taxes and penalties. Samsung would be expected to fully cooperate with the General's plans to build a new, wealthy, powerful nation. If B.C. refused to cooperate, a prison cell awaited.

The son of a peasant, President Park was an unconventional thinker. Thumbing his nose at Western wisdom, he closed the markets, imposed martial order, and coerced businesses to act under his decrees. A good deal of Park's worldview came from his training by the Japanese Imperial Army. Stationed in Manchuria, a center of wartime industry, he had bowed before a picture of Hitler every morning, astonishing other students, and watched the unfettered rise of the Japanese *zaibatsu* groups and their garish excesses that hurt the national interest.

"Their old, dated ways of managing the business do little to inspire progress. They do everything to concentrate and keep their properties within the family," the president once remarked of South Korea's *chaebol*. "This is standing in the way of the healthy development of companies."

"What is urgently required of us now," he said in a June 1965 speech, "is not to be envious of America's prosperity, or fearful of Japan's expansion, but rather to emulate the mental attitude and drive that made such things possible."

Under his watch, Korea retained parts of the prewar and postwar Japanese economic models while becoming more disciplined, militarized, state controlled, and bent on exporting high volumes of products at breakneck speeds. The Japanese miracle, in other words, was to be taken to an extreme in South Korea.

General Park had a powerful set of tools and levers to make this happen. He fired up South Koreans with the belief that they derived strength from a pure and unadulterated bloodline called the *minjok,* an idea first introduced to Koreans by their Japanese overlords earlier in the century.

The regime seized private banks owned by Samsung, Hyundai, and the like and barred them from future bank ownership. The banks gave too much financial power to the industrialists and were prone to creating financial bubbles and backdoor deals. President Park forced the barons to manufacture their way into his good graces. State-led consortiums pooled together capital and technology to help Korea race ahead in steel, shipbuilding, and petrochemicals. His regime set tough export quotas, pitting company against company. If one fell behind, its state loans would be cut, and the *chaebol* would collapse.

And finally there was the torrent of foreign money. American donor aid for military bases in South Korea and huge payments from the United States in exchange for Korean troops being sent to fight in Vietnam, as well as Japanese payments totaling $800 million—reparations intended to restore diplomatic ties—were put to use in building a sophisticated network of roads, infrastructure, and ports that fired up industry.

The chairman of the steelmaker Posco stood before employees in June of 1965 and told them, "This is a steel factory built with Japanese payments, the price of which was the blood of our ancestors. If we fail it will be an indelible sin against history and the South Korean people."

Chung Ju-yung, founder of Hyundai, was even more blunt, hurling an ashtray and slapping his senior executives if they failed to reach the rigorous financial goals the company had set.

B.C. Lee became the first chairman of the Federation of Korean Industries, a council of business elders who convened to align their goals with the regime's and to protect the *chaebol* groups from heavy-handed intervention.

Through this lobbying group, President Park prodded Samsung to build a fertilizer plant, a cornerstone of national development and a guarantor of the regime's legitimacy.

"The government will offer its full support," President Park told B.C.

In 1966, after eighteen months of construction and years of securing the needed Japanese capital, B.C. Lee's new plant was set to open.

"In the coastal city of Ulsan last week, old and new Korea came into symbolic confrontation," reported *Time*. "The spring mists filtering across the landscape were mixed for the first time with ammonia clouds, and Korean farmers wearing traditional costumes stood side by side with businessmen and government officials in trim, Western-style business suits. All had gathered for the dedication of the Korea Fertilizer Co.'s new urea plant, which, with an annual capacity of 330,000 tons of fertilizer, will be one of the world's largest."

But during construction, chemicals brought into the country for use in making fertilizer were sold to a saccharin-processing firm at a $40,000 profit.

It caused a huge scandal, as corruption was seen as undermining the country's efforts at economic development.

"Eat this saccharin!" an opposition lawmaker shouted as he threw a can of human excrement on the parliament floor, halting the day's proceedings.

Lee Chang-hee, B.C.'s second son and a manager at Samsung, was responsible for the sale of the chemicals. He was sentenced to five years in prison for corruption. B.C. was forced to step down and surrendered 51 percent of his fertilizer plant to the government in an attempt to get leniency for his son.

Following Korean tradition, he appointed his oldest son, Lee Maeng-hee, to head Samsung in his place.

But that son, Henry Cho told me, was a "troublemaker."

He was rumored to be violent and dissolute, unfit to lead the company. Maeng-hee later admitted to a South Korean journalist that he made the revered court whisperer Hong come into his office and kneel on the floor before him, a huge loss of face. B.C., unhappy with his son's erratic management style, returned to Samsung and eventually told him to resign.

When Chang-hee was released from prison, he had hoped for a warm welcome from his father.

"But his father refused to let him take over the company. He was a very good entrepreneur," Henry told me, "but according to my grandfather [B.C. Lee], he was [too] much into details and small in scale [in his thinking]."

So Chang-hee sent an anonymous tip to President Park Chung-hee detailing the real estate and other assets that his father was holding, prompting a grand schism between father and son. In the wake of that schism, the second son faded into obscurity and died of leukemia in 1991.

IN 1968, B.C. FORMALLY reinstated himself as chairman of Samsung. He was determined to pass the company to his youngest son, Lee Kun-hee, eventually.

"In the future, Kun-hee will be leading Samsung," he told the family in September 1976, according to his son Maeng-hee.

"I cannot forget the sudden shock I felt after hearing my father's words," Maeng-hee wrote in his memoir. "By that time, there was already a schism between my father and me, but I had still believed that he would someday pass the reins of Samsung to me."

Dejected and defeated, Maeng-hee later went into virtual exile in a bucolic fishing village, where the local fishermen called him by the nickname "The Chairman." Walking with a slight limp, a complication of a hereditary neurological disorder, he told those he met that his father had cut him off financially. At one point, he didn't even have enough money for bus fare.

The fallen prince became increasingly paranoid and was rumored to be violent and mentally ill.

"Actually, I was the one who had to flee, chased by my father swinging a golf club around like crazy," Maeng-hee told a visiting journalist. He was convinced he was being followed, that people within Samsung were plotting to have him institutionalized or kidnapped.

An unofficial South Korean biographer called him "Prince Sado of Samsung," recalling a historical tragedy of dynastic strife run amok in South Korea during the eighteenth century. Prince Sado, a twenty-seven-year-old feudal heir accused of being mentally ill, was locked up in a rice chest as punishment by order of his father; he died there from starvation in 1762.

WITH THE FERTILIZER BUSINESS gutted, B.C. turned Samsung to a much riskier, faster-paced endeavor: electronics. South Korea, on the precipice of a bonanza, defied a chorus of Western critics who believed such a rags-to-riches story in the global marketplace was impossible.

4

March of the Samsung Men

IN 1974 SAMSUNG CEO Jin-ku (J.K.) Kang got a call from a California entrepreneur named Joseph Sudduth, who told him about the financial troubles at Korea Semiconductor, a new joint venture set up by Sudduth and a Korean businessman.

The U.S. State Department was watching Korea Semiconductor's precarious financial situation closely, worried about the fallout of a default, since projects like this had loans from the U.S. government. They were believed to be in the national interests of both the United States and South Korea, a means to kick-start the sluggish Korean economy. With the OPEC oil embargo under way, it was clear that the United States and Korea needed to build new value-added industries that depended on highly skilled workers and not on natural resources like petroleum. Semiconductors were a good bet. They were an essential technology behind the Apollo space shuttles sent to the moon and in the laser-guided missiles deployed in the Vietnam War.

J.K. saw an opportunity. "I had been thinking that an electronics company without semiconductors was like a car without an engine," he wrote. With approval from B.C., Samsung bought Sudduth's half of the company.

Samsung managers warned against the acquisition. The chipsets were expensive to produce and incredibly risky. The semiconductor industry, notoriously unstable, required huge investments and years to recoup one's money, and it would force Samsung to grapple with chaotic price fluctuations. Success in semiconductors required a grand vision and long-term direction.

The industry leaders, especially the Japanese, had a remarkable ability to compete with, and even best, Intel and Fairchild Semiconductor, thanks to long-term government backing. Samsung would be a distant third mover in the market.

"The company made semiconductors only in name," B.C.'s third son, Lee Kun-hee, wrote later. "The reality was that they were essentially able to make little more than transistors," the basic parts used in microprocessors for calculators and computers, but with little added value of their own.

Samsung executives stepped in to reassure visiting American diplomats of the soundness of the operation. And the diplomats were convinced. "Everything appears well-organized and under control," a diplomat cabled back to Washington.

Unable to pull Korea Semiconductor out of the red, the remaining partner offered to sell his stake. Lee Kun-hee, convinced of the long-term importance of chipsets, agreed to finance part of the deal with his own money.

The shaky new partnership was renamed Samsung Semiconductor.

But the timing was ominous for a new venture. Two years later, in 1979, President Park's head of intelligence shot and killed the dictator after the two had a heated argument over dinner. In the resulting chaos, the military massacred hundreds of prodemocracy protesters in the city of Gwangju in May 1980. And then a brutish general named Chun Doo-hwan won a rigged election and seized B.C.'s broadcasting station, TBC.

B.C. looked for ways for Samsung to bounce back from the political turmoil. Hyundai founder Chung Ju-yung had won favor with the new government as a bastion of heavy industry and had received an honorary doctorate at George Washington University. B.C., eager to best him, lobbied successfully through a South Korean politician to receive an honorary doctorate at Boston University.

Visiting the United States in 1982 to receive his honorary degree, B.C. toured the semiconductor assembly lines of IBM, GE, and Hewlett-Packard. Their ingenuity dazzled him. And a grim realization hit home.

"We are too late," he told his son.

B.C. was prepared to abandon semiconductors. But under increasing pressure from Hyundai, which had begun manufacturing semiconductors, too, he decided Samsung needed to fight back. On the plane home from Boston, his son tried to persuade him that semiconductors were the solution.

After much deliberation, B.C. was convinced.

IN FEBRUARY 1983, B.C. woke up early in his suite at the Hotel Okura in Tokyo to call his right-hand man, Hong Jin-ki. He told him to publish a statement called the Tokyo Declaration in the *JoongAng* newspaper.

"Our nation has a large population in a small territory," he said, "three-fourths of which is covered with mountains but almost completely lacking in natural resources like oil or uranium.

"Thankfully, we have high levels of education and a rich supply of diligent and hardworking [people], which enabled us to see rapid economic growth through mass-exporting cheap goods. But with countries around the world experiencing recessions, and with the rise of trade protectionism, the export-based method of building national strength has reached its limit.

"We hope to advance into the semiconductor industry," he declared, "on the strength of our people's great mental fortitude and creativity."

AT THE INSISTENCE OF the chairman, Samsung began a series of physical and mental conditioning exercises for its managers.

"The people I'm going to call are going on the march," a manager announced to the room one morning.

Kim Nam-yoon's face lit up as he told me about the hike over Korean barbecue one night.

Nam-yoon, a semiconductor engineer, along with a hundred others, began his hike that night near Samsung's Everland theme park, traversing the mountains around Yongin—a distance of sixty-four kilometers (about forty miles)—through the day and overnight.

The hike was designed to test the team's mental toughness as they prepared to race forward making a 64K DRAM chipset, an early semiconductor used in calculators.

"It was the dead of winter," he said. A former infantryman in one of his country's toughest military units, Nam-yoon lined up in formation at the foot of the mountain.

As they paced through the frigid January night, a unit commander ambushed them several times, forcing them to solve games and puzzles on the spot.

"Find a living thing and bring it to me," the unit leader told the engineers. But there were almost no animals to be seen in the brown, barren mountainscape.

The men searched the area and turned up nothing. So Nam-yoon and his team pulled from their lunch box a dried anchovy, a common Korean side dish. They put it in a cup of water and handed it to their commander.

"It was alive a moment ago when we first caught it," they told him.

Clearly they had come up with a creative solution, which was the point of the test.

"All right, pass!" the manager said, laughing.

As the sun rose, the tired platoon descended the mountain into Samsung's new assembly line. Each of the hikers wrote a declaration and signed it.

"If Samsung does it, it'll be successful," Nam-yoon wrote, signing his name to the pledge.

At 7:00 A.M. the engineers clocked in, showered, and put on fresh clothes. Without having slept the night before, they worked until eleven o'clock that night.

"We treated it like any other workday," he said.

It was a historic moment. Samsung christened its first semiconductor fabrication plant in 1983. The company had built it in only six months, rather than the industry standard of three years. "A visitor to the Samsung plant must navigate the rough, narrow streets in the village of Giheung. Traffic is often clotted, as farmers driving rice cultivators vie for space with trucks carrying multimillion-dollar shipments of semiconductors," wrote journalist Mark Clifford, an early visitor to the site. It wasn't much to look at.

What drove them?

"The thought of going bankrupt," said former Samsung Electronics CEO Lee Yoon-woo, back then a manager at the fabrication plant.

The Samsung engineers were granted limited access to Micron, an Idaho-based chip maker, for technology training. Micron was in financial trouble and was willing to license its technologies for a fee. But the American engineers banned the Korean engineers from touching the computers or going in certain sensitive rooms. So the Korean engineers visited with eyes open, memorizing the diagrams they saw. They returned to the hotel room each night to re-create the diagrams, piecing together the incredibly complicated semiconductor charts from memory.

They gave gifts of pottery to Japanese executives at a joint Japanese-Taiwanese manufacturer called PSC, now defunct, in exchange for the opportunity to buy PSC's parts and chipset-making equipment, something the Japanese had a near monopoly on.

"Having had a lot of experience in Japan, I threw myself into traveling back and forth from our semiconductor plant to Japan in order to secure the technology," Lee Kun-hee, now promoted to vice chairman, wrote in his memoir. "Traveling to Japan to meet semiconductor experts almost every week, I tried to learn from them anything that might be of use. At the time, I would often secretly bring in Japanese experts on Saturday and have them teach my engineers overnight, before sending them back on Sunday."

"The Japanese brought this precision that we didn't have," Nam-yoon said. "It was a cultural difference."

B.C. knew Samsung couldn't afford to go on losing money forever. Semiconductors were expensive to manufacture, and government support was not guaranteed indefinitely. Japanese firms, meanwhile, were dumping their chipsets on the market, wreaking losses on Samsung.

B.C. diverted the profits from other Samsung businesses to prop up

the semiconductor projects for the time being. It would have been impossible to do in a publicly traded American company, forced to focus
on short-term profits. B.C., now in his seventies and diagnosed with
lung cancer, felt an increasing sense of urgency.

"He predicted that when he died, Samsung would just collapse,"
Henry Cho told me.

In December 1987, South Korea's Chun Doo-hwan stepped down
as the country's leader following mass protests, opening the way for
the country's first democratic elections. With the 1988 Seoul Olympics
opening in two months, the economy was bustling, fueling a patriotic
fervor that everyone hoped would bolster Samsung's standing in the
world.

A SAMSUNG ADVERTISEMENT FROM the period, created to celebrate
Samsung's fiftieth anniversary, showed two children gazing up at a picture of the solar system. The planets had been replaced with images of
a personal computer, a semiconductor, a DNA double helix, and an
orbiting satellite.

"Fifty years with our *minjok*," the ad declared, using the patriotic
word for the Korean race.

Samsung's success wasn't about sales, operations, economic policies, and bottom lines. It was a story of patriotism and spirit. Samsung
tapped into the feelings, emotion, and sense of belonging that South
Koreans experienced. It understood the human need to be a part of
something great, something bigger than oneself; it offered Koreans the
promise of glory.

In the ad, Samsung declared its mission: "To become a leading corporation to the age of humanity."

5

The Confucian and the Hippie

IN 1979 IRA MAGAZINER, a future adviser to President Bill Clinton, visited Samsung's Suwon campus as a British government consultant to consider the potential of a larger market for Samsung's black-and-white televisions. He was not impressed.

"Samsung's research lab"—with a bare concrete floor and people wheeling products around by hand—"reminded me of a dilapidated high school classroom," he wrote. He met a young chief engineer who was a recent graduate of an American university.

"I asked him about Samsung's color-television strategy, telling him I presumed the company planned to buy parts from overseas, [performing the] assembly in Korea.

"Not at all, he said. They were going to make everything themselves—even the color picture tube. They'd already identified the best foreign models, he said, and signed agreements for technical assistance." Soon, the engineer predicted, Samsung would be exporting televisions worldwide.

"I wasn't convinced," Magaziner wrote. Maybe in ten or fifteen years, he thought. But Samsung was starting out generations behind American and British technology.

"But the work going on there intrigued me," he wrote. "They'd gathered color televisions from every major company in the world—RCA, GE, Hitachi—and were using them to design a model of their own."

Magaziner returned five years later as a consultant for GE. And he was surprised to find that the South Koreans had done everything they promised. They were making the key parts for their TVs—from the tubes down to the glass—in a joint venture with Corning.

The plant was "as automated as any TV plant I'd seen in America," he wrote. As he toured the factory floors, he began to understand the mindset that fueled this minor miracle.

A woman who attached the serial numbers to microwaves said she treated her task as an exercise in discipline and integrity.

"They admit it's the same simple function, hour after hour, but neither thinks the days are dull," observed Magaziner of the workers he watched. One woman double-checked her work even after the inspector came around.

"I put my spirit, my soul into this product," she said.

The future of technology, people were realizing, was not West but East. And inconsequential South Korea was already inching toward the global market. The first PC had been released in America in 1981. Before long a torrent of early mobile phones and audio players gave way to the sudden dominance of Japanese and later South Korean corporations. By 1995, there would be no American-owned television manufacturer left.

IN NOVEMBER 1983 TWENTY-EIGHT-YEAR-OLD Steve Jobs arrived in South Korea. He was greeted by smokestacks and factory workers who wore Samsung company uniforms and lapel pins, employees who would not hesitate to salute their chairman if he so desired.

Jobs didn't visit South Korea out of romance or adventure—the motivations that had brought him to Japan and India. He was on a bold and prescient mission: to build a tablet computer, a full twenty-seven years before the introduction of the iPad, for his start-up company, Apple Computer.

"Steve knew the future was mobile. He was looking to build a Dynabook," said his colleague Jay Elliot, who accompanied him on the trip, as well as on subsequent Samsung visits. "He needed a supplier of memory and displays."

Skeptics were calling the Dynabook tablet concept, created by Xerox, a distant and fanciful idea. It resembled a prop in the movie *2001: A Space Odyssey*. It was thought to be too expensive to get to market and too small a niche in the marketplace. A decade earlier, Xerox's elite PARC laboratory had developed a prototype—"a personal computer for children of all ages"—but found the technology of the day far too primitive to produce it.

But that didn't deter Jobs. The personal computers and phones of the future, he believed, would need to be portable. If Apple didn't keep moving forward into this space, then the colossal technology giant next door, IBM—already unleashing an onslaught of PCs in an attempt to put Apple out of business—would gobble him up. So where could he get the parts he needed?

Jobs had traveled to Japan, where he had met Sony founder Akio Morita. He was keen to adapt Sony's management practices and sleek design ethos. But he felt Japan's neighbor, South Korea, which most of the industry saw as sloppy and backward, was starting to show promise, according to Elliot.

About an hour's drive south of Seoul, Jobs disembarked at the grimy, industrial entrepôt of Suwon. An entourage of solemn, bowing Samsung employees greeted him. At the time, Samsung was still an obscure family-run business manufacturing cheap microwaves for GE, as far as Jobs knew. It called itself Samsung Electronics, but the company's nickname among Western expatriates was "Sam-suck."

As Jobs entered the Samsung building, he was greeted by B.C. Lee. Leading Jobs to a majestic conference room with large regal chairs and Korean furnishings, B.C. unveiled his bold experiment: He intended to position Samsung Electronics, which at the time was easily a generation behind its American and Japanese rivals, as a massive supplier of the world's computer chips. With the PC revolution under way, the chairman wanted Samsung to be an engine of it, a driver of it, bringing global prestige and revenues to his country and company.

B.C. knew that the clock was ticking and that he needed to secure a path forward for Samsung. Jobs needed memory chips. Samsung was just getting under way in memory chips. But even at this stage, Samsung began supplying Apple some of the displays and components it needed for its PCs. B.C. Lee, the elderly Confucian who loved Chinese calligraphy and the Korean art of face reading, got on well with the mercurial, talkative, and occasionally obnoxious kid from California.

"Steve was boasting. He talked a lot. He wouldn't stop talking," said Elliot.

Jobs prattled on about his brainchild the Macintosh, which was slated for release the following year. B.C. Lee laid out some of his ideas for Samsung.

"That's gonna work!" Steve would exclaim, hearing of one idea, according to Elliot. That would be followed by "No, that's not gonna work!"

In Korea, a twenty-eight-year-old chattering away to his elder—especially the chairman of the country's representative company—would be a horrific insult by anyone else, worthy of banishment. But B.C. recognized Jobs's brilliance and overlooked his social solecisms.

"That was Steve. He threw his arms around, waving and pointing at everybody, speaking his mind," recalled Elliot. "I told him later, 'You gave away a few million dollars' worth of ideas!' He didn't care."

After Jobs departed, the Samsung founder, in his solemn, soft-spoken manner and with a few carefully chosen words, declared to his assistants: "Jobs is the figure who can stand against IBM."

Two years later, Jobs was fired by Apple's board, losing out in a power struggle with CEO John Sculley. The tablet, in its early form, fell into the dustbin of history, at least for the time being. And any plan for an Apple-Samsung alliance was scuttled.

It would be more than two decades later that Samsung and Apple, the component supplier and the computer manufacturer, would go to war.

ON AUGUST 4, 2010, a group of Apple executives arrived at Samsung's building in Seoul's glamorous Gangnam district. It had opened a few years earlier in what was the Beverly Hills of South Korea. All over Seoul, smokestacks had been replaced with glossy blue high-rises; cutting-edge technologies were on display everywhere. The previous spring, Samsung had released the Galaxy S, which some in the media had dubbed the "iPhone killer."

Jobs, back at Apple, was livid. He believed Samsung had ripped off the iPhone with a similar roster of icons, designs, and even packaging.

A group of Samsung executives, led by vice president Dr. Seungho Ahn, joined the Apple executives in a conference room. Chip Lutton, an Apple lawyer, launched into a presentation titled "Samsung's Use of Apple Patents in Smartphones."

When he finished, the Korean executives fell quiet.

"Galaxy copied the iPhone," Lutton said.

"What do you mean, copied?" Ahn asked.

"Exactly what I said," Lutton maintained. "You copied the iPhone. The similarities are completely beyond the possibility of coincidence."

"How dare you accuse us of that!" Ahn retorted. "We've been building cellphones forever. We have our own patents, and Apple is probably violating some of those." The meeting would mark the first salvo in one of the costliest, most sweeping lawsuits and business battles in history.

$$6$$

The Fifth Horseman

TWO MONTHS LATER, I got a call from an editor at *Fast Company* magazine who needed a story. "We're looking to do a big piece on Samsung," he said. "We have an invitation to visit the campus in Suwon. How about it?"

Like Jobs two decades before me, I disembarked from the black company car on a chilly autumn day and was greeted by similar platoons of bowing Samsung executives in suits. Bucking South Korean corporate tradition, I began to ask pointed questions of executives far older than me. We feasted on the same kinds of Korean food and had passionate conversations about the future of the tech industry and Samsung's role in it. The general personality of the company—its admiration of successful ideas and its drive to chase, learn, and adopt them—hadn't changed.

The "four horsemen" of Silicon Valley at the time, the four companies whose impact on technology overshadowed everyone else's, were Amazon, Apple, Google, and Facebook. In 2013 *TechCrunch*'s M. G. Siegler declared there should be a fifth horseman: Samsung.

"Not only is it bigger than Apple from a revenue standpoint," he wrote, "it's almost twice as large as the three other 'horsemen' combined ($190 billion in revenue versus what should be [in 2012] about $100 billion for Amazon, Facebook, and Google. And unlike Amazon and Facebook which make little or no profit, Samsung is hugely profitable."

B.C. Lee's far-fetched experiment in chipsets had been vindicated. Samsung, the butt of the technology world's jokes just two decades earlier, had leaped out of nowhere to become the designer of just about every type of premium electronic device in the world, as well as the parts inside them. With the company spending more than the entire economy of Iceland on marketing, Samsung's logo was everywhere: in Times Square, at the Olympics, slapped on televisions and handsets all over the world.

Before long, Samsung would supply one in three of the world's smartphones.

When you enter Samsung's headquarters, you enter a fortress. Few gain access to its inner sanctum. Its leaders avoid publicity and usually don't have Twitter accounts. But in my reporting, I had managed to build a degree of trust with its executives, and I was allowed inside its sleek, glossy office towers and able to tour its laboratories, where scientists and engineers tested ideas that most consumers could only imagine. Young designers told me about smartphones and TV concepts they were working on that I didn't believe could possibly land on the market in a few years. But they did. I met visionaries and dreamers, people whom the company billed as stars.

But there was something different about Samsung. As people there boasted about their achievements, they'd transition to something, well, unusual for a Fortune 100 technology giant.

"The founder had a vision in his mind to create a new world. We are moving into the dream and vision of the founder," declared Gil Young-joon, senior vice president at Samsung's ultrasecretive research institute, the Samsung Advanced Institute of Technology (SAIT). We were in the middle of a conversation about foldable smartphones, a far-off idea at the time but one unveiled almost nine years later, in February 2019.

"He wanted to make another miracle," Gordon Kim, human

resources director, told me as he went through the biography of B.C. Lee's son, Chairman Lee Kun-hee. "Three hundred and forty hours of speech he delivered. How can he talk for three hundred and forty hours? It's incredible." Employees, to my amazement, could recite the exact dates of his historic speeches.

I had already realized that Samsung didn't want me getting anywhere near its ruling family when I made the grave error of asking for an interview.

I was quickly told that this "will not be possible" by the rep from Samsung's PR agency, Weber Shandwick. "I am dealing with a variety of sensitivities that you would have no idea of." And yet it was Samsung that had invited me on the trip.

But as I continued to research Samsung, the stories of the chairman became ever more surprising.

When the chairman visited a Samsung manufacturing plant, employees were told to park behind the plant, as their cars were too ugly—apparently they offended his aesthetic sensibility. Mints were placed in the bathroom for employees to use, to sweeten their breath from the kimchi (Korean pickled cabbage) that they ate at mealtimes. Employees were cautioned not to gaze down from the windows at the chairman when he arrived. Security guards lined the road, and when the chairman's limousine pulled up, a long red carpet was rolled out.

"One employee was charged with trying all the local restaurants in a city where the Lee family was to visit," a former employee told me. "He'd write reports on their dishes and wines."

When the chairman and his family traveled to Germany, staying at the five-star Hotel Adlon in Berlin for a week of vacation in August 2004, Samsung booked the entire fourth and fifth floors and a full conference room, setting up a "situation room" where his aides could monitor the chairman's every move and ensure his well-being.

THE WAY SAMSUNG LOYALISTS (especially its older generation) talk about the chairman and his family—the way they venerate him and seem to regiment themselves as hard-fighting units inside the company in almost military fashion—reminded me oddly of the military-like culture of North Korea, which I had visited and reported on for almost a decade in my time in South Korea.

The odd similarities between the traditional culture of Samsung (and other South Korean companies) and the totalitarian dictatorship of North Korea are no coincidence. The Korea scholar B. R. Myers has written about North and South Koreans' belief in a shared, ancient bloodline that informs their politics and societies today. South Koreans, he argues, have identified strongly with the Korean race that transcends the border with North Korea, a far stronger identification than with their democratic system of government.

The result, he says, is that North Korea is the world's most nationalistic country, while "the second-most-nationalist country, in my view, is South Korea, which is completely open and completely wired, and still dominated by a very paranoid way of looking at the outside world."

For example, I watched a video that had been leaked by a Samsung recruit competing at the annual Samsung Summer Festival, a gathering featuring sporting events and a mass games ceremony, where recruits come together to form images using individual placards—a practice associated with communist and authoritarian governments. I saw the recruits create a formation of a punching fist and a formation of the word "victory," a martial symbol I had also seen in photos of a formation of North Korean patriots.

In the video, the recruits on the field also made a Pegasus formation. I lit up at the sight. It looked like a Chollima.

When I had traveled in North Korea a few months earlier, my government-assigned guide, Mr. Han, had taught me the significance of the Pegasus, or Chollima, in Korean culture. It was a symbol of speed and hard work used to inspire students and soldiers, who often double as construction workers building apartments, schools, and hospitals in grand nation-building campaigns.

"Charge forward at the speed of the Chollima!" Mr. Han told me, reciting North Korean propaganda phrases as we walked outside the train station in Chongjin.

Chongjin, a city with a heavy police presence even by North Korean standards, was located near the Chinese and Russian frontiers.

A Chollima statue in Pyongyang, North Korea.

Looking up, Mr. Han asked me to translate the phrase on the mountains overlooking the city. "One heart united," it read.

I saw an eerily similar slogan in the video of the Samsung recruiting event, at the front of an army of recruits in formation: "We who have become one."

"Can you see our fighting spirit?" Mr. Han would ask me during my trip to North Korea.

"Where is your fighting spirit?" Samsung executives would routinely ask their employees, demanding that they work harder and faster.

I will never forget the day Kim Jong Il, the dictator of North Korea, died on December 17, 2011. I was in London and about to return to Seoul. North Korea's news feeds broadcast videos of North Koreans crying and shrieking as Kim's corpse was carried by a black government limousine to his final resting place, where he was to be embalmed and preserved. Later I visited his mausoleum, where I saw his body resting behind a glass casing, the room illuminated with a spooky red glow to help preserve the corpse.

Samsung's employees, of course, didn't break down in tears when their leaders died—at least not to my knowledge. But Samsung was undergoing a similar cultural process as leadership was transferred within the family to the next generation.

"South Koreans cannot seem to get enough of the family behind Samsung, whose mystery-shrouded inner workings many liken to those of another famous Korean clan: the Kim family that rules North Korea," *The New York Times* reported.

Chairman Lee and North Korea's Kim Jong Un are both the third sons of their fathers but managed to inherit their leadership positions. Chairman Lee's older brother, like North Korea's Kim Jong Un's older half-brother, was effectively banished from the company and was alienated and unhappy.

The comparison, said Bruce Cumings, a legendary historian of Korea at the University of Chicago and the head of its history department, was "very apt." "It's a very basic Korean trait that trust rarely extends beyond one's family, and that includes the Samsung family," he said.

North and South Korea are two vastly different countries, but they have a shared history and culture of Japanese militarist and Confucian influences, as well as shared family links. North Korea's and Samsung's common heritage shows itself in five traditions: the extreme reverence for family dynasties; the belief that their strength is derived from an ethnic bloodline; the promulgation of military-like rituals, ceremonies, and slogans; nationalistic paranoia and distrust of outsiders; and the veneration of a supposedly wise, paternalistic emperor-like leader.

To illustrate this, try a brief thought experiment. Suppose the South had won the American Civil War, and seceded. Both the United States of America—formerly the Northern states—and the newly formed Confederate States of America would have kept their shared heritage, history, and family ties. Both would have essentially kept their "American" identity. The same is true in the real world of North Korea and South Korea today.

———

LET ME GIVE YOU another example of these similar cultural practices. In 2010 a Samsung employee tipped me off to a leaked video of Samsung recruits standing in formation before a scoreboard displaying a motto. In black-and-white rococo regalia, including a cravat—attire as ornate as the military dress of a French musketeer—a cheerleader uttered a battle cry: "Youth with boiling blood, conquer the summer season!"

The phrase PRIDE IN SAMSUNG was hoisted on a banner on a nearby hill amid the pine trees and fertile summer grass. Amid a sea of blue costumes and yellow capes, the recruits on the field smartly fell into formation in the shape of a trapezoid. Senior employees watched from the sidelines, behind the cheerleader, their company division identified by their color of dress.

"Victorious fighting spirit! Sensational telecommunications, team C!" the cheerleader shouted. She jumped in the air, then flung her right arm out, ruffles on her wrists, and snapped a forefinger in a white glove.

"Start!"

The day's recruits were the newest class to enter the gates of this silicon castle, at an event called the Samsung Summer Festival. They'd been preparing to join the knighthood of Samsung Men and Women for more than two weeks. They'd been through boot camps, hiked, and suffered sleep deprivation while learning to work together and treat each other as family.

Four Samsung recruiting divisions performed that day, each wearing its own distinct uniform, in what was meant to be a team-building exercise. It was meant to be fun. But the company elite were watching.

A trumpet blared over the stadium's loudspeakers, followed by the strike of a guitar chord and patriotic cries from the recruits. Choreographed to look like a human LED screen, they raised their colored placards in unison, creating a sea of red. A small group flipped their cards to yellow, creating the image of a digital watch display, counting down the seconds and milliseconds: *05, 04, 03, 02, 01, 00.* The regiment dropped to the grass and flipped their cards in flawless unison back to blue, then snapped up into a standing position with the precision of a Super Bowl halftime show. They formed an animation of a soccer player sprinting and kicking a ball, followed by the word "goal." The soccer player then leaped in victory.

Next the recruits formed the letters "DMB" (digital multimedia broadcasting). The technology behind the world's first service that allowed you to watch TV on your mobile phone, developed in South Korea and kicked off the previous year, one of the myriad inventions to come out of Korea.

The Samsung recruits had reason to be proud; their country and, even more, their company had created it. The recruits broke formation and sprinted outward to form a rectangle, picking up bags at their feet to create a checkered pattern using the Samsung colors of blue and white, spelling the word "victory" one letter at a time.

The recruits then formed the numeral "10,000,000," the target number of mobile phone sales that Samsung had set as the threshold for success. That year's Samsung D500 handset, a simple, compact slider phone, had rung in a bonanza. The recruits formed a picture of the phone, followed by the word "champ," before forming a digital watch with the word "hero," then another mobile phone with the word "star."

"It was amazing, scary, and weird," said a Samsung employee whose manager helped run the event. She and many others likened the pageantry to North Korea's mass games ceremony.

North Korea's Arirang Mass Games, August 4, 2012.

WIKIMEDIA COMMONS

———

A **SAMSUNG PUBLIC RELATIONS** executive, told of the comparison between Samsung and North Korea, told me over Korean barbecue one night, "That's offensive. We are a company. Don't compare us to North Korea. Compare us to Apple, IBM, HP.

"Yes, we're secretive," he admitted. "But so is Apple. The Samsung Man is just a stereotype," he said. "It's not the company I see."

Samsung hates it when journalists draw comparisons between Samsung company practices and North Korea, even though such comparisons are fairly common among Samsung employees themselves. Steve Jobs had his own cult at Apple, Samsung employees tell me. Samsung isn't the only electronics maker with cultish tendencies.

But Samsung's logic was a type of red herring, a way of diverting attention.

Pointing to similarities between Samsung and other companies doesn't dismiss the common culture and heritage between North and South Korea and, to some extent, Japan and China. The fact is that the South Korean *chaebol* have little in common with the more entrepreneurial and shareholder-driven firms in the United States.

Even the biggest companies in the United States do not enjoy the privileges of companies in South Korea today. More than half of the family leaders of the ten biggest *chaebol* groups are convicted criminals. All have been pardoned by the president, often without serving prison time. Three of them, including Samsung chairman Lee Kun-hee, have been pardoned twice. From January 2015 to February 2016, the outside members of Samsung Electronics' board of directors—who were supposed to be independent, as a check on corporate governance—unanimously approved every proposal put forward by the company, except the two times a director was absent.

Imagine the heirs of the Carnegies and Rockefellers being so powerful and revered that *The New York Times* would self-censor its coverage out of deference. Imagine a White House pardoning the heirs of Sam Walton or Ray Kroc, as they ran the operations of Walmart or McDonald's from their prison cells. Or seasoned journalists turning their eyes away when confronted with Donald Trump's conflicts of interest between his presidential duties and his businesses.

Because of the outsized privileges of Samsung and the Lee family, South Koreans tell me Samsung has grown too big to fail.

But Sangin Park, an economics professor at Seoul National University, puts that "too big to fail" label in perspective. The continued success of South Korea's economy hinges not just on the Samsung Group as a whole but on a single company *within* the Samsung Group:

According to crisis simulations I have carried out, if Electronics stocks fall by 70%, both Samsung Insurance and Samsung C&T will become bankrupt. It is an inevitable domino game. If Samsung Life Insurance and Samsung Fire Insurance become bankrupt, the entire insurance industry in South Korea goes into crisis. If employees at Samsung Group and all its supplier firms (whose exact number is unknown to the public) lose their jobs, unemployment rate in the country is estimated to rise by 7.1%. (Its current rate is 3.5%.)

If the Samsung Group were to fall, he goes on, the country's National Pension Service, a major shareholder, would lose an estimated 19 trillion won (about $16.7 billion) in investment. And corporate taxes would decrease by an estimated 4 trillion won (about $3.5 billion). "If the entire Group falls, and with it multitudes of suppliers that depend solely on Samsung, South Korea's biggest banks are at risk of insolvency."

It is a startling admission. South Koreans refer to it as the "Samsung risk." And many see it as an urgent concern.

IN FACT, SAMSUNG HAS no equivalent among its Silicon Valley peers. Nor does the company have Silicon Valley's rebellious, counterculture origins. There is no marijuana-smoking college-dropout equivalent of Steve Jobs; there is no mischievous Mark Zuckerberg, ranking co-eds on his dorm room website. There is no flamboyant engineer like Sony's Akio Morita, who survived World War II, co-founded the company in a bombed-out department store, and drove its success.

Sitting down for interviews in Seoul, I was often met with looks of suspicion and distrust—even terror. A number of South Koreans told me that I might come to harm for writing a book on Samsung.

I treated such concerns as groundless conspiracy theories. But they did show me how much fear the company can strike in the minds of South Koreans.

Everywhere I went in South Korea, the "Republic of Samsung" and its imprint were inescapable.

"Do you plan to criticize Samsung?" one former senior employee asked me over lunch.

"Samsung helped me with my magazine reporting," I told him, "granting me official access. But I intend to ensure that my coverage is independent and credible. Samsung has no say over this project."

But Samsung did not know I intended to write a book.

"Key contacts," I explained, "would be coached on what to say, sources would go quiet, and my hard-earned access through unofficial channels would go cold."

Nonetheless, I couldn't keep the book secret forever. In September 2015 I woke up to an email from one of my PR contacts at Samsung, "Nam," who had arranged a number of my earlier interviews.

"I hope you aren't annoyed or alarmed by us knowing about your media inquiry," he wrote.

A former Samsung employee had forwarded him my request for an unofficial interview. Now that Samsung knew of my intentions, we agreed to meet over coffee, a gesture meant to smooth out any future interactions. In our subsequent meetings, another public relations executive told me that a suggestion that Samsung cooperate on the book had been flatly rejected.

Nevertheless, through my magazine work, the company's official channels, and my own unofficial sources, I interviewed more than four hundred current and former Samsung employees, executives, politicians, businesspeople, board members, journalists, activists, and analysts, as well as a member of the founding family. The interviews took place in South Korea, Japan, China, New York, New Jersey, Texas, and California. Some interviewees were afraid of being named. Others did not want to go on the record at all. I have respected their wishes. Many graciously offered to help me, knowing their careers were at great risk, and with little benefit to themselves.

7

The Scion

B.C. LEE HAD BEEN nonreligious his entire life. But now, at age seventy-eight, on his deathbed, he found himself asking existential questions of a Catholic priest who remained at his side. "If God loves humans, why does he allow pain, misery, and death?" He died of lung cancer in his home at 5:05 P.M. on November 18, 1987.

Twenty-five minutes after his death, the presidents of Samsung's thirty-seven affiliates convened at the Samsung Group headquarters. They unanimously voted in Lee Kun-hee, age forty-five, the third son of B.C., as their next chairman. Known as Lee II, he'd been named heir by his father in 1976.

The rest of the Samsung empire was split into four parts.

The new chairman controlled the crown jewel of Samsung, Samsung Electronics. Samsung's food and snacks company, Cheil Jedang (CJ), went to B.C.'s oldest son, Lee Maeng-hee, fifty-six. The VHS tape company, Saehan, went to B.C.'s second son, Chang-hee Lee, fifty-four. The paper and chemicals company, Jeonju (now Hansol), went to B.C.'s eldest daughter, In-hee Lee, fifty-eight. And the department store chain, Shinsegae, went to B.C.'s younger daughter, Myung-hee Lee, forty-four.

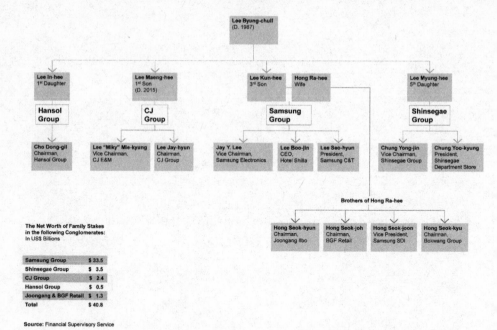

The family tree of Samsung's founding Lee family, one of the wealthiest families and the world that controls much of South Korean business. When Samsung founder B.C. Lee died in 1987, his business empire was separated into five parts among his children. Today, those business groups are known as Samsung, CJ, Shinsegae, Saehan, and Hansol.

FAMILY TREE DESIGNED BY GORDON BRUCE AND JON CRAINE. USED WITH PERMISSION.

These five companies touched almost every facet of South Korean life. Much of the South Korean economy, in fact, was centered around Samsung and B.C. Lee's clan.

In a seemingly royal tradition, purges of potential rivals within the family began following B.C.'s death; members of the family were let go from Chairman Lee II's domain.

Henry Cho, my contact, had long been close to Chairman Lee II; he had lived with the young scion in Japan during his days as an undergraduate economics student at Waseda University. He was playing golf with his colleagues at Samsung's Anyang course after Lee was named chairman, when the manager of the club approached the green.

"You guys are no longer members," the manager told the group. They were asked to leave. The fact that Henry and the chairman were close friends didn't matter.

And that was how Henry knew he was out.

Stories of intrigue within the family emerged, owing to the less fortunate siblings' resentment over the inheritance. In 1995 a video camera was set up on the roof of Chairman Lee's home, many assumed to monitor the home of Chairman Lee's estranged elder brother's family nearby. Almost two decades later, in 2012, the brother's company, Cheil Jedang, claimed to police that it had footage of a Samsung C&T (which stands for Construction and Trading) employee tailing a rival family member, Chairman Lee's nephew. The incident happened after Chairman Lee's estranged elder brother filed an inheritance lawsuit against Chairman Lee, attempting to get what he believed was his fair share of the Samsung empire. Four Samsung employees were indicted and fined without being convicted; Samsung denied the allegation.

Twelve days after B.C.'s death, the new chairman made his first major public appearance at the company, with a speech under the slogan the "Second Foundation," waving a company flag before the crowd.

"I am going to found the second establishment with my young ambitions and enterprising spirit," he proclaimed. In an act of filial piety, he named his home Seungjiwon, a portmanteau that means "respecting and following my father's wishes."

But the senior executives at Samsung seemed to pay little attention to him. They considered him the "emperor's son" and a "lucky heir," as they put it, referring to his older brother Maeng-hee's effective banishment two decades earlier, which had resulted in Lee being elevated.

And in fact the young chairman was an unproven leader.

"I was feeling bleak," he wrote later in his 1997 book, *Lee Kun-hee Essays*. "I had been partly involved in the management of the company since I became Vice Chairman in 1979, but I'd always had the safety net of my father."

Three months after B.C.'s death, in February 1988, Samsung released its greatest success to date in the electronics industry, the four-megabit DRAM. It was the result of half a decade of work. It still left the company roughly six months behind its Japanese competitors, but when Samsung had begun to work on producing chips in 1983, it had been a generation behind, and few people had thought it could catch up. It had made huge strides in closing the gap.

Samsung executives took the first semiconductor off the assembly line and laid it on B.C. Lee's grave, in a sign of gratitude and respect.

"MY NICKNAME IS 'THE silent one,'" the new chairman of Samsung admitted. "At home, I'm known as the person who's no fun." He had originally wanted to be a film director or start a movie production company.

Chairman Lee was something of a recluse who spent much of his time in his home, where he lived a life of contemplation. He refused to take calls from Samsung executives. Stacks of videotapes littered his bedroom. His interest in movies was a passion that he had developed as a lonely child raised for a few years in Japan, with few school friends, hiding away at the local movie theater on weekends or locking himself in his room and watching as many as eight movies a day. His favorite genre was wildlife documentaries, and he also loved the Academy Award–winning film *Ben-Hur*. He would also take apart radios, televisions, and other electronic devices to better understand how they worked.

"He digs into one issue for hours and hours, days and days, years and years," former aide Hwang Young-key, a banker who advised Lee on finance and translated texts into English for him during his tenure, told me in his office over tea.

Lee Uh-ryeong, a professor at Ehwa Womans University, called him *Homo pictor*, or "man the artist." Uh-ryeong, who knew Lee personally, described him as an abstract thinker and a rather awkward, brooding fellow, but with broad insight into Korean culture and civilization (as opposed to *Homo faber*, the maker of tools, the hands-on craftsman of the workshop). In any event, he was a man brimming with ideas.

"Even in idle conversation, I could not help but marvel at Chairman Lee Kun-hee's sharp insight into 21st century civilization and his steadfast grasp of Korean culture," Uh-ryeong wrote in Lee's book, *Lee Kun-hee Essays*.

"He did not strike me as particularly lively," the Korean novelist Park Gyeong-ri wrote about meeting Lee, on the other hand, in the same book of essays. "He was neither nimble nor elegant. But he was unique because he had an air of fineness, meticulousness and a sense of reticence. A creative sensibility, that is what I felt."

An erudite loner lacking the loquaciousness of so many modern CEOs, he spoke little, and sometimes not even in full sentences. He seemed to prefer spending time with his dogs to human contact. He had more than two hundred pups in the magisterial family kennel and was credited with the first pure-breeding of South Korea's national dog, the Jindo.

"After returning to Korea [from Japan], I enrolled in middle school," Lee wrote about his childhood. "At the time, anti-Japanese sentiment was high and for someone who had just come from Japan like myself, adjusting to school life wasn't easy. It was in those circumstances that I grew closer to dogs, and I have had dogs ever since."

But the chairman needed frequent stimulation as well. He spent his free time blazing around Samsung's private racetrack—his favorite car was a Porsche 911. His hobby led to more than one high-profile crash.

"Driving at 200 mph puts your life at risk," he told *BusinessWeek* in 1994. "It makes you fully alert and relieves most of your stress."

"A very interesting man," Henry Cho, his nephew and former friend, said of him. Henry followed that with a qualifier. "He was the laziest guy I've known."

Manic and mercurial, the chairman also had a dark streak. He could be, in the words of Cho, "cruel."

"In the morning, he would come out of the shower, and his orange juice was on the table, and without looking," said Henry, mimicking how the chairman would cover his eyes as he fended off his morning grogginess, "he would use his right hand" to get his orange juice.

"Where's my orange juice?!" he'd shout if it wasn't there, berating his assistants.

In another example of his quirkiness, Henry said, "he would instruct his driver, say from the Hyatt, to use the car's brake only five times."

As the chairman himself wrote in his book of essays, "Several years ago I walked from my office on the 28th floor of the Samsung headquarters all the way to the lower level employee offices in order to measure the travel time and to see what the quickest course was. I have also changed the location of drawer handles to see if there's a more convenient way to use it, and rearranged furniture."

South Koreans, understandably, didn't appreciate this image of the mean rich kid. Rumors spread that he had a prescription drug addiction—no evidence was offered—as well as a tryst with an elevator operator in the late 1970s, resulting in a secret child. He was rumored to have a lover in Los Angeles and to have fathered dozens of other children outside marriage—stories that created an aura of power and mystery around him.

The chairman denied tales like these to the media. On the legend of his ninety-five children—a number worthy of Genghis Khan—he claimed he had gotten a vasectomy after his fourth child with his wife, making such claims impossible.

"He was on drugs," Henry told me. Henry said that the chairman got surgery after a car crash in the 1980s. "He [the chairman] was hurt so bad that he had to take Demerol. It got him hooked." Demerol is a potent opioid painkiller.

Henry was chairman of the board of Samsung's Koryo General Hospital Foundation from 1981 to 1994. He said that the Korean government inquired about an alleged supply of Demerol being stored at the hospital for the Samsung chairman. Henry said the allegations were true, even though the government didn't press charges. "We had to supply [Demerol]," he said. "It's totally illegal for us."

These kinds of claims about the Samsung chairman were familiar to South Koreans. Kim Yong-chul, a former prosecutor who later went to work for Samsung, wrote in his 2010 memoir, *Thinking of Samsung,* about a strange case he brought in court against a drug dealer who set up a cocaine ring, because the drug dealer thought it would help him land the business of Chairman Lee.

Then there were the hookers.

In a series of videos that were made public, Chairman Lee sits back in a recliner chair, the window shades closed, wearing a ragged white T-shirt, watching a television screen, and pulling from his trademark stack of DVDs next to him, while he proceeds to pay four women for their sexual services. One of the women secretly filmed him from a hidden camera attached to her handbag. In April 2018, three people were convicted for attempting to blackmail Samsung with the video before it was leaked to the media. They were each given a prison sentence of three to four years.

One of the journalists who broke the story works for the investigative website *Newstapa*. He told me, "I always heard [that] the chairman [was] this deity." Recollecting his thoughts when he first saw the leaked video, he said, "And now we get to see him in the flesh. This is it. This is the emperor. It's disillusioning."

WHILE LEE II WAS slow in taking the reins of Samsung Electronics, he gradually began to rise to the occasion.

"He didn't come really to the office for about a year and a half," said his aide Hwang Young-key. But he began showing up more often in 1990.

"Do you know," he asked a group of his executives, attempting to make a point, "how many kernels of rice there are in a single piece of sushi?"

They had no idea.

"Two hundred and fifty," he said. A master sushi chef, explained the chairman, reaches into his rice bowl and takes out exactly two hundred and fifty kernels of rice, "perhaps eight out of ten times," he said, according to Hwang.

The chairman was referring to the mastery of the *shokunin,* a master craftsman of Japan who seeks erudition through repetition. Lee urged his executives to think like that.

"Seeing the same movie ten times, you think that's not possible," he told his executives. "I think that it's interesting. If you see the movie for the first time, you know the story and you know the leading actor, maybe the leading actress. The second time, you see the extras or supporting actors, maybe three or five supporting actors' roles. Then you see it three times, you see the decorations. The cars on the streets, the buildings on the back side, you see other sides."

But even ten times is not enough to get inside the head of the director, the chairman explained. And that should be the ultimate goal of the viewer who wants to truly understand the film.

"Whenever [Lee] said something, he had a message," said Hwang.

In 1991 the Cold War ended. The Berlin Wall fell in 1989. And Lee realized, in the isolation of his home, that Samsung was failing to keep up with the emerging global boom.

"Even in such a situation, there was no sense of urgency at Samsung," Lee II wrote. The company was floundering under the mistaken belief that Samsung was number one in its field.

The chairman knew that was ominous. According to a booklet Samsung published about his thinking, a century earlier Korea's leaders had tried to hunker down in their isolated feudal kingdom, resisting the arrival of foreigners (and even executing some), only to have their kingdom pried open like oysters by gunboats and then annexed by Japan into a half century of colonialism.

The world was reliving such a moment, Lee was convinced. Borders were opening, trade was increasing, and South Korea, still on an isolated peninsula, was once again vulnerable to getting squashed by foreigners—this time by the likes of global companies like IBM, Microsoft, Motorola, and Sony.

Samsung chairman Lee Kun-hee, also known as Chairman Lee II. After inheriting the company from his father, he began a movement to transform Samsung from a manufacturer of third-rate microwaves and TVs into a global powerhouse.

Lee knew that while Samsung may have been the number one company in South Korea, it was still weak by global standards.

"From the summer of 1992 until that winter, I suffered from insomnia," Lee wrote. "I was feeling desperate, as though Samsung, beyond simply having to give up a business or two, might completely wither away. I never slept more than four hours in those days. Instead of my usual big appetite, which would only be satisfied with three portions of *bulgogi* [marinated slices of beef or pork], I barely managed to eat a meal a day. That year, I lost more than 10 kilograms [20 pounds]."

And then he came out of his hermitage.

8

Glorious Chairman!

IN FEBRUARY 1993, CHAIRMAN Lee II gathered his executives for a conference in Los Angeles at the Century Plaza Hotel.

"Come and see for yourself how our products [have become] worthless in the American market," the chairman told his executives. He led them to a nearby department store, where they peered at Japanese and American televisions on shelves at eye level, while Samsung products were gathering dust at the bottom.

The chairman barked that these products did not deserve the name of Samsung. He later told *BusinessWeek,* "But they didn't listen."

Eager to get around his executives, who told him little about what was going on in the company, the chairman summoned reporters from the Samsung Broadcasting Center, the company's internal television news channel, and ordered them to place hidden cameras in a plant in South Korea. He would watch the video of his employees later.

That Samsung's business was in trouble was no mystery to Lee. But he didn't realize how bad things were. Lee wanted an outside appraisal. As Samsung's journalists set out on their investigation of the company's business practices, Lee consulted the Japanese—who had been faithful teachers and tough competitors—for wisdom and guidance. On June 4 of that year, he traveled to Japan, where he invited his Japanese advisers to his suite at the Hotel Okura, where his father used to stay.

Leading the Japanese delegation was a soft-spoken industrial designer named Tameo Fukuda, who'd been helping Samsung for the past three years out of an office in Osaka. It was his first meeting with the chairman.

"Tonight you should tell me straightforwardly how you feel about Samsung Electronics," the chairman requested.

"The chairman [asked] some incisive questions and the Japanese designers frantically replied to all of them," Fukuda told the *JoongAng* newspaper. "We stayed up all night talking."

Up to this point, Samsung managers had been ignoring Fukuda's advice, to Fukuda's displeasure. Offices were siloed, refusing to share the most basic information. Standards were lackluster. The parts inside electronics products like microwaves and cellphones were sloppily arranged, and quality control was abysmal. An expensive piece of testing equipment at Samsung Electronics' research and development center had gone unused for several days because of a broken electrical socket.

The Japanese delegation continued talking with Lee until 5:00 A.M., leaving him with a report on their findings from their visits to the factories and offices of Samsung Electronics. The chairman put the report in his briefcase, preparing to fly to Frankfurt the next morning.

Official histories of Samsung don't mention that the chairman's trip to Germany was about cars: He was scheduled to meet with the heads of German automakers. He was eager to learn from them, as he was interested in entering the auto industry, fulfilling a little-known pillar of his father's vision.

"Automobiles were the most important part of Lee Kun-hee's tour," Hwang, his adviser, acknowledged.

On the flight, however, the chairman opened his briefcase to read the report that the Japanese consultants had prepared.

Samsung's designers, Fukuda said in his report, needed a stronger mandate to work *with* the engineers who traditionally ran the company, rather than *under* them.

He wrote that the true beauty of a product is on the inside. "The outer shape is important," the report stated, "but the inner design is also important." It is "not simply creating a form or color of the product, but rather forming or tapping cultural activities to create a new kind of user lifestyle by increasing added value, starting from the study of the convenience of the product." Design, in other words, is not about making things pretty. It is about making things work.

As he read on, the chairman grew furious. Samsung's management was ignoring Fukuda's much-needed advice. He called his five direct reports over from the back of the plane.

"Why is this happening?" he asked them, repeating his question several more times. He handed his executives the report and told them to read it and answer his question by the time they landed in Germany.

But the "vague" conclusion the executives came up with—that the education, mindset, and attitude of Samsung employees were lacking—didn't satisfy Chairman Lee.

In his hotel room in Frankfurt, the chairman next watched the secret surveillance tape made by his in-house investigative reporters.

What he saw appalled him. The doors of washing machines were so poorly manufactured that they scratched against the mold when shut. Instead of restructuring their assembly-line processes to reduce errors, the engineers chose quick fixes, such as shaving off the excess plastic with a utility knife. How had things come to such a pass?

He called Samsung executives in Seoul and yelled at them for an hour, demanding that his call be recorded and distributed within the executive group.

"What have you been doing all this time?" he shouted at them. "How little you have changed, despite my repeated emphasis on quality management. From now on, I will be in charge."

Samsung would replay recordings of the chairman's voice for years in company buildings, as a continuing reminder of the mandate of every Samsung employee.

Next the livid chairman ordered two hundred of Samsung's executives in Seoul to fly to Frankfurt for an impromptu management session immediately.

———

ON APRIL 29, 2016, I visited the chairman's presidential suite at the Kempinski Hotel Frankfurt Gravenbruch—room 312. It was a majestic duplex, furnished with a Samsung television, overlooking a pond near Frankfurt. Over the years, it had become the closest thing to a holy site for Samsung. The chairman chose the hotel for a reason—Frankfurt for its industry, and the Kempinski for its quietude. With fresh air and forests nearby, it was a refuge for self-reflection.

"Samsung sent a film team here a few years ago, to make a documentary," my hotel guide told me on my tour. Samsung's executives, I later learned, returned occasionally for pilgrimages.

We made our way downstairs through the lobby, decorated in the fashion of a Bavarian hunting lodge. Our guide opened the doors to the room where what would become known as Samsung's revolution took place: the main conference room.

The hotel was under renovation when I visited, so the site wasn't completely historically accurate. But I stood in the conference room in awe. This was the site of the speech that kick-started a radical managerial transformation within Samsung. It was a moment that would help redefine the world of tech.

On the morning of June 7, 1993, the assembled Samsung executives were seated around the table with notebooks, wearing identical white shirts and blue or black suits. At the front of the room stood a speaker's table with a bed of pink flowers—a South Korean tradition. Behind it was a lavish oil painting of the canals and townspeople of Venice.

That morning, as the chairman entered the conference room, the Samsung executives stood up and clapped. The air in the room was heavy with tension.

The chairman wasn't sleeping more than two or three hours a day, Hwang recounted. He sat down, adjusted the microphone, and, without a script or preamble, unleashed his fury upon them for the next eight hours. Samsung has never released the full transcript of the speech. But through interviews with eyewitnesses, videos, media reports, and official company materials, I was able to piece together its key elements.

"I have felt a cold sweat running down my back at the thought of this crisis," the chairman said, as paraphrased in an internal booklet called *Samsung's New Management.* "We are standing on the edge of a cliff facing a life-or-death situation.

"The Cold War has ended, but a more intense economic war has begun. In this new war, a country's firepower will be determined by the level of its technology," he warned. "Many people at Samsung don't realize how cold and cruel this technological warfare can be."

He was determined to cut defects to raise quality, a strategy that Sony and Samsung's other Japanese rivals had mastered. "At Samsung, we must adhere to three credos: Faulty products are our enemy, faulty products are the root of all evil, and if we produce a faulty product three times, we must take it upon ourselves to resign."

Over the next three days, in sessions that ranged from eight to ten hours a day, he expounded on his new philosophy and strategy, barely pausing to use the restroom. Executives were given box lunches and sandwiches to get through the day.

"Change everything except your wife and children," he said, using a phrase that became his motto. Samsung called his philosophy "perpetual crisis."

The "Frankfurt Declaration," as these three days came to be called by those inside Samsung, became part of company lore, invoking his father's "Tokyo Declaration" a decade earlier.

The speeches were "absolutely unexpected by anybody, including myself," said the chairman's aide Hwang Young-key. "It was quite a well-aged message. He'd been thinking about it for years and years."

But not everyone was convinced by the previously awkward heir, now suddenly bursting with charisma. Some thought they were being collectively reprimanded, when personally they had done nothing wrong. One executive, Oh Jung-hwan, said that Lee Kun-hee's rebuke of his executives included "expletives and vulgar words."

On the third day, the chairman asked his executives over a meal what they thought of the previous three days.

"I'm sorry, sir, but quantity also matters," said one man at the table. "I think quality and quantity are just like two sides of the same coin."

Chairman Lee, holding a spoon, threw it on the table and stormed out of the room.

Clearly the speeches in Frankfurt alone were not enough to turn the company around and transform the culture. Over the next three months the chairman went on a whirlwind tour to London, Osaka, and Tokyo, giving forty-eight talks, amounting to 350 hours of lectures. Some of these lectures went on for as long as 16 hours.

"After two weeks, we had to start washing our underwear ourselves and hang them on the hotel veranda, to the protests of the hotel management," said Hyung Myung Kwan, head of the chairman's secretariat, who accompanied the chairman on the tour.

"I don't know how he did it," his nephew Henry admitted.

In the end, his speeches came to 8,500 pages of transcripts. The company began calling Lee's speeches the "New Management Initiative."

9

Church of Samsung

THE CHAIRMAN WAS BUILDING a quasi-religious corporate culture, what South Koreans called "emperor management." But there was a problem.

"These were not, at the time, very well organized messages," said Hwang. The chairman gave Hwang the job of disseminating his philosophy throughout the company. "He just poured out what he had in mind."

With a team from the chairman's office, Hwang was tasked with designing a book about the chairman's philosophy, a comic book, and films and pamphlets to ensure Samsung's hundred thousand managers and employees understood the chairman's message.

Samsung began distributing the book of proverbs, called *Change Begins with Me: Samsung's New Management,* and a raft of other educational materials to every recruit.

"It was kind of like Chairman Mao's red book," joked former vice president for sales and marketing Peter Skarzynski, "but it was Chairman Lee's blue book."

Executives would read the chairman's proverbs and some hung them up on their walls:

- "Neglecting to nurture talent is a kind of sin." (February 1989)
- "A genius is one in 100,000. South Koreans alone won't fulfill our need for genius, so we need to expand our horizons overseas." (June 1993)
- "In the future, one person will be feeding thousands more." (June 1993)

As I sifted through these internal documents and old news reports, five points leaped out at me that seemed to define the chairman's philosophy:

1. Healthy paranoia and a disdain for complacency, to survive in a cold, cruel, Darwinian technology industry.
2. A sense of perpetual crisis and a need to find opportunities in crisis.
3. The importance of quality control and the reduction of waste.
4. Human resources, talent, and training as the pillars of a strong workforce.
5. The urgency of building a flexible, long-term, globally minded corporate culture, rather than an inward, short-term, bureaucratic one.

A comic strip called *Let's Change Ourselves First: A Comic Book About Samsung's New Management Story* was published in 1994 by the Samsung Economics Research Institute (or SERI). The cartoons were professionally drawn by famous South Korean cartoonist Lee Won-bok. In it, Chairman Lee chided Samsung's executives for being arrogant and petty and lacking manners. One cartoon showed a Samsung Man getting drunk, sticking his face in a noodle bowl, and declaring, "This is my country," in front of a Western businessman who, sipping wine, calls him "uncivilized."

"How miserable it is to become a country which is economically subordinate to other countries," the book went on. The statement was accompanied by an illustration of American, Japanese, Russian, and Chinese villains ganging up on a placid, scholarly Korean aristocrat.

"If you know your enemy, you yourself can win every battle," the comic book continued, calling on Samsung employees to plan their strategy properly and work together for the coming fight. Their success, the chairman declared, required the trinity of government, people, and corporation, unified under a grand cause—with business at the vanguard.

"To make myself, my country, the Korean people, my children, and my descendants successful," the comic book claimed, "we should make a new leap forward."

In 1993 Samsung employees watched thirty-minute sermons by the chairman every morning for several months on the in-house broadcasting service, where Lee ordered his employees to examine their morality and rediscover their pasts.

The chairman instituted a 7:00 A.M. to 4:00 P.M. workday to improve his employees' quality of life—rather than having them work late nights every day. Of course, that meant the actual working hours were "more like 7:00 A.M. to 10:00 P.M.," as Kim Nam-yoon, an engineer, told me.

In 1995 the chairman's New Year's gift to family and friends was Samsung's new black mobile phones. But he was embarrassed to discover that some were returned because they were faulty. Nearly one in eight of Samsung's mobile phones, in fact, that year were defective. Clearly the chairman's grand proclamations weren't getting through.

In March of that year, he ordered his employees to prepare a giant bonfire, a sort of purification ritual, near the mobile handset factory in Gumi, an industrial city in the south-central part of the country.

The chairman summoned factory workers and engineers to a courtyard, assembling them in phalanxes against the barren, wheat-colored mountains. They were made to don headbands that read QUALITY FIRST. A banner over the courtyard read QUALITY IS MY PRIDE. A virtual mountain of cellphones, fax machines, and whatever else was deemed junk—over 140,000 devices worth $50 million—stood before them.

A few employees at the front approached a microphone, raised their right hand, and read pledges that they would treat quality control with the utmost seriousness. The chairman and his board of directors listened from a row of seats nearby.

At a prearranged signal, nine employees rummaged through the mounds of metal and plastic, hammering each phone or device into pieces and throwing the shattered remains into a pile.

Then they "covered the pile with a net and poured petrol on them," Gordon Kim, human resources director, told me, and set them on fire. After they had melted and burned, a bulldozer razed the remains.

"If you continue to make poor-quality products like these," the chairman announced to the workers before him, "I'll come back and do the same thing."

Some of the people who had designed and built the phones cried. It was "as if their babies had died," Kim Seon-jeong, a former financial executive, told me. Moreover, to be humiliated like that before the Samsung "emperor" was the ultimate loss of face.

BUT WHILE THE CHAIRMAN promulgated an image within the company as a globally minded leader, behind the scenes he perpetuated a dynastic Korean tradition. He made it clear he eventually wanted to pass the entire Samsung empire to his first and only son, Jay Y. Lee (in Korean, Lee Jae-yong), a dapper and well-spoken doctoral candidate at Harvard Business School.

The former financial executive, Kim Seon-jeong, who was involved in some of the succession planning, told me that a tranche of high-level executives were involved in drawing up plans for the sensitive inheritance process, exploiting legal loopholes, cash gifts, and financial tools like convertible bonds and bonds with warrants.

"That was the perfect tool for the succession plan," corporate governance lawyer Jisoo Lee told me. The plans got under way in early 1995, when Chairman Lee gave a gift of money to his children, which they used to buy shares in unlisted Samsung holdings at what seemed like bargain prices.

The companies went public a few months later; share prices jumped on the open market. The heirs sold them off for a small fortune and used the lucrative earnings to buy shares in the Samsung Everland theme park. The succession plan established Everland as the de facto holding company of the Samsung empire, effectively giving the chairman's children a vehicle through which to gain control of the empire's cross-shareholdings with a minimal financial stake.

"We knew that this transaction was problematic," said Cho Seung-hyeon, a lawyer who challenged the deal with a group of corporate

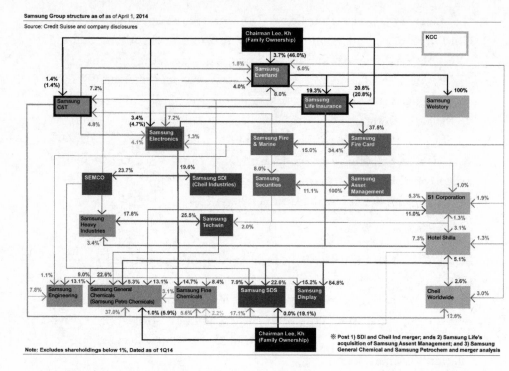

Samsung Group structure as of as of April 1, 2014
Source: Credit Suisse and company disclosures

Note: Excludes shareholdings below 1%, Dated as of 1Q14

Samsung's ownership structure. Samsung's theme park, Everland, was the de facto holding company through which the Lee family held control. In 1996, Samsung issued a shareholding mechanism called convertible bonds to Chairman Lee Kun-hee's children, at a below-market rate. The transfer turned Jay Lee into a major shareholder, the first of many financial chess moves to keep control within the Lee family. It also sparked a massive controversy in South Korea over corporate governance. Two Samsung executives were criminally convicted for the share sale.

DIAGRAM DESIGNED BY GORDON BRUCE AND JON CRAINE. USED WITH PERMISSION.

governance activists. But he failed to get it reversed in the courts. "It was designed to keep power in the family at any cost."

A South Korean court later ruled that two Samsung executives had illegally sold controlling shares in Everland to Jay Lee at a below-market rate. Although convicted, the two executives did not serve prison time. Executives at Samsung were extremely loyal to the Lee family, eager to win the patronage of the chairman.

"They were all sort of falling over themselves saying, 'It's my fault,'" Kim Seon-jeong said. "People were lining up to go to jail for [the chairman]."

What was curious was the fact that the shareholders supported the succession plan, acting against their own immediate interests, forfeiting their right to buy the unlisted shares at a market price and therefore allowing the chairman's family to take them over.

The chairman, meanwhile, instituted the next phase of Samsung's global expansion, eager to make his mark.

"On June 7th, 1993, a meeting was convened in Frankfurt, Germany," the chairman wrote. "This historic day will long be remembered as the turning point in Samsung's evolution. Let us commemorate this day together for the many years to come."

10

Go West, Young Heir

CHAIRMAN LEE II NEEDED an emissary to the rest of the world on Samsung's behalf. He turned to his beloved niece Miky Lee—her Korean name is Mie-kyung—making her Samsung's chief ambassador to the United States.

Miky's aesthetic sense, her people skills, and her relaxed demeanor made her an ideal candidate to represent Samsung in New York and Hollywood. A short, spunky thirty-something with a flair for film and the arts, she enjoyed befriending the likes of Quincy Jones and Steven Spielberg. She was a far cry from the typical Samsung Man (or Woman)—a dapper employee in a formal white shirt who would have had trouble adapting to Hollywood's self-promoting business culture.

"She is not like Korean men, giving orders and expecting them to be followed," her high school English teacher told *The New York Times*. "To do anything creative, you have to have a soft touch."

For Miky, representing Samsung was nothing short of a patriotic mission. As an assistant instructor of Korean at Harvard, where she attended graduate school, she couldn't help but notice how students ignored Korean-language offerings because of Korea's dismal international standing. Japanese, on the other hand, was seen as cool, and Chinese as useful.

"My lifetime obsession to promote the Korean culture started then," she told *Bloomberg Markets*.

Miky wanted to find a marketer who could help build Samsung's brand and presence in the minds of American consumers. Her family used another of the chairman's Japanese advisers, a fashion entrepreneur named Tomio Taki.

"[Tomio] had been at Donna Karan, where I was doing advertising work," advertising guru Peter Arnell told me from his scenic villa near Rome, showing off the lush view and cylindrical green topiaries. "So Tomio recommended me for the job with Samsung. Miky and I, we became good friends."

Peter was a rambunctious Madison Avenue advertising executive known for his eccentric ideas and aggrandizing demeanor, who'd branded Donna Karan's wildly successful DKNY, among other companies. Arnell had a booming presence and an unconventional portfolio. He had grown up in relative poverty and had worked at the fringes of boutique advertising before joining the ranks of Madison Avenue.

Despite weighing three hundred pounds, he dashed from meeting to meeting with a Rolodex of high-powered contacts, from Martha Stewart to Celine Dion, on speed dial. Sometimes he would show up for meetings at his office, which was filled with model spaceships and Star Wars toys, in a tennis outfit, barely stopping to catch his breath.

"I did all the ad work myself," Arnell said. "I did the concepts and I shot my own photos."

For fashion shoots, he made his mark with a perennially favorite subject: well-sculpted human bodies in black and white.

In our chats, Arnell came across as thoughtful and wise, if chatty. But I heard from many former colleagues about Arnell's sudden swings between charm and explosive bouts of anger. In 2007 *Gawker* named him one of "New York's worst bosses." *Newsweek* wrote that he made employees do push-ups in front of clients.

"But, in his favor," a reader wrote to *Gawker*, "when he's not pretending to kiss the ass of insipid rich, famous and powerful, he shows a refreshing contempt for authority and takes an anarchistic delight in creative destruction."

Recalled Pete Skarzynski, former senior vice president for sales and marketing at Samsung, "Not all of his ideas are good, but there's a ton of ideas." He once charged Pepsi $1 million for a document that connected its circular blue-and-red logo to the *Mona Lisa*, the Parthenon, and Hinduism. "The universe expands exponentially with $f(x)=e^\wedge x$," read the Arnell Group report, attempting to link the speed of light to a cosmic transcendence that included the Pepsi logo. "The Pepsi Orbits Dimensionalize Exponentially."

"I think that electronics today are more and more like fashion," Arnell told *Adweek* in 1996, referring to his early creative thinking for Samsung. Speaking to me more than twenty years later, he was proud of this insight: "It was before aesthetics were big in technology, before Apple released the iMac. This sort of thinking was still new, and Chairman Lee had the vision to see it coming.

"We decided that since Motorola controlled the space for mobile phones, and Sony had televisions, we were going to enter the American market with microwave ovens."

But this posed a marketing problem. Microwaving food, it was rumored, was a cause of cancer and other maladies.

"One night after a bad marketing meeting, I was so depressed. I took a taxi home and looked at the driver, a young Russian migrant. He was perfect for what I needed." Arnell invited him back to his studio to pose shirtless for a test shoot. "He probably thought I was coming on to him," Arnell joked.

He gave the nineteen-year-old driver a Samsung microwave, told him to hold it under his arm, and started taking photographs of the microwave and his model's sculpted abs.

Developing the black-and-white photos in his darkroom, Arnell immediately sensed a hit. He called his newly discovered model "Microwave Man" and ran a series of magazine and television ads and billboards featuring the sculpted taxi driver.

"Simply healthy," the writing called out in a magazine ad. "Vegetables cooked in a microwave retain more vitamins than any other form of cooking."

"It was a clear standout," said Thomas Rhee, former senior vice president of new business development at Samsung. "We were sensing a new thing at the time"—a growing movement to package the product in a simple, neat, aesthetic way, rather than just telling consumers about its hardware features. Samsung was connecting to Americans and their everyday needs.

But Arnell's vision for Samsung had its detractors as well.

"People said this is strange," Thomas said. One problem was that Samsung's products still fell short of pristine quality. The marketing campaigns—which screamed luxury and "premium"—risked disappointing customers.

"Let's put it this way," a Samsung insider told *Advertising Age* during the company's ad agency review years later. "Six years ago we had a small share [of the U.S. electronics market], and six years later, we still have the same share."

AS A FILM BUFF, Chairman Lee was looking to invest in Hollywood; Sony had done the same six years earlier, buying Columbia Pictures in 1989 for $3.4 billion, in a much-watched marquee acquisition. In 1995 and 1996, three partners—Steven Spielberg, Jeffrey Katzenberg, and David Geffen—founded a start-up company called DreamWorks but were having trouble finding investors. They needed $900 million in investment money. They hoped two or three suitors would pool together to come up with the money.

Miky Lee heard from a Los Angeles attorney that Spielberg was looking for investors. After meeting with Spielberg's Amblin Entertainment, she came away convinced that DreamWorks would be a smart and fun business. She convinced her uncle, Chairman Lee, to invest.

One night in 1996, the DreamWorks trio welcomed Chairman Lee and more than a dozen other Samsung guests to Spielberg's Beverly Hills home for dinner. During a meal of Chilean sea bass and white wine, the Samsung executives presented the details of their offer. They were willing to invest the full $900 million, far more than the amount DreamWorks was requesting from a single patron. But Spielberg got a funny feeling in his stomach at the offer.

"The word 'semiconductor' must have been used 20 times during that two-and-a-half-hour encounter," Spielberg told *Time*. "I thought to myself, 'How are they going to know anything about the film business when they're so obsessed with semiconductors?'"

His partner David Geffen concurred.

"I kept on saying to them, 'If we were only interested in making money we'd also build semiconductor plants.'"

Yet Samsung insisted on being the only outside investor, giving it effective creative control. Such deals were standard in South Korea.

Ultimately, Samsung walked away from the talks, feeling that DreamWorks was "asking for too much freedom and not enough accountability," said one executive. And Spielberg didn't pursue it. The DreamWorks founders instead went with Microsoft co-founder Paul Allen, who invested $700 million.

Instead, Samsung invested a much smaller amount, $60 million, in New Regency, the production company that had produced the hit *Pretty Woman* in 1990. The chairman also created a short-lived film studio called Samsung Entertainment.

Meanwhile, Samsung executives smelled another opportunity in California when shares in AST Research fell 27 percent after posting a loss in 1994.

AST was a classic California success story. Once the third-largest PC maker in the United States, it was founded in 1980 by three immigrants in a garage with $50,000 from mortgaging their homes. It focused on high-end computing. Nimble and freewheeling, it had an Irvine campus with a Wild West culture that allowed it to beat the market hegemon IBM to the latest technologies.

"The best technology they have," went an internal joke about cheaper competitors, "is a screwdriver."

But the PC market became volatile over time, with profit margins thinning. And AST was having trouble securing a stable supply of memory chips. That meant crippling delays in getting its latest high-end PCs to the market. By 1994, the losses were draining its balance sheets.

AST desperately needed capital, and Samsung executives jumped in with a win-win proposal: Samsung would acquire part of the struggling PC maker, inject it with the necessary resources, and bring AST into its ruthlessly efficient supply chain, giving Samsung a company to sell its hardware to and a respected brand name and distribution network.

The deal guaranteed AST "a committed source of supply for critical components," AST chairman Safi Qureshey told the *Los Angeles Times*.

But longtime AST employees weren't taken with the deal. They feared Samsung would strip them of their rebellious culture, where they were able to challenge AST bosses, and turn them into just another bureaucratic corporation.

"My biggest concern was with upper management, [that] they're going to do the same thing that they do in Korea," said Dan Sheppard, vice president for global product marketing. Visiting the Samsung offices, he was unsettled by the sight of groups of employees entering together, dining together, and leaving together at precise twenty-two-minute intervals.

Samsung executives attempted to reassure AST executives during acquisition talks.

"We know we have to stand back and let you guys do what you do best," an executive on Samsung's side of the table said, according to Sheppard.

In April 1995, the first stage of the deal went through; Samsung bought just under half the company for $377 million. But chief executive Kim Soon-taek (S.T.) told the *Los Angeles Times* that Samsung, with its 49 percent stake, quickly felt hamstrung from pursuing its own long-term vision, since it had to deal with other shareholders. It was an early sign of a culture clash.

"Public companies are dictated by short-term results, which is not the reason Samsung acquired AST," he later said. Samsung, he said, was in for the long haul to build a major PC manufacturer.

In 1996 the DRAM market, the bread and butter of Samsung's profits, collapsed, draining balance sheets. Still, Samsung followed with a full acquisition of AST for another $469 million the following year.

As Samsung tried a slew of strategies to take AST out of the red, it asserted a stronger and stronger grip on the company.

"That's when our worst fears came true," Sheppard said.

More Samsung employees began trickling into the California office, sitting in on team meetings and taking notes, and filing reports with Samsung headquarters. Samsung called these people dispatchers. Their role was to bridge the AST management culture more closely to the South Korean mother ship, smoothing the cultural and language barriers.

"Why don't you do it like this?" they started asking about various PC or notebook computer designs. "We like this idea. We don't like that idea."

To the chagrin of AST employees, Samsung engineers insisted on bulkier designs teeming with more features, even though Samsung had publicly insisted on cost-cutting measures.

"You need to put this LCD on the front of the mini tower," an AST manager, taking orders from Samsung, told his team. "It'll show the music playing on the CD track."

"A mini tower goes under the desk," an employee told his boss, according to Sheppard, who was present in the meeting. "It's not on your desktop, so people won't see it. That's taking a five-dollar hit, which means on the street that's going to be about twenty dollars, for a feature nobody wants. We're not going to do it."

The boss called the team two months later with further instructions from Samsung.

"Put the LCD on the computer."

The battle lines were hardening, and the hostilities were beginning to flare.

As AST's finances continued to slide further into the red, senior executives were summoned to the conference room and asked to retire.

"Who's next?" they joked with one another.

By December 1997, after six years of losses, three chief executives in four years, and wide-ranging layoffs, Samsung had whittled down AST from 6,500 people to 1,900. Even Korean Samsung employees, seeing the writing on the wall, defected to companies like HP.

"The very good ones in there saw the opportunity and they jumped ship," Sheppard said.

By the end, Samsung had sunk $1 billion into the $2.1 billion manufacturer, before deciding to shut down AST in 1999 and sell the name and patents to Packard Bell.

The chairman's lofty ambitions for Samsung were faltering thanks to its colossal, high-risk expenditures. Samsung's management practices, international standing, and unimpressive credit rating (Moody's gave it Baa2) were still too backward and archaic for the kind of global expansion the company envisioned.

Focusing on improving the quality of Samsung's products with its own in-house designers seemed like the only viable way forward. Already the chairman and Miky Lee had yet another idea.

11

Seoul Searching

GORDON BRUCE PICKED UP the phone. It was his boss, college president David Brown. "Ever heard of Samsung?"

"Yeah!" Bruce laughed. "But they make really terrible stuff."

"Well, they're coming here next week, and I have no idea what they want."

A philosophical, introspective man with wire-rimmed spectacles, Bruce was vice president at the renowned ArtCenter College of Design in Pasadena, a feeder school for large corporations and design houses. He was the school's point man for companies in Asia, and his job entailed setting up a campus in Japan.

But *Samsung*? Samsung was known as a cheap follower. The phrase "Made in Korea" was an international laughingstock.

"The request puzzled me. Maybe they wanted corporate training," Bruce told me.

Gordon Bruce was a master designer, a protégé of the descendants of the renowned school of Bauhaus designers, the German movement that unified arts, craft, and technology with its simplicity, and that influenced Steve Jobs and Jony Ive, among others. Gordon had spent years creating IBM's famed Watson Center, along with sleek products and buildings for Siemens and Mobil Oil.

The following week he saw four limousines pull up outside. Gordon recalled, "Seventeen executives and designers disembarked; I counted them."

The employees acted with all the deference and discipline of a secret service unit. The woman at their head was Miky Lee. She was joined by Peter Arnell, the larger-than-life Madison Avenue advertiser.

Bruce brought along a team of Asia hands at ArtCenter to help. By his side was James Miho, who had been interned in a U.S. detention camp as a child during World War II and had served as a tank commander in the Korean War. Miho went on to have a legendary career as a consultant for Chrysler and Xerox.

The school president, David Brown, was also in attendance. As a former marketing and public relations executive, he had led ArtCenter's projects in Japan. At six foot four, he towered over his Korean guests.

"We gave Miky a tour," Bruce said. "The meeting was cordial."

They strolled through the bridge structure spanning an arroyo, getting a look at the building, a midcentury work of art with glass panes and dramatic lines and diagonal steel pillars. As they walked, Miky talked about Samsung's mission. The chairman wanted to build an in-house design school where Samsung's best craftsmen could forge their own corporate identity. Samsung felt that Pasadena's ArtCenter, renowned for its curriculum in automotive design and industrial design, was an ideal partner.

"Chairman Lee is hosting a Samsung Day, renting out Olympic Stadium," she explained. "He would like you to attend as a guest."

"We wanted to emphasize that we weren't for sale," recalled David Brown, "that we were an educational institution first and foremost, that we weren't really a consulting arm that would do work for hire. It had to have a larger purpose for us."

"Samsung's designers need to be different," Miky pressed on. "We want them to think different."

"It's not enough to be different," Gordon Bruce retorted. "You have to be good."

It looked like the project was a nonstarter, but the ArtCenter trio reconvened after Miky departed.

"What persuaded me that this was a good idea was a little booklet that Chairman Lee published," said Brown, describing the Frankfurt Declaration. "There was some silly stuff in there, but the chairman was partially educated in Tokyo and had a pretty deep understanding of the Japanese rise to an exporting power and the role that design played in that."

The pair looked to James Miho for insight on whether Samsung could pull off a similar feat. He felt the prospects for Samsung's designers were good.

"In Asia," he told the group, "form is everything."

"We agreed to set up an educational institute at Samsung, but only as short-term consultants," Bruce said. It would be called the Innovative Design Lab of Samsung, or IDS. It was set to open in 1995. "But first, we needed to do some fact-finding, to diagnose the problems at Samsung. [Only] then could we help the Korean designers find inspiration."

Their mandate would prove much harder than they thought.

AN ENTOURAGE OF LIMOUSINES picked up Gordon Bruce, James Miho, and David Brown at Seoul's Gimpo International Airport to begin a fact-finding tour, and later drove them to a helicopter pad, where they'd embark on a tour of the country. Miky Lee, Peter Arnell, and a group of executives joined them.

"Is that ours?" Bruce asked.

A flying fortress of a helicopter befitting presidents and dignitaries emerged above them from the clouds.

Samsung's VIP helicopter landed, and the group boarded.

"I pulled out my video camera and started recording," Bruce said.

Bruce could see that South Korea was a nation in the throes of change, at the dawn of a national awakening. The chopper flew over rolling mountains and the autumn riverine palette, where shamanistic burial mounds and quiet farming villages were interrupted by factories and military bases.

After they landed, they traveled across the country by car and on

foot. Bruce, Miho, and Brown walked around marketplaces and bucolic villages for aesthetic inspiration. They wandered into a Buddhist
temple with a stone turtle at the front and an obelisk on top, and took
notes on its meaning: the turtle as a symbol of longevity, the pattern
and shape of its shell used in ancient fortune-telling. Bruce walked
around markets and saw traditional garments sitting there unsold that
had modern applications.

"My favorite was the *bojagi* cloth," Bruce later recalled, "a patch-
work quilt with pockets sewn together. It had all the design principles
of economy, quality, user-friendliness, and ecology."

His enthusiasm was met with disinterest from his Samsung guides.

"That's just Korean stuff," one of them said dismissively.

When Bruce met Samsung's internal design staff, he quickly
learned about their mindset.

"Management tells us Sony is successful," another American designer was told. "So if we want to be successful, we have to do things
like Sony."

Because they didn't trust their own designers, Samsung design offices were hiring foreign studios like Porsche Design to import their
house styles to Samsung. As a result, rather than develop a house style
of their own, a phone might look like Motorola's, a television remote
control like Sony's, and a car prototype like Nissan. This scattered
Samsung's identity across product lines, confusing customers.

"I saw the same effect in Seoul, a city built to show it had overcome
hardship," Bruce told me. "Seoul was a charmless city of gray and
white buildings—almost Soviet in its architecture; it was sacrificing its
heritage and identity in the race to modernity."

Old architectural treasures—traditional wooden homes and
Victorian-style colonial structures built by the Japanese—were being
razed at an alarming rate, replaced by corporate housing develop-
ments. The city itself was becoming an island of luxury—Louis Vuit-
ton shops and plastic surgery clinics—as Seoul's rising middle class
began to seek rounder eyes and sculpted noses, choosing from what
seemed like a catalog of Western traits.

Samsung lacked identity, Gordon could see, but perhaps its identity
crisis was a national one.

On one of his cross-country tours, the helicopter touched down on

Jeju Island at Samsung's Shilla Hotel, a honeymoon resort where the consultants would start planning the new design institute.

"The paparazzi rushed up when we landed at the hotel," Bruce told me. "They wanted to snap [pictures of] Miky with the three of us. It was quite a sight. And Chairman Lee [wrote] articles about the importance of design in the *JoongAng* [Samsung's newspaper] while we were there."

Dining in a private room in the hotel, where Samsung had hosted former Soviet premier Mikhail Gorbachev the previous night, Peter Arnell and Gordon Bruce got acquainted. They were asked to work together on a two-pronged strategy to build both Samsung's products and brand.

Though their relationship was cordial on the surface, tensions between the marketer and the product designer would stretch its bonds.

Gordon Bruce, as an industrial designer, homed in on the need for Samsung to win over customers with products of beauty and quality. Peter, as a marketer, focused on drawing customers in by stitching together a coherent image that would encompass all of Samsung's product catalog.

"Samsung needs to exceed customer expectations," Bruce flatly said in one of the meetings.

"Did you hear what Gordon just said? Exceed customer expectations!" Peter proclaimed at the table, acting as a sort of cheerleader.

Gordon was uncomfortable; he felt that Arnell was trying to take credit for his work and use his product portfolio for his own advertising.

"I refused to work with him after a while," the designer told me. He was turned off by Peter's dominating personality.

Peter had a different recollection of events. He called the ArtCenter designers "heroes."

One day on a tour bus, Miky Lee's hotline rang.

"It's the chairman!" she exclaimed.

She listened closely, then hung up and turned back to her American guests.

"We are summoned," she announced. "You are going to be received by the chairman at his home in Seoul."

Returning to Seoul, the team was driven by an escort up the aerie of hillside homes, approaching the chairman's home overlooking the helter-skelter cityscape from atop a quiet peak. Security guards opened the gates.

"We were immediately [struck] by the chairman's very, very refined taste," Gordon told me. "What he had in Korea was nothing like the showy mansions and gold-plated bathrooms [ultra-wealthy Americans had] back home."

This was something altogether different. The chairman had embraced subtle, understated displays of beauty and wealth, devoid of garishness. Every screen door, wooden fixture, and piece of furniture seemed to have been carefully selected and put in just the right place. Clearly the chairman had spent a lot of thought and time designing his home. Bruce and Miho saw that he knew the principles of good design.

The team walked by the chairman's monument to technology, a towering sculpture of dozens of televisions by a famous modern South Korean artist, Nam June Paik, before they came face-to-face with the chairman, his wife, and his inner circle in the chairman's living room.

They encountered a penetrating and reverent silence. Cigarette smoke drifted through the room. The chairman sat calmly, his gaze studious. A video camera recorded the meeting, which would be broadcast across the Samsung empire the next morning.

"I want you to do three things," the chairman explained. "First, make my designers creative. Second, make them globally aware. Finally, I expect a three percent return."

Bruce wasn't sure about the last request, but he suspected Chairman Lee was setting his expectations low.

"We realized that the chairman was seeking a *gaiatsu*," Bruce told me. "It's an outside force, the Japanese say, that brings about sweeping change."

BRUCE AND MIHO NEXT toured Samsung's Corporate Design Center, documenting what they saw and writing a report on Samsung's best and worst practices. They met with resistance almost immediately.

"The place was just shark-infested," Bruce said.

"What do you know about Korea?" a Samsung executive asked Bruce and Miho. Samsung's team leaders were especially hostile to Miho because of his Japanese heritage.

"I've been all the way up to the Yalu River," Miho retorted, reminding the Korean businessmen that he was a tank commander in the American army who had been shot and injured in the Korean War. He'd lost his brother to wounds sustained on the battlefield, defending South Korea from a communist invasion.

The Samsung executives fell quiet on this point. *A Japanese fighting for Korea?* The pair could sense their hosts easing up a little on them. But they knew there was still a hard road ahead.

"I was appalled about the way they viewed Miho," Bruce said later.

On February 15, 1995, Bruce submitted his critique and recommendations to Chairman Lee. But there was immediate pushback.

"The initial suggestions for department chairs in the report were rejected," he told me.

The chairman had told them that people don't change easily. And he was right. Gordon and James were scheduled to depart, but Miky asked them to stay longer to help implement the chairman's vision.

"Miky insisted that Miho and I become chairs of the new in-house design school," Bruce said, "which was now under construction and would be called the Innovative Design Laboratory of Samsung (IDS)."

At first they resisted. It would mean running an institute in a country far from home where they didn't know the language.

"We thought about it. We realized that Samsung was an influential company," Bruce said. "If we were patient, we knew Samsung could make a big ripple."

So the two went back to Miky and accepted her offer to run IDS.

BRUCE SETTLED INTO HIS new office inside the half-constructed building, set to open in the coming months, and peered out the window into the urban expanse. The autumn sky was grimy, with a grayish-yellow color, the "yellow dust," as Koreans call it, that swept in from the Mongolian desert each spring.

He daydreamed about riding on the steppes with the hordes of

Kublai Khan as they advanced southward through Asia, and then westward to eastern Europe, redefining the known world. Daydreaming was a designer's inspiration.

The cracking and clawing of machines jolted him back to reality. A construction worker was spraying lacquer onto the metal-paneled walls. A sickening, pungent odor drifted through the room, where it was breathed in by the staff and hovered over a small fortune's worth of computer equipment.

This was the Samsung Way, Bruce thought. Move fast, cut corners.

His mandate was to teach good design principles to Samsung. But the confusion, chaos, and last-minute construction of IDS was deafening his thoughts as Samsung prepared to open the lab. It was hardly a place for contemplation, solitude, and exercising the creative imagination.

Bruce decided to work from his apartment in Seoul. He had promised to give Samsung his best shot, but as he prepared to start design classes, he realized that reforming Samsung was going to be far more challenging than he had anticipated.

GORDON BRUCE INTRODUCED ME to Ted Shin, one of his best Korean designers. Ted had approached Bruce and Miho when they had started the lab, handing over a stack of paper filled with hundreds of design concepts for wristwatches that he had created. In the mid-1990s, Samsung made analog watches, mostly for Middle Eastern markets.

"They interviewed every designer who applied," Ted told me. Of the company's seven hundred designers, twelve got acceptance letters. "I was one of the twelve."

On September 1, 1995, classes began at the newly finished IDS. Bruce entered the classroom and introduced himself and his curriculum. Then, on day two, he pulled out a banana.

"Nature is the best designer," he explained. "The banana fits in your pocket. It comes in its own sanitary package. It's biodegradable. And the color indicates when the fruit is ripe."

"It's intuitive. I know you don't eat it this way!" he joked, pretending to eat the banana like an ear of corn. "The peel protects the inside.

That's sanitation." He peeled it open and took a bite. "Potassium! Delicious! Easily manufactured! It's the biggest market in the world. Everybody loves bananas!

"Now imagine if you could design a product that uses these same principles."

A befuddled silence fell over the classroom.

"You mean," asked a student, "you want us to design a cellphone in the shape of a banana?"

When Bruce and Miho received their end-of-term reviews, the Samsung designers marked them lower than South Korean lecturers, complaining that the two Americans didn't tell them what to do. They were accustomed to the rote memorization and top-down leadership style taught by South Korea's educational system.

The solution, Bruce and Miho decided, was to get them out of the classroom. Only by relieving them of the presence of their bosses and bringing them on a journey around the world could they provide Samsung's designers with inspiration.

Gordon Bruce leads an industrial design exercise with his students at the
Innovative Design Laboratory of Samsung, or IDS, in the mid-1990s.

Chairman Lee liked the idea. He approved the curriculum of the Global Design Workshop, a plan to take Samsung's design students for nine weeks each year to Europe, the Americas, or other Asian countries, where Bruce and Miho would teach them about different cultures, peoples, and their tools.

"A mobile phone, after all, is not just a device but an artifact, reflective of the society that made it," Bruce told his students.

For weeks at a time, he and Miho brought the Samsung designers to see Apple's first computers at the Smithsonian, the graceful temples of the old capital of Kyoto in Japan, the palaces of France's Bourbon dynasty, and the cars of Germany's world-class autobahn.

"Mercedes-Benz prospered with its design," Bruce recalled telling them. "But Mercedes had been through hard times. They'd been through the destruction of World War II and the division of East and West Germany."

The message? If Germany could progress this far, then Korean employees—another people divided into two countries that had experienced the devastation of war—could do it at Samsung.

Perhaps Bruce's landmark trip was to the Taj Mahal.

"I took off my shoes and gazed at the majesty around me. I'd seen the pictures and on TV so many times. But being there was completely different. It's immersive learning."

There were things you couldn't absorb from a photo, like the smell, Ted Shin realized. *It stinks,* he thought. "I learned something I couldn't really describe," Ted said, abandoning the South Korean practice of rote memorization he had grown up with. "It's not memorizing it."

Bruce explained to the students that the drape of sunlight reflected differently off the Taj Mahal's bulbous dome and marble minarets, shifting with the time of day; and this shower of sun was a symbol of god's presence. The imitation of nature was one of the goals of the Mughals who built the Taj Mahal. Marble was in short supply, so cutting waste and carving with precision were key. It was a story that was similar to the experience of those from South Korea, with its lack of resources.

The Taj Mahal became the inspiration for new product ideas that could be developed only through experiencing new environments.

One student told me he had been videotaping the inside of the Taj Mahal and realized that he couldn't capture everything around him with his flat, two-dimensional camera. So he went back to Samsung and designed a five-lensed camcorder that could capture angles not just in front of the cameraman but all around him. He won all sorts of in-house design awards.

"Hey, why don't we design something for ourselves?" someone in the classroom suggested when they returned to Seoul. "What kind of workstation do we want?"

It would be the capstone to their class. The students had been designing products for other people their entire careers. Now they'd have to pass the ultimate test: designing something good enough for themselves.

After days of toiling away at sketches of lamps and tables and chair placements, Ted approached the front of the class and nervously handed Gordon Bruce his blueprints. He was afraid Bruce was going to respond like his Samsung bosses, who picked their favorite ideas without consulting others.

Bruce thumbed through the papers and looked up at Ted.

"So what do you like?" Bruce asked him.

Ted froze. *What do I like?*

"I just couldn't answer," he said. "I never thought that my opinion mattered."

Ultimately, he made the difficult choice.

"The left side was a workspace for sketching and light model making," he recalled. "The right side was for a computer, and it had a pull-out keyboard tray. The area in the center was for pinups like sketches and research boards. Everything was ergonomically designed, in arm-reach distance."

Though many workspaces have this sort of layout today, back then it was novel.

12

Design Revolution

"THE UPCOMING TWENTY-FIRST CENTURY marks the Age of Culture," the chairman of Samsung proclaimed in his New Year's address in 1996, "an era in which intellectual assets will determine a company's worth.

"No longer is a company selling products. Instead they have to sell their philosophy and culture." He went on to declare 1996 the Year of the Design Revolution. Gordon Bruce had launched his first design class at Samsung four months earlier.

In the mid-1990s, the forces of "digital convergence" were leveling the giants of Silicon Valley, turning computers into televisions and phones into music players and music players into obsolete junk. Tech giants like Sony and IBM, which had grown up through hardware sales, were pushed back to the starting line along with Samsung. Every device was now being connected through design, software, content, and user experience.

At the chairman's proclamation, Samsung went into a frenzy. The team tasked with disseminating the chairman's messages, the Samsung Economics Research Institute (SERI), published a handbook of the chairman's tenets, called the "design scrapbook." The clear message? Samsung was to forge an identity of minimalism and sleekness. Among the benchmark products he identified were the Sony Walkman, the Coke bottle, the Braun coffeemaker, the Sydney National Opera House, the Minneapolis Zoo sign system (one of *Time* magazine's best designs of 1981), and the PowerBook, Apple's early laptop.

Samsung was eager to study Apple and its intuitive simplicity. The PowerBook, the chairman's design scrapbook noted, had to sacrifice features for portability in an era of boxy, large computer components, back when people bought PCs for their hardware features rather than their design. Apple designers threw out the internal floppy drive—a much-criticized decision—and installed a trackball mouse. It was new territory. Yet even at a $2,500 price point, this stripped-down Power-Book was a hit.

"Apple Computer realized that their goal was not to design a 'small computer,'" the booklet claimed. Apple learned "a lot about portability and the value of the product during the design process of the PowerBook." This was at odds with Samsung's affinity for feature-heavy devices.

Samsung was learning more about itself through the study of others. And ironically, a group of American foreigners at the lab were closing in on a South Korean design ethos.

Gordon Bruce wasn't satisfied with Samsung's design slogan, "Smart and Soft." Its products were highly logical and counterintuitive. The designers began pondering the concept of yin and yang, the juxtaposition of light and dark, east and west, heaven and earth, representing the simultaneous unity and duality of all things.

Bruce convinced Samsung to hire former IBM designer Tom Hardy, an outgoing and likable southerner from Georgia who bore a physical resemblance to Ernest Hemingway.

The Tao was Chinese; Samsung needed to draw on its Korean roots.

"I decided to use the *taegeuk*," Hardy told me. The *taegeuk*, which means "supreme ultimate," is the Korean variation of the emblem that appears on the national flag, the circle of blue and red interlocking spirals. It's placed on the center of the national flag, surrounded by

The *taegeuk* symbol on the South Korean flag.

four trigrams used in divination and symbolizing quatrains like spring, summer, autumn, and winter.

"I thought that might be something that could really resonate with them, and it would be an easier sell [to Samsung's designers] because they would appreciate what we're talking about," Hardy said. "Every Samsung product should be harmonious with a certain philosophy to include both the rational and emotional. . . . It has to work. It also has to look good. It has to resonate with people."

Hardy also zeroed in on a new design slogan for Samsung: "Balance of Reason and Feeling."

Reason and feeling, in his view, were a *taegeuk* for the technology world.

The slogan stuck; it reflected Samsung's identity seamlessly. The company embraced a two-pronged strategy, strengthening its consumer products while simultaneously churning out more and better chips, screens, and components behind the scenes.

Reason versus feeling, product versus brand, tradition versus modernity, a dynastic chairman versus the freewheeling world of technology. Samsung became an empire of dualities, in a constant struggle to find unity between its opposites and among its disparate offices, people, and products. The Silicon Valley firms, focused on narrower product lines, avoided such intense internal conflict.

Samsung's design office, looking to globalize its products, was sending designers to work at its young office in Palo Alto, California, which opened in 1994, and set up design centers in London in 2000, Shanghai in 2004, Milan in 2005, and New Delhi in 2008.

The Samsung designers in Northern California, in an office called Samsung Design America, began embracing a practice called perceptual mapping to identify their *taegeuk* of design. A new logo, used across the company, had one axis consisting of the opposites "simplicity" and "complexity" and another axis featuring the words "feeling" and "reason."

The designers put Apple in the quadrant of "simplicity/feeling" and Sony, the hardware maker, in the quadrant marked "complexity/reason."

Samsung's electronics were complex and rational, filled with square angles, placing Samsung closer to Sony's domain.

But being in this quadrant was undesirable, as it placed Samsung in narrow competition with Sony without differentiating its product. So the designers targeted their assault in the middle ground between Sony and Apple—which they felt gave the company its best market opening, overshooting slightly more toward Apple's soft and curvy designs.

The effort was a needed kick start to Chairman Lee's vision. One student developed Samsung's first "smart home" concept, uniting every Samsung appliance, washer, phone, and computer in a conceptual ecosystem. He won a design award from the chairman himself.

Lee Min-hyouk, who later designed the early Galaxy smartphones, used his new knowledge to convince Samsung to manufacture the first flip-cover phone, without an external antenna. It was called the "Benz phone" after a Norwegian newspaper likened its look and feel to a Mercedes-Benz car. At ten million units sold, it was a bestseller of its time.

In the late 1990s, Ted Shin, having graduated from Gordon Bruce's course, was working as a mobile phone designer in an elite in-house group, tasked with thinking up futuristic, far-off concepts.

"I called it the I-Phone," he said. "Information phone."

"The cellphone was becoming a hub of everything, a computer," he said. "My proposal was a color screen in which you can connect to the web, you can send messages, you can write with a pen, you can connect to whoever you want to."

Novel for its time, the I-Phone had a keyboard and made use of software and hardware advances that resulted in larger and sharper displays, allowing users to more easily surf the Web and text one another.

When he showed the concept to his managers, his boss said, "You can talk to people on a phone. Why do you need to send a text message?" When the project was quickly shut down, Ted quit.

"You can't quit," his boss told him. "We invested millions of dollars in you."

But quit he did. He went on to become a design professor at Metropolitan State University of Denver, a role that gave him the freedom to create his own design philosophy.

In fact, lifelong Samsung managers did not take kindly to this new generation of freer spirits Samsung had hired and trained.

"Miho and I were becoming disliked because we were changing the culture," Gordon Bruce said.

As the pair continued to lecture at IDS, Miky Lee made an ominous move affecting their design operation.

After Samsung's deal with Steven Spielberg broke down, Miky invested $300 million in DreamWorks on behalf of the CJ Group, the food conglomerate run by her rival arm of the Samsung family. She would use this access to turn CJ into a serious movie producer in South Korea. Employees received training from DreamWorks filmmakers, and CJ distributed Spielberg's films in South Korea.

CJ's food business was a far cry from the film business, but this was how South Korea's sprawling *chaebol* groups often worked, expanding into whatever industries interested their ruling families and might grow their profits.

CJ was still tied to the greater Samsung Group through a complex web of cross-shareholdings. But Samsung and CJ were completing a break from each other through a series of share swaps and sales, giving Miky and her side of the Lee family (descendants of Chairman Lee II's older, estranged brother) more freedom to act without Chairman Lee's approval, completing the inheritance process. Now CJ would stand on its own.

Miky's DreamWorks deal "was like slapping the chairman in the face," Gordon Bruce said. "Our connection to Chairman Lee evaporated as Miky—our conduit to the chairman—was banished from the IDS picture."

On her own, Miky would live out her dreams to promulgate Korean culture globally, going on to enormous success in her new production house, CJ Entertainment, releasing in 2000 one of South Korea's first megablockbusters, *Joint Security Area,* a story about North and South Korean soldiers who secretly befriend each other at the demilitarized zone (DMZ). It was a social commentary on the tragedy of the national division, and it resonated with South Koreans.

Miky's cinematic success helped fuel a cultural renaissance known as the Korean wave, in response to the growing popularity of Korean music, cinema, and food around the world. She never returned to the Samsung empire—which had to close its struggling film company, Samsung Entertainment, in January 1999 after three and a half years of operation.

At IDS the growing neglect from Samsung's highest leaders, and burgeoning resistance from the bureaucracy of Samsung's middle managers, led Bruce and Miho to leave the company in September 1998. Their legacy, however, would continue to define Samsung's design ethos. The company continued their practice of sending its designers on world tours to find inspiration for televisions and smartphones.

In 2001 IDS was merged with another Samsung institute, the Samsung Art and Design Institute (or SADI).

"I call it Spartan creativity," Rich Park, chair of SADI's product design department, told me on an official company trip in 2010. "The train's coming and you better get off the railroad fast."

AS CHAIRMAN LEE PUT on a progressive face for his American guests in the mid-1990s, he was being enmeshed in an investigation into money he and eight other business leaders had paid to a former Korean dictator, Roh Tae-woo, in the 1980s.

Prosecutors announced they were trying to end the "Korean tradition of collusive ties between politics and business."

The chairman was called in for ten hours of questioning; he insisted that he had given about $10.5 million to Roh Tae-woo, strictly as a "donation."

But prosecutors noted that as Samsung's cash made its way to the government's coffers, his father's business seemed to get the government licenses it needed to expand.

"Back in the old days, everything was done that way," Henry Cho, the chairman's nephew, told me.

As the investigation unfolded, formal accusations followed.

"From Samsung Group Chairman Lee Kun-hee (53 years old), [President Roh Tae-woo] received 2.5 billion won [$32.6 million] in nine installments"—more than three times the chairman's claim of $10.5 million.

On January 29, 1996, with an entourage of lawyers and assistants, Chairman Lee showed up at the courthouse. A deluge of reporters and cameras followed him, shouting questions.

Inside, the accused businessmen and the state officials who were accused of receiving funds were seated before the judge.

"Why did you give the president such a large sum of money?" the prosecutor asked Chairman Lee during his testimony.

He answered, "We customarily gave such money since the days of the Third Republic [the name of the dictatorship of Park Chung-hee from 1963 to 1972]. Generally, we called them political funds. Following precedent, we paid political funds once or twice a year to the Blue House. For business owners like us, we came to think of this as obvious and a kind of custom."

The prosecutor replied, "Can you just hand over that kind of money without any kind of interest relationship? Was there not an implied request for favors in Samsung's pursuit of the commercial vehicle business?"

"The purpose was that the company not be harassed by political power or sustain unfair losses [of government intervention]. . . . Samsung has been the biggest victim [of such power abuse] since the Third Republic. We've come to think of these things as a kind of tax."

It was time for the judge to ask Chairman Lee some questions.

"Why launder the money when giving it?"

"If the money isn't laundered it isn't accepted, so it is for mutual convenience."

"From [an average citizen's] perspective, it's a tremendous sum of money. Did you hand it over without so much as a receipt?"

"In the Third Republic, when the Blue House [the President's house] called, it meant they would ask for money. In the Fifth Republic, they gave a receipt. In the Sixth Republic, it was done with tacit understanding."

When it came time for the defendants to respond, one lawyer signed off with an unusual defense.

"The only reason this trial is happening is because [President] Kim Young-sam refused to take political funds. If that were not the case, this trial would not have even opened."

And with that, the hearing ended.

The judge later found the chairman guilty of bribery and sentenced him to two years in prison. But he went on to commute the chairman's sentence.

"[The court] has taken into account the fact that Lee Kun-hee, as Samsung Group Chairman, contributed a considerable deal . . . with his business activities; the fact that he continues to contribute greatly to the advancement of the nation's sports, culture and arts; the fact that he is deeply repentant of his offenses; the fact that he has pledged to strengthen and disseminate sound business practices following this event; and the fact that he is a first-time offender, among others."

He exited the courtroom and was driven off by his chauffeur a free man.

But the Korean business community reacted with disgust to the verdict.

"The verdicts are shocking," a spokesman for the Federation of Korean Industries, the lobbying group first chaired by Chairman Lee's father, told the *Los Angeles Times*. "This will gravely affect Korean business activities overseas. It will also hurt the morale of the business community."

In October 1997, Korea's president, Kim Young-sam, pardoned Chairman Lee and the other convicted businesspeople.

The minister of justice told the *Korea Economic Daily* newspaper they had been freed "out of consideration for the fact that their imprisonment would lead to a restriction of their business activities and negatively impact their credibility overseas."

"My uncle's very lucky," Henry Cho, the chairman's nephew, told me. "He didn't go to jail. A lot of people did."

———

AS THE CHAIRMAN WAS awaiting his pardon in the summer of 1997, Nam S. Lee, one of his aides in the Tower, noticed the tense silence on the twenty-eighth floor of the Samsung headquarters.

"We knew something was wrong," Nam S. Lee told me in a coffee shop in central Seoul.

The stock tickers and news headlines across Asia were ominous. "The Thai Gamble: Devaluing Currency to Revive Economy," read a *New York Times* headline in July 1997.

Runaway currency speculation and high-risk, high-debt economic policies across Asia led to financial collapse. One by one, currencies were falling off a cliff, from Thailand to Indonesia, wiping out the savings of farmers and workers and later prompting Indonesian students to rise up and topple their longtime dictator, Suharto. Harried American officials suspended trading on the New York Stock Exchange, reacting to its worst decline ever in a single day until then, as the contagion spread to Japan, Brazil, and Russia.

Despite the "Asian financial crisis," as it was called, Chairman Lee stubbornly insisted that Samsung become an automaker.

"I've studied the automobile industry more than anyone else. . . . There are concerns that we made a mistake in starting the automobile business, but I am confident that Samsung Motors, which will be launched in March 1998, will put these misunderstandings and fears to bed."

The Samsung Group had been loading up on debt, as a result of this and other aggressive expansions. Interest payments alone were crippling Samsung Electronics' respectable operating profit in 1996 of $1.8 billion—which after deductions came to a net profit of $202 million.

In March 1998, as Korean companies were defaulting on their loans, Samsung's first car rolled off the $3 billion assembly line in the southern port city of Busan.

Samsung employees rushed out to buy up their employer's cars. In doing so, they saved the company, since these sales accounted for most of Samsung's 45,000 automobiles sold that year—low for a plant that could manufacture 80,000. Worse, Samsung was losing $6,000 per vehicle. The next year, car sales fell precipitously to fewer than 7,000.

It was a painful moment—the worst loss of face for the chairman since he became the head of Samsung. Samsung Motors went under creditor protection and was later bought by Renault in a fire sale, which created Renault Samsung Motors. Samsung's name was kept, since the company held on to a minority ownership.

Samsung instituted emergency financial measures unheard of in South Korea.

"There was a sense that this entire company could go down," Samsung Electronics CEO Yun Jong-yong later told *Fortune*. "It was that extreme."

South Koreans donated their gold rings and trophies to help the government pay its colossal bill to the International Monetary Fund, which had just approved the largest bailout in its history up to then— $55 billion. Samsung Electronics laid off one-third of its workforce, or about thirty thousand people, replaced half of its senior managers, and sold off $2 billion in non-core businesses, including an executive jet and an entire semiconductor division. Senior managers took a 10 percent pay cut.

"Workers can no longer count on such perks as preschool tuition," *Fortune* reported. "At headquarters, the thermostat is now set so low that execs wear thermal underwear."

In the most devastating act to loyal Samsung employees, South Koreans abandoned the expectation of lifetime employment, driving a dagger into the tradition of filial piety.

The chairman had been preaching crisis with little success until now. Having slashed Samsung's bureaucracy, the chairman's empire was emerging nimbler, ready to focus on the businesses that mattered: mobile phones, televisions, LCD displays, lithium ion batteries for cellphones, and NAND flash memories—Samsung's last investment, which later made advances in the iPod possible.

"Think about it this way," the chairman's aide Hwang Young-key told me. "The Frankfurt Declaration, in 1993, was when the chairman laid down his vision. But the Asian financial crisis was his chance to execute it."

13

My Boss the Shit Kicker

THREE FRANTIC SOUTH KOREAN executives showed up at Pete Skarzynski's office in Richardson, Texas, in January 1998. "What do you need to double your plan for the year?" one asked. "We want you to grow six times."

A gregarious and talkative veteran of AT&T and Lucent Technologies who once did a stint in London debriefing teams of CIA officers on the telecommunications industry, Pete had joined Samsung a year earlier as senior vice president for sales and marketing at Samsung Telecommunications America.

"In essence," the tall and always smiling redhead told me over sushi and sake in Dallas, "I ran the American cellphone business."

Samsung's first employees in Richardson hoisted the biggest logo in the neighborhood up onto their new six-floor building, where Samsung occupied four floors, despite the fact that only seven employees worked there at the beginning. Chairman Lee had given Skarzynski a blank check to work with Peter Arnell in New York and others to put Samsung phones in the hands of Americans. The timing was perfect, and not just because of the economic crisis. In 1996 the South Korean government had finished building a Code Division Multiple Access (CDMA) cell tower network—American carriers were using a different network to this point—and this smart government policy immediately made Samsung an early mover in CDMA phones.

Sprint was going to be the first U.S. carrier to adopt the same network.

"I need a cheap Asian company in here, and you guys are doing something," a Sprint executive told Skarzynski, inking a $600 million deal to sell 1.7 million bottom-shelf Samsung phones.

Before the economic crisis in Asia, Samsung had been reluctant to sell phones in the United States, because it could sell them at a significantly higher margin in South Korea's protected market.

But with the economic crisis, Skarzynski was told, "You're the most important guy in the company because you can bring in hard U.S. cash currency and we need that."

The plan? The fast-rising and soon-to-be CEO of Samsung's mobile division, a former factory floor manager named Lee Ki-tae (K.T.), flew out to Texas—and Skarzynski occasionally flew to Korea—as the two plotted how to win carrier contracts.

Their strategy was simple: They would bend over backward for the carriers, charm them, massage them, buy them drinks. They'd win them over by elevating Samsung's product quality. They'd release novel designs and give them lots of marketing money. They'd position Samsung phones as the best value for what the carriers offered, not simply the cheapest.

"In my view," Skarzynski said, "I had ten customers. We'd go see the CEOs of the ten customers."

Back then, carriers were the gatekeepers of the cellphone industry, the channel through which cellphones were sold and marketed.

Skarzynski and K.T. quickly got acquainted on the Samsung corporate jet. On one long-distance trip from Texas to Seoul, Pete was about to put on his headphones and watch a movie when a clamor arose in the back of the airplane.

"You could hear K.T. yelling in the back of the plane," he told me.

For about three hours, K.T. shouted and berated his dozen or so staffers, rattling off the Korean equivalent of "sons of bitches" and worse. Their transgressions? To K.T., they were not working hard enough. *Work harder, go faster,* he yelled at them.

I'm glad I'm not a part of that, Skarzynski thought.

When they landed at Seoul's Gimpo International Airport, K.T. gathered his staff in the airport lounge, where he continued to let loose as he chain-smoked. Nor did he stop at the company lunch. His staffers solemnly bowed their heads at the table, not daring to touch their meals.

Unsure of what to do, Skarzynski started eating, along with a non-Korean colleague. No one seemed to care.

"It was a couple hundred dollars' worth of food no one ate other than two of us," he told me, "because we didn't understand Korean."

As they prepared to get on the plane again, K.T. glared at a vice president.

"You can get on the corporate jet if you think you've done your job."

The vice president declined to board.

"Well, I'm sure we can get you a ticket on a commercial flight," someone told him.

Later, Pete and two hundred other executives had flown in for Samsung's annual meeting of global executives. The first foreigner to be promoted to such a high level, Skarzynski was the lone Westerner in the room, unable to speak the language or understand the cultural nuances.

Little did he know the "global meeting" was going to be more like a Communist Party self-criticism session. He took his seat next to an interpreter, who whispered translations into his ear and explained how things worked. The translator informed him that everyone in the room was seated in order of rank and performance. Top global executives sat up front at the bigger tables, country managers behind them. About forty rows back in the auditorium, near the exit sign, were the heads of the lowest-performing subsidiaries.

"I was in row seven or eight," Skarzynski said, relieved.

The chief financial officer from the feared financial unit that was close to the chairman took the stage bedecked in a flower bouquet and the Korean flag.

"I didn't really prepare anything to talk to you about. So you know what? If you have any questions, why don't you ask me some questions?"

Silence.

"You could hear people breathe," Pete said. "You could hear someone try to hold their cough in. Everyone had their heads down a little. This went on for three minutes. Dead silence for three minutes."

Finally the CFO broke the silence. "If you don't have any questions for me, I have a few questions for you."

He called on an executive. The man stood up, terrified.

"Could you tell your colleagues in the room why you decided to buy all that material? Since you weren't able to sell it, you had to discount it, didn't you?" the CFO said. "Please explain to everybody your strategy and what your thoughts were on this."

The executive, too terrified to speak, bowed.

"I apologize to all my colleagues."

"For the next ten, fifteen minutes, the CFO talked about whatever he did and how bad it was and how damaging it was," Skarzynski said.

The executive, shamed, scorned, and suffering a devastating loss of face, sat down. The CFO then called on another executive. And he proceeded with a similar self-criticism exchange.

"It went on for the most painful two hours," Skarzynski said. Luckily, he was never called on.

But the yelling and shaming of employees—this ingrained fear of failure and loss of face—paid off for Samsung in its own way.

In his meetings with America's carriers, K.T. had a chance to show off the progress that he'd made with his phone developers.

K.T. and Skarzynski made a pitch to Sprint in 1999 over dinner in Kansas City. K.T. began with a quality-control demonstration. He held up Samsung's latest flip phone for the Sprint executives. He attacked it, pulled at it, tried to squeeze and twist the flip cover in every direction, struggling to bend and break it. But he failed to do any serious damage. He even dropped the phone on the floor and stomped on it. Then he picked it up and made a call. And it worked. Then he turned to a Sprint engineer at the table.

"Can I borrow your phone for a second?" he asked, reaching for the engineer's handset.

K.T. opened the flip, bent it, and broke it apart with the mere twist of a hand before giving it back to the engineer.

"Can I see your Motorola phone?"

"You're not gonna break this one, are you?"

K.T. took the phone and held it up to his face. The table was nervous.

"Motorola, get outta of here!" he shouted at the phone in his Korean accent. He gave the phone back fully intact.

K.T. was an instant hit at Sprint; his relationship with Sprint soon bordered on a bromance. Samsung invited Sprint CEO Ron LeMay and Sprint PCS COO Len Lauer to South Korea for a grand tour of its factory lines. After breakfast at the Shilla Hotel, they joined Pete in an entourage of three limousines. With emergency lights flashing, "We had police escorts, during rush hour, to stop traffic across the bridge and the streets and let us through," Skarzynski said.

They arrived at the helicopter pad by the river and boarded the company helicopter, flying to the handset factory at Gumi. Upon landing, they were greeted by cheering crowds of factory workers and women in traditional garb offering giant flower bouquets. Each factory line hoisted conspicuous signs that said SPRINT NUMBER ONE, SPRINT NUMBER TWO, SPRINT NUMBER THREE.

"So we take the helicopter back and hit the five-o'clock rush hour," Pete said. "It's brutal. Worse than New York City." But the police escort showed up again for the dignitaries.

"What's nice about this is I enjoy having a customer who owns a country," a Sprint executive remarked. "How about one more spin around the city?"

The nights were even crazier, full of *heung,* the Korean word for carefree joyousness. The Samsung Men took drinking, merriment, and toasts to an extreme as they entertained the foreign executives during their visit. After shows by national musical troupes dancing to traditional drums, dinners served up by Korea's finest chefs, and endless shots of *soju,* the company brought out a giant cake that the CEO cut with a sword. Samsung employees popped open aged whiskey worth hundreds of dollars and, to the shock of their foreign guests, dropped shots of it into glasses of cheap beer, which Samsung employees called "atomic shots," for a special kind of Samsung toast.

"Colleagues, we had a tremendous day," a Samsung manager proclaimed. "This collaboration was fantastic, and it's going to make us number one in market share. We're going to beat Motorola by 2000. When I say Samsung is number one, you say 2000, 2000, 2000!" Everyone raised their glass and drank, in a noble Samsung tradition. All fifty people at the table then serenaded their employer and finished with a drink.

Samsung was receiving more and more American orders, winning over contracts with the mobile carriers. In 1999 its stock rose from its crisis-stricken discount rate by 233 percent, hitting $227 a share. The strategy was paying off. Samsung was exporting its way out of the financial crisis.

These were still me-too phones, mimicking existing phones. Pete still needed to appeal to Sprint's bottom line, which was to deliver to its executives something more profitable, with brand-building power. So he stopped by for a chat with the product designers at Samsung. They introduced him to a new phone called the SCH-3500—a clamshell model inspired by the Motorola StarTAC, the dominant handset at the time. But the SCH-3500 had a unique curvy flip-up earpiece clipping to the display.

"What's that?" Skarzynski exclaimed. "I have to have that."

The SCH-3500 gave the flip phone of the day a stroke of flair. But the engineers explained that the phone was a bust in South Korea. Pete thought there might be a market in the States. He carried a prototype home and showed it to his Sprint counterparts.

"This is great. We're going to do an exclusive with you guys," a Sprint executive told Skarzynski, admiring the clever design.

They signed a deal to buy roughly three million SCH-3500s from Samsung. It was a huge order at the time.

"We developed the phone with a big launch. We had all these big promotions planned," coordinating with Sprint, Skarzynski said.

But when Samsung launched the phone in Europe, it flopped there, the same way it had in South Korea.

It was not a good sign. Samsung engineers informed Pete, "The factory has decided not to ship this phone. It failed and we don't want to be stuck with inventory," Pete recalled.

Skarzynski panicked. It was two weeks to launch, and he was getting inundated with calls from Sprint.

So Skarzynski jumped on a plane to South Korea.

"Here is all of our research," he told the assembly-line managers in frantic presentations. American consumers, research showed, liked the unusual design. "You're in a real danger zone," he explained, in canceling this phone with Sprint. "Trust me," he pleaded.

In the end, with the data to back Skarzynski's order, headquarters decided it was better to preserve its new and uncertain relationship with Sprint. Skarzynski got his phones, and the launch went forward as planned.

"The SCH-3500 became the flagship product in the early days for Sprint," Skarzynski said. "It was a great product and sold more than six million units. Generally products last about a year, but this product lasted about two years."

And Samsung's Texas office began getting more inquiries from the other carriers.

K.T. AND SKARZYNSKI TRAVELED to Verizon's headquarters in New York City, where they hoped to land more orders. But Verizon had a long-standing partnership with Samsung's crosstown rival in South Korea, LG.

As K.T. prepared to discuss the latest Samsung models, one of the Verizon executives handed over his business card. To the surprise of everyone in the room, K.T. rejected it.

"You buy LG," he said, scowling. He threw the business card back at the Verizon executive.

"That hurt us big for quite some time," Pete said. "It was like, all right, okay. Nice meeting you, K.T. Now cancel those orders."

Keeping an eye on Samsung's Korean executives—the ones who had never traveled before to the United States—became a matter of survival for Skarzynski, to ensure Samsung wasn't embarrassed in front of customers or in the press. After a night of getting smashed with carrier executives in the United States, he learned from a sales employee, the visiting South Korean executives had left a bag with cash inside at the front desk, with instructions to give it to a Samsung buyer staying at the hotel. Perhaps it was a mistake made in the previous night's drunken haze, or maybe it was a bribe to win the favor of the carrier—Skarzynski thought the latter was more likely. In South Korea, getting drunk with business partners is the time when real business gets done. After bonding over booze and karaoke, it's an accepted practice to roll out bags of cash and other gifts for your partners. Samsung executives, Pete said, preferred rounds of golf and drinking to entertain executives, not escorts and bribes—though he admitted Samsung made the occasional slip.

The buyer took one look, called someone at Samsung, and delivered the cash back to the company. Pete raced into the office of his boss, Jeong Han Kim, whom Pete knew was steadfastly honest.

"I know we can't stop all of this," Skarzynski said to him. "But please. This is really embarrassing."

Samsung still hadn't won over Cingular, a newcomer among carriers, founded in 2000, and a big target. Skarzynski struggled to get a meeting for K.T. with Cingular's brash CEO, Stan Sigman. Finally Cingular agreed to talk.

Skarzynski and K.T. arrived at Cingular's headquarters in Atlanta. The meeting was set for Friday at 4:00 P.M.—a sign of its low priority. It was the worst possible time to pitch a sale, as everyone was heading home for the weekend.

K.T. and an unusually large delegation of Samsung executives—a dozen or so handlers, as per Korean custom for a man of his rank—waited in the lobby. Skarzynski watched as the clock hit four, and still Sigman didn't appear, an insult in South Korean business. The wait lasted an excruciating thirty minutes.

Skarzynski, fearing K.T. would perceive this as a "big insult," pleaded with Sigman's assistant to at least get K.T. out of the lobby.

"Come on up. The meeting's really running late," said the secretary.

Unfortunately, there was another problem: K.T.'s entourage could barely squeeze into the small conference room. In the end, K.T. picked three of them. Then they waited for another half hour.

Finally Sigman entered. A portly Texan with a cutthroat demeanor—"a shit-kicker kind of guy," as Skarzynski described him.

There was nothing friendly about Sigman's expression. Skarzynski started his spiel on behalf of K.T., only to be interrupted by Sigman.

"I don't really ever meet with vendors," Sigman said. "Unless there's something interesting or unique you can tell me, I have no reason to work with another vendor."

As the interpreter began translating Sigman's response into Korean, K.T., who understood English, looked Sigman in the eye.

"Stan, I like you. You get right to the point."

As it turned out, the pair clicked.

"Okay, K.T. Show me what you got."

From one son of a bitch to another, Skarzynski thought.

But no major deal came out of the meeting, and Cingular remained a hard sell for years to come. Samsung's break came much later, in 2005, when Hurricane Katrina wrought devastation in Louisiana.

"Stan is creating a recovery fund, and he would like everyone to chip in," a Cingular executive told Samsung executives. "Everyone's giving around fifty thousand dollars."

"We've got a half million dollars in our budget," Skarzynski told his boss. "Let's give it to Sigman and fund a half-million-dollar contribution for the victims of Katrina."

At ten times the pledge of the other phone manufacturers, Skarzynski was worried the gift would come off as a pay-for-play deal. That's certainly what it looked like. Skarzynski said he just wanted to show a commitment to Cingular. So he told Cingular, "We don't want you to advertise that we gave this amount to you. Simple as that."

The strategy worked. Cingular needed a manufacturer that could be the engine of its 3G initiative, and Samsung was the first to offer it, with Samsung's BlackJack phone, an early smartphone designed to compete with BlackBerry.

But the name "BlackJack" was too close to the name of the Black-Berry. Research in Motion (RIM), the maker of the BlackBerry, filed a lawsuit. So Samsung settled the lawsuit and eventually renamed the phone "Jack."

But bottom line, Samsung became a big supplier to Cingular for its 3G push.

14

Sony Wars

PETE SKARZYNSKI, K.T., AND Peter Arnell were given enormous freedom by Samsung in America and were pretty much running their own show. That is, until a Korean American named Eric Kim was hired by Samsung Electronics as its executive vice president for marketing in 1999, and then became its global chief marketing officer.

Samsung had never before hired an outsider into a position of such authority at its headquarters. Before this, executives needed to be Korean and rise through a lifetime of employment as Samsung employees. Kim took the stage in front of four hundred Samsung employees for his introductory speech. CEO Yun Jong-yong, who had hired Eric, sensed he would be resented and preceded him with an introduction true to Samsung's heart.

If you mess with him, "I will kill you."

Kim's first appearance wasn't a blistering success. A Korean who'd lived overseas for decades, Kim wasn't seen as a "true" Korean by this tight-knit company, even though he felt his colleagues went out of their way to make him feel accepted. Nor was he able to muster the fluency in Korean the audience expected. One Samsung vice president called Kim a "kimchi-eating American."

"Was it hard being a Korean American there?" I asked Kim over dim sum in Berkeley, California, where he's now retired.

"I'm Korean," Eric insisted in a soft tone, disagreeing with how his colleagues viewed him and the press covered him. "I was born in Korea. I moved to California as a very poor immigrant when I was twelve. My Korean-language skills never improved beyond the twelve-year-old level."

"What about your unconventional background as a marketer?" I asked.

"My first job was as an engineer," he told me. He had a graduate degree in engineering from UCLA. His former colleagues would repeat those words, with either favor or disdain: "Eric was an engineer." But he also had degrees in physics and business from California's Harvey Mudd College and Harvard Business School.

Before Samsung, he'd been CTO of the Dun & Bradstreet Corporation, a firm that crunched data on people's credit history, and CEO of Pilot Software, a business intelligence vendor.

"Now, the CEO wanted me to unify the entire brand under one umbrella," Eric said. "Samsung's brand was scattered. Each regional office was doing its own thing. If we wanted to become a global company, we needed to direct these efforts from headquarters."

Samsung needed one giant marketing cannon with which to blast its messages out; right now, the brand sprayed its messages more like a BB gun, spewing little metallic balls one after another, without overarching purpose, direction, and vision. Samsung had fifty-five advertising agencies under contract. American offices had three minuscule accounts of $20 million each.

"Eric wanted to start fresh," said the vice president for new business development, Thomas Rhee.

Kim and his new marketing office began traveling throughout the company, convincing Samsung's top officers of the merits of global marketing.

"Marketing was viewed as a kind of sales-promotion function," he said. "My job was to change the mindset." As he consolidated marketing under his watch—the whole point of hiring him—he ran into resistance from the C-suite down to the manufacturing plants. That resistance was especially strong from the product developers and the finance people.

"K.T. hated his guts," Pete Skarzynski said. "He didn't want anyone messing in his sandbox." But the resistance came not just from K.T. but from a generation of factory floor managers. They'd built their company from the rough-handed factory floor. To them, Eric Kim was a silver-spooned dilettante who knew little of the hardships Samsung had gone through, and in their minds he was preparing to spend too much of their hard-earned revenues on marketing.

"He got a reputation for wanting to dine at the finest restaurants," Skarzynski said.

"There's no substitute for a constant stream of bigger and better new products," a Samsung line manager later told him. "Money spent on R&D is always a better investment than money spent on advertising."

Kim stayed the course, bringing together a global marketing budget of $1 billion under his Global Marketing Operations (GMO) unit and a $400 million advertising budget—a big account but not an endless amount, considering all the products and divisions he had to advertise—and he ordered an ad agency review. He also ended the practice of setting marketing budgets at a percentage of *current* sales—which encouraged employees to work for the here and now—and began setting the marketing budget to "potential sales." That would force them to think of the long view.

Next he had to forge an overarching Samsung marketing identity.

"Products don't sell themselves," he told me.

He enlisted Peter Arnell to help.

"We had a brainstorming [session] at this office, and then we had a dinner," said Kim's marketer Thomas Rhee. "Sometimes Peter said, 'Oh, I got an idea!' and scribbled it on a napkin at a restaurant." He recalled that Kim and Arnell wanted a slogan with a sense of novelty, excitement, inclusion, and optimism as the millennium approached.

"Digital" was the buzzword of the era. Arnell scribbled variations on his napkin. He added an *L* to digital. The result? "DigitAll."

"A lot of ideas started on the back of a napkin," Thomas Rhee recalled.

Eric Kim was pleased. He settled on the new campaign, which would last for years to come: *DigitAll: Everyone's Invited.*

———

ARNELL DIDN'T CONTINUE WORKING with Samsung much longer. Samsung was looking for a break from what one consultant jokingly called his "soft porno." One television ad showed an almost-nude woman caressing a sculpted male model on her television set to the music of a seductive electric guitar.

"I love my TV. It's Samsung," she said in an erotic embrace.

Kim's DigitAll campaign was unleashed on November 10, 1999, steamrolling over Arnell's earlier campaigns, blaring out Samsung's new mission: "to integrate digital home, mobile and personal multimedia on a global scale."

For Arnell, Kim's much-anticipated ad agency review in 2000, which was getting headlines in *Advertising Age* and *Adweek,* was the end.

"Peter kept Eric waiting for almost an hour or so. He blamed the traffic," Thomas recalled. "Eric was really pissed off."

The two parted ways. Eric brought much of the advertising work back in-house to Samsung's agency, Cheil.

BUT THERE WAS A deeper purpose to branding campaigns like this. Behind the scenes, Samsung executives were benchmarking themselves against Sony. But no one wanted to say it publicly and risk angering their Japanese rival, which happened to be a massive buyer of Samsung parts for its PlayStation and Walkman CD player.

But in April 2001, Kim gave an interview to *Forbes*'s Heidi Brown that was the first public acknowledgment of the conflict.

"We want to beat Sony," he told the reporter. "Sony has the strongest brand awareness. We want to be stronger than Sony by 2005."

The *Forbes* article made headlines in Japan, where Japanese business readers reacted with great interest. Korea, after all, had a long and troubled history with Japan, and few people thought a Samsung coup against Sony—the Apple of its day—was possible.

Samsung executives, on the other hand, were freaked out about Kim's slip. Sony was Samsung's customer, not just a competitor. How could Eric Kim, an outsider, challenge a partner so boldly and publicly? And so Kim was reprimanded. "The article almost got me fired," he told me. "But it rallied the troops!"

This was a dilemma that Samsung executives would continue to face with companies like Sony, Apple, Motorola, and whomever else they supplied but also competed with.

"Can you please not publish that?" Samsung's PR representative once asked me when, after an official interview with my recorder out, an executive vice president for strategy began talking about how Samsung was in a "weird situation" by supplying parts to HP and Nokia, two customers-cum-competitors. I didn't include the quote in my reporting, thinking at the time that it wasn't that interesting.

But if Sony felt Samsung was pushing too hard, it could cut back Samsung as a supplier and potentially cripple Samsung's memory chip business. And memory chips were the bread and butter of the empire. It could be the end of Samsung's global ambitions.

There were other reservations in the executive suite. "Let's not overspend on advertising and promotion just for the sake of beating Sony in some *BusinessWeek* ranking," an assembly-line manager later told Kim. "The brand will gain consumer credibility naturally, at its own pace. You can't force it."

Two months later, Sony CEO Nobuyuki Idei offered his perspective from the Japanese side, gazing down at Samsung, to *The Wall Street Journal*.

"The product design and the product planning—they're learning from us. So Sony is a very good target for them." But, he dismissively added, "We still believe that Samsung is basically a components company."

Then Sony retaliated in a way that sent Kim up the wall, in a product placement in the new movie *Spider-Man*, starring Toby Maguire.

"Once upon a time," the judge wrote in his poetic ruling in a court case on the use of fictional Times Square logos in the film, "at a gathering of many thousands in New York's Times Square for the 'World Unity Festival,' the crowd was murderously attacked by the jet-powered Green Goblin, who was, however, eventually put to flight by the timely arrival of Spider-Man."

As the Green Goblin flung little green grenade-like devices at the parade of innocents, the redheaded Mary Jane Watson, Peter Parker's love interest, was helpless, trapped on a balcony with the villain.

"It's Spider-Man!" shouted a bystander.

Slinging his web through New York's concrete jungle, the superhero swooped down to catch a child in danger and flung himself in front of Two Times Square, the iconic high-rise of flashing corporate advertisements. *Spider-Man* was set to hit national theaters in May 2002 with this scene.

Samsung employees watching the trailer noticed a peculiar logo on the iconic Two Times Square building, where Samsung's signage should have been. Instead of SAMSUNG it said USA TODAY.

"I . . . went ballistic," Kim told me. "This was too blatant an affront to Samsung, and I felt that we could not accept the situation."

In the real-life Times Square, Samsung had spent truckloads of cash for that top-shelf ad space; its blue oval logo was in the second-to-highest spot. Columbia Pictures, the studio behind *Spider-Man* and owned by Sony, was clearly playing foul.

Samsung's legal team prepared to join a lawsuit whose plaintiffs included NBC—which, as a Sony Pictures competitor, had also had its logo snipped out—and property owners in Times Square who felt their rights were violated.

But before going to court, Samsung executives convinced Sony to reinstate the logo, using the leverage of being a parts supplier.

"At the end of the day, Sony needed Samsung for its components as much as Samsung needed Sony," a lawyer familiar with the case told me.

Samsung's decision to settle was undoubtedly the right call. For the other plaintiffs, the judge ruled in Sony's favor on free-speech grounds. Since *Spider-Man* was a fictional work of cinema, the film studio was free to alter the real-life Times Square as it saw fit.

"They tried to remove us from Times Square. But Samsung is still alive," Chin Dae Je, president of the digital media division, bragged to the audience at a Manhattan party.

Eric Kim became a hero within Samsung.

———

IN 2002 WARNER BROS. approached Kim with an offer: It was planning a sequel to the iconic film *The Matrix,* and the filmmakers needed someone who could design the green-and-black futuristic handsets for protagonists Neo, Trinity, and Morpheus.

On the surface, *The Matrix Reloaded* looked like an extraordinary product placement opportunity. The sci-fi franchise had incredible pop-culture appeal and a fanatical following. Nokia had broken through with a product placement of its own in *The Matrix,* released in 1999. The Finnish company designed a sleek, minimalist phone for the movie's main character, Neo, played by Keanu Reeves, who receives a surprise call from Morpheus warning him that Agent Smith and his goons were searching for him in his office. The Nokia phone was a hit. To this day, *Matrix* fans buy and sell Nokia's imitations of the phone on eBay.

During the 2002 Consumer Electronics Show in Las Vegas, Kim and his marketing team dined with a delegation of about two dozen executives from Warner Bros.

Some of the people on Samsung's side of the table were skeptical.

"I knew immediately what was happening," Peter Weedfald, CMO of Samsung North America, who reported to Kim, told me over lunch near his new workplace in New Jersey. "Warner Bros. smelled huge opportunity and was surrounding our [global] CMO with executive muscle, smiles, and hunger."

Eric Kim called Peter over.

"We'll have a Samsung product placement in *The Matrix!*" Kim exclaimed at the table. Samsung, in return, could use *Matrix* footage and references in its phone advertising campaigns.

Introducing himself to me as "the Sledgehammer," Weedfald was not the kind of person you could win over easily. "Brand is the refuge of the ignorant" was one of his sayings. He believed that products, not brands, were the core of any marketing effort.

Weedfald whispered to his boss that Samsung did not have to give Warner Bros. any money up front. All it needed to do was promote the movie through its advertising channels.

Kim, to Weedfald's surprise, became angry and told him to go back to his seat.

Weedfald, K.T., and Pete Skarzynski accompanied Kim to follow-up meetings with Warner Bros. where they planned the look, feel, and build of Neo's new phone.

"You could tell pretty quick the bullshit and the rigmarole that was going on, and how [Eric] wanted to deal so bad he would do anything," Skarzynski told me.

The film's producer, Joel Silver, was demanding and opinionated, "notorious for squashing people at a moment's notice," according to one advertising executive. Warner Bros. was also secretive and difficult to work with, allowing only one representative of Samsung's ad agency to read the script behind a locked door guarded by security officers. Samsung, it seemed, didn't have much leverage in the deal.

K.T.'s team was annoyed at the imposition of brand marketing on his product unit. Weedfald and K.T. needed to protect their carrier relationships. A Samsung engineer can't simply create a phone at the behest of a Hollywood filmmaker. The agreements with Sprint and others mandated a process of reviews and approvals for projects like this.

Watching him in these meetings, Skarzynski described Kim as "a neophyte."

"He didn't understand anything about products. . . . [He] had no clue." As Skarzynski made his presentations on what Samsung could design for Warner Bros., he sensed the Warner Bros. executives nodding off. He got the impression it was not a high priority.

Warner Bros. went with K.T.'s suggestion of using an existing Samsung design. The mobile CEO fastened on a neon-green keyboard and an unusual spring-loaded earpiece that snapped up to reveal the display.

"It sounded to me like a gun cocking," Skarzynski said. "It made me incredibly uncomfortable."

In May 2003, Kim and his marketers watched Warner Bros.' screening with excitement. Neo's female sidekick, Trinity, roared out of a parking garage, sending her car hurtling in the air, landing, and swerving to the right. Morpheus, in the passenger seat during this daring escape, pulled out his Samsung phone.

"Operator," Link, the operator of their mother ship, answered.

"Get us out of here, Link," ordered Morpheus.

The product placement appeared eight times, sometimes for a split second, sometimes for a few seconds at a time. Kim's marketers celebrated victory, the South Korean press went crazy, and Samsung wrote the product placement into its official history. Kim kicked off an ad campaign using scenes and characters from *The Matrix Reloaded*.

"That was a big success," he told me. "It elevated the brand to something more unified and noticeable."

Though the product placement itself didn't upend everything, it showed that Samsung was starting to graduate in the public's mind from a cheap knockoff manufacturer to a global multinational.

Kim, reading *BusinessWeek* in August 2003, saw Samsung racing to the top of the world's hundred most valuable brands. It was the only Korean company on the list. Samsung came in at number twenty-five with a valuation of $10.8 billion, shooting up from number thirty-four the previous year.

Samsung was closing in fast on Sony, which was stuck around number twenty each year. When Kim first started at Samsung in 1999, the company wasn't even on the list.

SONY, MEANWHILE, WAS IN trouble for a deluge of reasons.

Four years after the death of Sony's visionary co-founder, Akio Morita, in 1999, Japanese firms were afflicted by the scourge of complacency and "Not Invented Here" syndrome, as management expert Chang Sea-jin, a business professor at the National University of Singapore, put it. Their aging workforces clung to obsolete technologies that they had invented two to three decades earlier. Too comfortable with their own success, they were unwilling to engage in acts of "creative destruction" and start afresh.

It mirrored Samsung chairman Lee's prophecy of "perpetual crisis." In 2003 Sony president Kunitake Ando asked "for a report on what Samsung is doing every week." Its CEO, Nobuyuki Idei, admitted that Sony's revenues and profits were tumbling, forcing him to slash jobs and shut down plants.

In the chaos of Sony's decline, the Japanese rival was struggling with unstable LCDs for its televisions, unable to make all the parts it needed on its own. Samsung executives seized on this weakness. They approached Sony in a bid to open a joint venture to manufacture LCDs, giving Sony and Samsung a stable supply, but also giving Samsung a chance to learn from Sony's stellar operations.

In October 2003, CEO Nobuyuki Idei announced a move that would have been unthinkable ten years earlier. Sony, he announced, would join Samsung in a joint venture that was being named S-LCD. The plant was opened in South Korea the following year, and Samsung was its majority owner. Sony engineers continued to question the quality of Samsung's hardware on their visits to the new plant. But Samsung suddenly had the upper hand.

In 2004 *BusinessWeek* declared Sony's Idei one of its worst managers of the year. It was a horrific fall from five years earlier, when the magazine had named him one of the best.

Now Samsung just needed to drive a stake in the Sony brand.

15

Bordeaux

FOR ITS FINAL ASSAULT on Sony, the chairman of Samsung—now in wavering health—summoned a rising star named Choi Gee-sung (G.S.), who, over two decades, had proven himself one of Samsung's most loyal knights. Born into a poor family during the Korean War, G.S. was the fourth son of a government civil servant. He joined Samsung in 1977 "to put food on the table," he said. It was his chance to make a difference. He would quickly rise to the task.

Promoted to overseas semiconductor salesman in the 1980s, G.S. became a legend in this company of factory floor engineers.

"To overcome his weakness as a non-techie, he read and memorized a 1000-page book about the VLSI [vertical large-scale integration] manufacturing process," business consultant Eun Y. Kim wrote. "His first-year sale of $1 million was record-breaking."

"I was often impressed at his capacity to understand and recite minute details of my U.S. TV business," said James Sanduski, former senior vice president for TV marketing. "He seemed to know everything."

In 2004, after a steady rise as a television marketer, the chairman appointed him chief design officer, an unusual title that gave him responsibility for overseeing the creation of a new line of televisions designed to go head-to-head with Sony. But G.S. was also the man behind the throne, the kingmaker and regent to the crown prince. The chairman, as his health deteriorated, prepared to hand the Samsung empire to his only son, Jay Y. Lee. G.S. was expected to mentor Jay in the ways of business and technology.

Those around Jay treated him with near-absolute reverence. Everything was done to ensure that he looked good and was able to ascend to the role of chairman without blundering.

"The whole room would stand up and clap and bow and things like that as he would make his way to his seat," Pete Skarzynski told me. "He's the chairman's son. He would be your future boss someday."

An eager learner, Jay rarely made direct, substantive statements; instead, he chose his words carefully and asked a lot of questions.

"He's sort of like a god inside the company," said former senior vice president for content and services Daren Tsui, who met with him four times. "Every little thing that he does is analyzed, and it becomes basically an edict right away." *Fortune* reported that when he doled out mild criticism about Samsung's television marketing efforts in early 2015, "thirty-year veteran Park Gwang-gi, an executive vice president and the unit's head of strategic marketing, was said to be so rattled by Lee's criticism that he immediately resigned his position."

Reserved and dapper, Jay sat on the board of directors of the new television display venture with Sony. He spoke eloquently and, in public, almost always appeared in a conservative suit and thin-rimmed spectacles. "He's exactly the kind of guy you would want running your company," said Peter Weedfald, who made presentations to the future chairman in New York. "He handled himself with poise in front of people like Colin Powell and Rudy Giuliani. It was impressive."

Kang Tae-in (T.J.), former senior vice president at Samsung's Media Solution Center, said of Jay: "I had the impression that Jay Y. is probably more of an introvert than an extrovert. . . . But I think through training and education he's able to turn on and display leadership quality."

"He's nice," another South Korean mobile employee at Samsung told me, "but he doesn't have the charisma of his father."

It was a sentiment that was repeated to me by a dozen South Koreans who met him.

Growing up under the tutelage of Japanese and English teachers, Lee was a sportsman, equestrian, and polyglot. He studied East Asian history at Seoul National University, South Korea's equivalent of Harvard, which gave him the foundation in arts and letters favored by Samsung's ruling family. Jay continued his studies and received an MBA at Japan's Keio University, where he studied Japanese business history and practices. There he wrote his thesis on the Japanese currency, the yen, and the role of its sharply rising value in the 1980s boom and the subsequent decline of Japan.

In 1995 he began a doctorate at Harvard Business School, where he cultivated an interest in the dot-com world of eBay and Amazon. Three years later he married (and later divorced) the granddaughter of the founder of Daesang, a packaged-food company, and left Harvard without completing his doctorate, looking to get more hands-on business experience.

"He was not involved in daily operations of any business of Samsung Group," the chairman's aide Nam S. Lee, who accompanied Jay on trips, told me. "But he wanted to make an investment in new, promising business areas."

Jay had everything a Samsung prince and future leader needed: personality, likability, curiosity. But he lacked one prerequisite. He had no balance-sheet success story with which to prove himself.

"Jay Y. Lee was eager to participate in management not a day later than possible," wrote one of Samsung's lead lawyers, Kim Yong-chul, in his memoir *Thinking of Samsung*. "The result of this anxiousness was 'e-Samsung.'" In May 2000, Jay became the chief shareholder of a group of more than a dozen new ventures under the name "e-Samsung," consisting of an online financial services aggregator, a network security firm, and other dot-com investments popular at the turn of the millennium.

"But a Samsung style of management in a venture business was a recipe for disaster," Kim Yong-chul wrote. "And true it was."

"e-Samsung was supposed to be the crown prince's sensational debut," the lawyer wrote. "It didn't sit well with me at all. . . . Saying that it would turn out well because it had Lee Jae-yong's name on it, they offered [me] an investment opportunity as though doing me a great favor. Though I didn't feel good about it, they said it was looking so good that I invested 20 million won [almost $17,000 at the time]."

Jay, the lawyer recalled, didn't show up to many e-Samsung meetings, instead treating his executives to barbecue and drinks after work.

In 2000 the dot-com boom went bust. And now a similar effect was taking hold in South Korea. Shareholders cried foul that e-Samsung was benefiting unfairly from the Lee family's favoritism, getting millions of dollars in investments from another Samsung affiliate, even while the future looked grim. Soon, insolvent and with $20.4 million in losses, e-Samsung was shut down.

But Jay Lee, protected by the Samsung empire, wasn't the one to take the brunt of the losses, despite having invested his own money. Samsung ordered nine Samsung affiliates to buy up the failing e-Samsung shares, relieving Jay Lee of much of the financial damage—a decision that sparked a government investigation. Merrill Lynch claimed that the e-Samsung shares were overvalued.

"I dispatched two Restructuring HQ legal team lawyers . . . to the investigation site," Kim wrote, ". . . to destroy the relevant documents. This effort was successful, and the Fair Trade Commission found us not guilty."

Kim explained how he personally instructed employees, using a training manual, to destroy evidence and protect their leaders in the case of an FTC investigation by the Korean government, a tactic he called the "FTC checkpoint."

"FTC investigators cannot show up with a warrant—they need to have a certification," he told the employees. "So first, buy time by checking their certification. During that time, download all the documents from the computer and hide them, and delete everything on the computer. If there isn't enough time, deleting the files is the priority."

Over the years, South Korea's FTC investigated and fined Samsung Group affiliates six times after they were caught destroying evidence, blocking the investigators from entering, and obstructing justice.

With the failure of e-Samsung coming on the heels of the high-profile failure of Samsung Motors in 1999, the mood within the company was glum.

"Failing in cars, failing in ventures," employees said.

Minority shareholders complained that the debacle showed that Jay Lee was more entitled than he was competent. In the aftermath, Jay took on a lower profile, one carefully managed by company aides. He focused on learning the ropes under G.S. and others, and serving as a liaison with Apple and other companies that needed Samsung components.

"The only people who really knew him or what he was doing were his 'guides,'" recalled Park Sang-keun, a former senior vice president at Samsung Electronics' wireless terminals division.

Alan Plumb, the regional director for Rolls-Royce, was having lunch at the Seasons restaurant in the Millennium Seoul Hilton with his South Korean partner Hyun Hong-choo, a Rolls-Royce adviser and partner at the well-connected Korean law firm Kim & Chang, when he saw Jay dining nearby with his mother. Hyun told Plumb that he was Jay's mentor, "teaching him how to engage with people internationally." He then beckoned him to their table. "Hyun introduced us, saying to Jay that he should get to know me, as I was a big customer of his."

Rolls-Royce had just signed an agreement with Samsung in which Samsung would supply the components for a new jet engine, the Rolls-Royce Trent 900, used in the Airbus A830.

"Jay said he didn't understand, as Samsung no longer made cars," Alan recounted. Told that Samsung made the core of the engine for the A380 aircraft for Rolls-Royce, Jay still seemed mystified. He said he didn't know about his family's jet engine company, Samsung Techwin, and said he had never been to its plant in Sacheon.

"He said that his father doesn't have a Rolls-Royce but a Bentley," Alan said. It was an awkward and unsettling moment. The two bade each other goodbye without discussing their supplier relationship.

"That was my only interchange with Jay," Plumb told me, "someone I marked down as unsuitable to run a group, since he didn't have basic knowledge of the businesses he would one day be managing."

In 2004 Jay asked a Japanese engineer at Samsung why a device that played both VHS tapes and DVDs was selling well. He didn't seem to realize that people were still watching their old videocassette tapes. All of this underscored the importance of educating Samsung's heir.

"Jay came into design meetings a lot," a Samsung TV designer told me. "He was very interested in design and learning about it, but the big decisions on design came from G.S. Choi."

"WHICH DEVICE IS GOING to be the center of multimedia, the PC or the TV? We believe TV is the winner," Kim Gaeyoun, director of visual display product planning, told me in November 2010. Mobile executives disagreed and insisted to me that mobile phones were the center of tech.

"That became a fun rivalry," said Bill Ogle, former CMO for Samsung's mobile marketing unit in Texas. The competition between Samsung's TV and mobile divisions extended to who could build the higher office tower next to the other in Suwon.

"The TV side added four floors to their building," he said. "Then the mobile phone side added five floors to their building. So the TV side added three floors to their building. . . . Finally one of the top executives who were running Samsung Electronics went in and said, 'Stop!' "

PETER WEEDFALD, SAMSUNG NORTH America's CMO, who was based in Cherry Hill, New Jersey, was a pianist and a television enthusiast. One evening he sat down at the grand Steinway at a Seoul restaurant and played a piece for a delegation of Korean executives. When he finished, he recalled later, "We started thinking about this black ebony or white ebony and how beautiful it is and how the TVs were sort of flat at that time in terms of color."

What if the TV itself could be designed like a beautiful piece of furniture, almost like a grand piano or a painting hung on a wall—rather than an ugly black box you threw in the corner? On orders from G.S. Choi, designer Yunje Kang and his team scoured the furniture stores and hangouts of Seoul, where they talked to shoppers who were buying furniture based on its look, rather than just its function. It was a way of thinking that was just gaining momentum in South Korea. After a great deal of debate, they settled on a thin, glossy television design.

When they presented their concept to G.S. Choi, he exclaimed appreciatively, "It looks like a wineglass!" He gave the project the code name "Bordeaux."

Samsung selected an elite team of eleven engineers and designers to work on the TV. The division heads had to sign a confidentiality document and vow not to come out of the studios until the top-secret Bordeaux project was completed.

The team put in grueling days and nights at the company's Value Innovation Program (VIP) Center, which had been set up by CEO Yun Jong-yong. The facility was open 24-7; it had thirty-eight bedrooms, a kitchen, a gym, traditional baths, and pool tables. For Samsung's product developers, it was like training for a Navy SEAL mission.

The teams of designers and engineers fought each other daily; the engineers wanted a bulkier, bigger TV with more hardware features, while the designers were focused on a sleeker, thinner unit. Both presented their ideas and designs to G.S.

"The design appears to be by designers from Milan, but the engineering by engineers in China," he wrote in an email.

The final design, however, achieved what the chairman had been pushing for: a minimalist work devoid of feature bloat. In April 2006, the Bordeaux TV was released to the public at a price of $1,300, for what was a relatively small twenty-six-inch set. Yunje Kang, the designer, worried that the market wasn't ready. But the innovative TV flew off the shelves, eventually selling more than three million units.

"The selling point is very clear—it's slim and sleek," Shin Sangheung, senior vice president of Samsung Electronics Visual Display Division, told reporters at a press conference. The new television was so successful that in 2006 Samsung eclipsed Sony as the number one manufacturer of LCD TVs.

As Samsung products rose in prominence, so too did the career of Jay Lee, even though he played no obvious role in product development. He was now the chief customer officer (CCO) of Samsung Electronics, a position created especially for him, giving him greater responsibilities and allowing him to travel the world meeting executives from Samsung's supply chain to ink supplier deals.

In 2005 Sony's CEO, Idei, stepped down. In response, the Japanese, mired in a recession, did the unthinkable. They hired a Westerner to run the company out of their headquarters in Tokyo: Howard Stringer, a Brit–turned–American citizen and former CBS executive.

Stringer met internal resistance and a language barrier, and over the years of his tenure, he struggled to help Sony regain its footing. The company experienced losses nearly every year from 2009 to 2015. In 2011 Sony exited the Samsung-led displays joint venture, and Stringer, unable to repair the broken company, would step down in 2012. Once the world's admired product, Sony Walkmans had disappeared from the marketplace with the rise of the iPhone. The Sony Ericsson phones had failed to catch on. Sony Trinitron televisions had been passed by in the marketplace. Once the world's most innovative consumer company, Sony, the Japanese miracle, had been reduced to a whimper.

"Running a big company is like running a cemetery: there are thousands of people beneath you, but no one is listening," Stringer later remarked. "It was a bit like that at Sony."

16

Unholy Alliance

IN 2005 HWANG CHANG-GYU, the president of Samsung's semiconductor and memory business, traveled with two fellow executives to Palo Alto, to the home of Steve Jobs.

"I met him with the solution to Apple's life-or-death problem hidden deep in my pocket," Hwang wrote.

In the course of their meeting, he pulled out the NAND flash memory, as it was called, and put it on the table. He called it "my trump card."

His pitch? Flash memory was a much more lightweight and efficient storage device than the traditional hard disk. And Samsung was one of few companies that could guarantee a rock-solid supply. But as Jobs presented his vision for Apple on a whiteboard, Hwang knew this was not going to be an easy meeting.

"At the time he had achieved tremendous success with the mp3 player and the iPod. But he had concerns about the drawbacks of the hard disk storage device, which consumed a lot of battery power and was [susceptible] to shock" from being dropped.

Hwang, who hailed from the southern port town of Busan, had

gotten his engineering doctorate from MIT and completed his post-doctoral studies at Stanford. His fascination with semiconductors had been kindled thirty years earlier, when he picked up a book called *Physics and Technology of Semiconductor Devices,* by Intel's legendary co-founder Andy Grove.

In 1999, three years after the collapse of the DRAM semiconductor market—the original engine of Samsung's rise—Chairman Lee II was looking to invest in NAND flash memory. Hwang, at the time a senior vice president, told the chairman, "Just leave it to me."

In 2002 Hwang rolled out his new system, called "Hwang's law," paying homage to "Moore's law," a term coined by Intel co-founder Gordon Moore in 1965. Moore had predicted that semiconductors would double their computing power every two years, bringing down the costs of any business model that relies on computing and powering the economy of the future.

Hwang's law was Samsung's steroids-injected version of this industry dictum. Memory density, Hwang predicted, would double every year. Samsung engineers did just that, applying Hwang's law to innovate at a pace unseen in the industry. By 2004 Samsung flash memory was advanced enough to meet Apple's needs. Hwang just needed to convince the fastidious, finicky Jobs of that.

"This is exactly what I wanted," Jobs said of Samsung's flash memory, according to Hwang. He agreed to make Samsung the sole supplier of flash memory for the iPod.

"It was the moment that marked the beginning of our dominance in the U.S. semiconductor market," Hwang wrote. With that, Samsung had a launchpad from which to eventually get into smartphones, when they came out. They would go from supplier to competitor.

But soon there were signs of trouble in the new, fragile relationship. Samsung executives had already been gathering with competitors at hotels and golf courses for what were called "Glass Meetings," "Green Meetings," and "Crystal Meetings," colluding to fix the prices of LCDs and DRAM chips—components that Apple and others depended on. Samsung's executives heard rumors that a powerful customer, which they had code-named NYer, was onto them. NYer, in reality, was Apple. "Samsung ran to the Justice Department [in 2006] under an anti-trust leniency program and ratted out its co-conspirators," wrote Kurt Eich-enwald in *Vanity Fair*. "But that didn't lessen the pain much—the com-pany was still forced to pay hundreds of millions of dollars to settle claims against it by state attorneys general and direct purchasers of L.C.D.'s."

In September of that year, one of the Samsung executives who had been present at a meeting with Jobs, Tom Quinn, a vice president for memory marketing in the United States, pleaded guilty and agreed to serve eight months in prison for his role in the DRAM price-fixing con-spiracy. He was joined by five other Samsung executives.

Apple, however, kept Samsung as a supplier. The relationship was simply too valuable to the company to sever it. It turned out to be a wise choice. In the lead-up to the launch of the iPhone, Samsung saved Apple from calamity.

In February 2006, David Tupman, Apple's vice president for iPhone and iPod engineering, realized that he was a year away from shipping the first iPhone—but he didn't have the main processor ready, or even a time line for creating one. Fortunately, Apple's hardware engineers were already getting their chips for the iPod from Samsung. Tony Fadell, one of the leaders of the iPod team, asked Hwang's team if they could produce a chip for Apple to the iPhone's specifications.

Hwang's employees could—by modifying a chip currently used for a cable box. Apple gave Samsung an impossible deadline: to create a chip for the iPhone in five months. It did not tell Samsung what the chip was for.

Hwang sent a team of engineers to Cupertino to work beside Ap-ple's people. Samsung built a chip "as fast as they've ever built a chip in a [fabrication lab]. Normally it takes days per layer, and it's twenty or thirty layers of silicon you're trying to build," Tupman said. "Normally it's months and months to get your prototypes. And they were turning this around in six weeks. Crazy."

When the iPhone's semiconductor chip—its nerve center—was completed, the transistor count in each chip was a remarkable 137,500,000, a number possible due to the breakneck pace of Hwang's law. (The first Intel microchip, in 1971, had only 2,300 transistors.)

Meeting their impossible deadline, Hwang's team shipped the components two months before the product announcement. The new iPhone booted up fine. But when the engineers tried to push the system, it crashed due to an unknown flaw. Steve Jobs declared an emergency. Apple's engineers sat down with Samsung's team to look for ways to get more bandwidth from the chips.

In the end, the iPhone was completed on time. On January 9, 2007, Jobs walked onto the stage at the Macworld convention and unveiled his beloved product.

"This is a day I've been looking forward to for two and a half years," he said.

Without Hwang's vision for mobile chips, the iPhone would not have come into existence as quickly as it did—an Apple engineer admitted it wouldn't have been released on the time frame Apple had set if it hadn't been for Samsung's scorching speed in creating the semiconductor chip for the phone.

Steve Jobs was grateful to his South Korean partners. For now.

17

The Emperor Has No Clothes

THOUGH THEY SUPPLIED APPLE'S chips, Samsung's executives, com-placent and successful, paid little heed to the new iPhone. In fact, many companies failed to foresee its significance. The South Korean govern-ment banned sales of Apple's new device in 2007, citing a regulation that smartphones in South Korea had to support a technology called the Wireless Internet Platform for Interoperability, or WIPI. In reality, it was a protectionist trade measure to keep the iPhone out of the country.

"You don't understand! Koreans will never browse the Internet on their phones!" an American Google employee heard from his local staff in Seoul. Samsung didn't have to worry about competition from the iPhone.

But a crisis was brewing, and Samsung would soon be in a tough position.

A STAFFER BASED IN the chairman's office at Samsung told me that in late 2007 or early 2008, "we were given a directive. 'Someone's going to come and wipe your hard drive. Sorry.'"

Another employee in the chairman's office objected to what seemed like the destruction of evidence and refused to have his hard drive wiped.

"I understand the position and your pain," a more senior employee told him. "Why don't you and I go down and get coffee and talk about it?"

When they returned, his hard drive had been wiped.

I FIRST MET KIM Yong-chul in September 2016, at a modest municipal government office in his hometown of Gwangju, a city on South Korea's southeastern rice bowl. It's a region known for its political protests and its resentment of Samsung and the company's conservative political allies.

Yong-chul didn't ordinarily grant interviews. He was one of the country's most famous prosecutors. As a prosecutor, before working for Samsung, he got the former dictator Chun Doo-hwan and President Roh Tae-woo convicted of corruption in 1997. In South Korea prosecutors enjoy a great deal of prestige and have enormous legal power. They have the ability to open investigations without the involvement of the police. But South Korean prosecutors often have political ambitions as well. Corporate executives have access to them, offering them lucrative positions at *chaebol* groups when they leave office.

Yong-chul greeted me in a conference room and gave me a wheat tea. As a foreign journalist, I was suspect, he made clear to me. For all his public challenges against the nation's most powerful company, Mr. Kim was still a Korean patriot who didn't want to smear his country's good name.

"I live in Korea. I like Korea," I reassured him. "I've been here for four years."

Hired in 1997 as a top legal aide at Samsung, where he worked for seven years, Yong-chul was living the dream in the Republic of Samsung. He drove an Audi and had a posh apartment in the upscale Gangnam neighborhood. Prospective political allies clamored to meet him, setting up appointments with him as much as two years in advance, he claimed.

But he'd been unable to stomach the illegal activities that he found at Samsung, and ultimately quit. In October 2007 Yong-chol teamed up with a group of activist Catholic priests, who had credibility for their role in Korea's prodemocracy movement in the 1980s, for a series of press conferences in Seoul, at which he made explosive allegations against his former employer.

At those press gatherings over two months, he claimed that Chairman Lee had stolen billions of dollars from Samsung affiliates and stashed it in bank accounts under the names of Samsung employees, including Kim's own name. Kim claimed his former employer was managing a $215 million slush fund to bribe influential figures.

This was a far more damaging allegation than the price-fixing scandal in 2005 or Chairman Lee II's criminal conviction in 1997. Samsung forcefully denied the allegations and later described Kim Yong-chul in this way to *The New York Times:* "When you see a pile of excrement, you avoid it, not because you fear it but because it's dirty."

Kim told an audience at the Seoul Foreign Correspondents' Club that he had come forward when Samsung fired his law firm.

KIM'S ALLEGATIONS CAUSED ENOUGH of a ruckus to merit an investigation, approved by the South Korean president himself. South Korean prosecutors often respond to the shifting political winds of the moment, rather than the merits of the evidence alone. Investigators rummaged through the chairman's office and, a day later, on January 14, 2008, swooped in on the Samsung headquarters and his mountainside home.

Small bands of elderly protesters gathered around Seoul holding pickets, complaining of how Samsung was being treated. But it's unclear how many of these demonstrators were authentic. An investigation by South Korean news channel JTBC later revealed that one of these protest groups, the Korea Parent Federation, had been receiving money transfers of more than $400,000 from a probusiness lobbying group, the Federation of Korean Industries (FKI).

South Korean newspapers called Kim Yong-chul a "traitor" and "no different from a gigolo or a gold-digger." One angry protester torched a photograph of the whistleblower with a makeshift flamethrower.

Nonetheless, Korean prosecutors indicted Chairman Lee for tax evasion and breach of trust, claiming he damaged his company's interests and profits by forcing Samsung subsidiaries to sell shares at unfairly low prices to his son, damaging the interests of shareholders. The share sales made it easier for Jay to raise his stake and get the throne to the Samsung empire.

The chairman denied the allegations in April 2008 but resigned as chairman of the Samsung Group that month. Prosecutors didn't pursue the bribery charge, citing a lack of evidence, and complained that much evidence had been destroyed.

"I sincerely apologize and will do my best to take full legal and moral responsibility," he told reporters. "It grieves me, for I still have many things to do." Samsung still didn't have an iconic product, nor had it completed its generational succession.

"It was just a show," Yong-chul told me bitterly. "An interesting piece of theater. . . . It was like watching a fire at a rich person's house," he explained. "The public doesn't want the fire to spread to their houses. They still want their kids to get a job in Samsung."

Nine others were implicated in the charges: vice chairman and financier Lee Hak-soo, the second-most-powerful man at Samsung, another top executive, Kim In-ju, and seven more executives.

ON JUNE 13, 2008, Chairman Lee and his aides entered the courthouse in Seoul. The grim silence was punctuated by the flashes of cameras. The chairman, walking very slowly and obviously in ill health, spoke in a frail, weak voice.

"I apologize for causing concern," the chairman said in his opening statement to the court. "Everything was my fault, and I take full responsibility. Those who stand trial with me are people for whom I was responsible. I sincerely ask for leniency on their behalf."

Two months later a Seoul district court found Chairman Lee guilty of tax evasion but cleared him of Samsung's decision to sell stocks to his son at unfairly low prices. "I am sorry," he told reporters as he left the courthouse.

He had been stripped of his imperial aura, shamed, and disgraced.

He was sentenced to three years in prison, but the judge commuted his sentence, sparing him prison time.

YET WITHIN SAMSUNG, BIG changes were already taking place. A generation of key figures had left the company. Eric Kim's contract ended in 2004; K.T. Lee left in 2007 and CEO Yun Jong-yong in 2007.

Despite the chairman's guilty verdict, Kim Yong-chul, the whistle-blower, felt little sense of triumph. Old friends and colleagues stopped calling him. Once Mr. Popularity for his Samsung connections and for the celebrity cases he had tried as a prosecutor, he was now persona non grata. No longer a prosecutor, he couldn't bring the case against the chairman and Samsung himself. He was helpless.

"A state of panic," as he put it, "overtook me."

He retreated to the countryside east of Seoul, where he locked himself indoors, listening to Beethoven piano sonatas and caring for his kennel of dogs.

"Beethoven is the only musician who could express his rage through his music," Kim explained. He called his children in the United States and told them never to come back to South Korea—because there wouldn't be any jobs for them. "I am not even a father," Kim lamented.

"TELL ME ABOUT THE X-File," I told Kim.

The X-File was a secret batch of recordings, named the "Samsung X-File" by the press, that was leaked and splashed all over the media, posing a threat to the rulers of the Samsung empire. A month before my meeting with Yong-chul, I'd met Roh Hoe-chan, a former parliamentary lawmaker, at his office at the outskirts of Seoul, in a gray industrial zone where we sat outside and ate barbecued pork and drank soju.

Roh, a balding middle-aged man with a giant smile and booming voice, bore the brunt of what it means to be a Samsung critic. I was surprised he maintained such an upbeat, optimistic outlook.

"It was a search for justice," he told me. "We could not allow Samsung to treat our country like it's a Republic of Samsung."

A former activist and journalist who covered labor movements before entering politics, Roh recounts a story similar to Kim's.

"I found a package in my mailbox at the National Assembly," he told me. "It was 2005. The sender was anonymous. Inside it was a CD."

Roh put the disc in his computer and up popped a recording. It was a top-secret tape recorded by the nation's spy agency and leaked by an unknown source.

"Ah, and will there be any greetings for *Chuseok?*" asked one person on the recording. The man was referring to gifts—sometimes a cover for bribes in South Korea—during South Korea's equivalent of Thanksgiving.

"The ones that are worth it, yes."

"I'm already doing the prosecutors. And the K1s," he said, referring to alumni of the influential Kyunggi High School, an academy for the Korean elite on the level of Phillips Exeter Academy.

"What about the others?"

"I think there will be some overlap there."

"I'd also like to get to Kim _____." (The full name of the individual was redacted from a transcript of the recordings that was later broadcast on national television.)

"We will send something over if you set up a budget for us."

Neither participant knew he was being recorded, but it sounded like the two men, speaking that day in September 1997, were plotting to bribe prosecutors and other people of influence.

Roh hired audio experts to analyze the voices.

"Our conclusions were groundbreaking," he said after a shot of soju and a slice of barbecued pork. "The first speaker was probably Hong Seok-hyun, the South Korean ambassador to the U.S."—who happened to be the Samsung chairman's brother-in-law and was publisher of the widely read *JoongAng* newspaper, owned by Samsung until 1999.

"Hong Seok-hyun was a big figure," Roh told me. "There was talk that he wanted to be the next UN secretary general, that he had political ambitions."

"The other man," he said, "was Vice Chairman Lee Hak-soo," the chairman's confidant, whom Kim Yong-chul told me was a key figure in the company.

And why?

"The purpose of these gifts, we believed, was to get the government's goodwill. Chairman Lee had to get his son into the company leadership." It was yet another scandal arising from an attempt to hand over control of the company in Samsung's troubled generational succession.

It was also an indictment of the company's tentacle-like reach, showing how deep its connections went in the South Korean government. Roh knew the grave consequences of publishing this tape and naming judges, prosecutors, politicians. And who exactly, he wondered, were the recipients of the holiday "gifts"? And there were additional "X-Files" still kept confidential by the spy agency but waiting to be released. This was just one of hundreds of recordings, he said.

"I couldn't trust the mainstream media," Roh told me. "I had to go online, to my blog."

Roh concluded that seven people, including some prosecutors, were the targets of Samsung's alleged payments or attempted payments. He held a parliamentary hearing. After lawmakers proposed a summons for Chairman Lee, he flew to Texas's Anderson Cancer Center for treatment—a move widely criticized as a ruse. (Medical excuses were common among *chaebol* leaders under scrutiny or facing trial.)

National Assemblyman Roh read the transcript before the parliament in August 2005. Then he turned to a vice justice minister who was present.

"You're on the list," Roh asked. "Did you take the money?"

"No," responded the vice minister.

The vice minister resigned a few days later, although he never was found guilty of wrongdoing. Chairman Lee's brother-in-law, Hong, the ambassador to the United States and someone with political ambitions, also resigned.

Prosecutors opened an investigation into . . . well . . . the prosecutors; ultimately they declined to press charges against themselves or against Samsung, citing the statute of limitations, which had expired.

Roh, however, continued his campaign against Samsung. So loud and boisterous was his activism that two people he was indebted to approached him—at the request of Samsung, he believed—and asked him to stop attacking the company.

A year had gone by with no state investigation into Samsung, when Roh got a call from the prosecutor's office.

"Roh Hoe-chan," said the bureaucrat on the other end of the line, "we're going to have to charge you."

Because Roh had published the transcripts online, a team of prosecutors had concluded that the politician, not Samsung, was guilty of a crime: The tapes were made from illegal wiretaps conducted by the spy agency; releasing them was in violation of the country's anti-wiretapping law. Samsung, in other words, was the real victim.

Suddenly the wheels of government turned against Roh. Prosecutors also charged him with defaming the prosecutors—a verdict that can carry a prison sentence in addition to a hefty fine. The broadcaster who first revealed the existence of the X-file transcript, Lee Sang-ho, stood accused of the same crimes: defamation and breaking the wiretapping law.

Found guilty at the conclusion of his first trial, Roh Hoe-chan appealed. In South Korea both prosecutors and defendants can appeal up to two times. In 2009 an appeals court found in his favor and chastised the prosecutors for not bothering to investigate the content on the tapes adequately.

"Any person who possessed an ordinary and rational intellect would naturally make the strong assumption that the money was paid according to the content of the conversation," said the judge.

The court believed that Roh was telling the truth. It believed that the person who received money from Samsung was indeed—as the lawmaker claimed—the top prosecutor whom Roh had named in his own, private investigation.

Prosecutors appealed to the Supreme Court, which in 2013 overturned the ruling in a sensational and much-watched trial.

In 2013 Supreme Court judges ruled that Roh had indeed violated the anti-wiretapping law when he released the transcripts. Moreover, the recordings, the court ruled, were not in the public interest and therefore shouldn't have been published.

Lawyers and pundits were puzzled by this part of the argument.

How could the recordings not be in the public interest if a top justice official and an ambassador who was the chairman's brother-in-law, though never criminally charged, stepped down?

The court also ruled that parliamentary immunity did not protect Roh. He was stripped of his seat in the National Assembly. He became a public example of what happens when you challenge the Republic of Samsung.

"I don't regret it," he told me. "I stood up for what is right."

CHAIRMAN LEE'S GUILTY VERDICT was problematic for the International Olympic Committee (IOC), where he was a member and played a prominent role in lobbying on behalf of South Korea.

Following the verdict, Chairman Lee voluntarily gave up his rights as an IOC member. He risked losing his membership—and, therefore, losing South Korea future bids to host the Olympics.

Rumors swirled of a pardon for the chairman, in the classic South Korean tradition. Meanwhile, an auto parts maker that was connected to the South Korean president was involved in a separate U.S. lawsuit. According to later court testimony by Lee Hak-soo, the chairman's former right-hand leader, Samsung agreed behind the scenes to pay that company's steep legal fees.

Samsung paid the money hoping to get a presidential pardon for the chairman, the former executive testified. In December 2008, five months after Chairman Lee's conviction, the president announced he was granting special amnesty to Chairman Lee. With his exoneration, Lee was able to remain a member of the IOC, leading a campaign to bid for the 2018 Winter Olympics, to be held in the pristine mountain county of Pyeongchang, South Korea. The bid was successful.

"The latest pardon reconfirms a common saying in South Korea that Samsung lies above the law and the government," claimed economics professor Kim Sang-jo, who would later head Korea's government financial watchdog, to The New York Times. "President Lee [Myung-bak] talked about national interest, but a criminal convict traveling around the world campaigning for South Korea's Olympic bid will only hurt our national interest and image."

The political winds had shifted back in Samsung's favor.

A day after the chairman's pardon, on Christmas in 2009, former *Guardian* correspondent Michael Breen—a noted British expert and author on Korea and an honorary citizen of the city of Seoul—published a satire piece in the *Korea Times* called "What People Got for Christmas."

"Samsung," he wrote, "the world's largest conglomerate, and the rock upon which the Korean economy rests, sent traditional year-end cards offering best wishes for 2010 to the country's politicians, prosecutors and journalists, along with 50 million won [$43,000] in gift certificates." He went on, "Employees received two framed photographs of Lee Jae-yong, the new Chief Operating Officer at Samsung Electronics Co."—alluding to the portraits of the father-and-son dictators that every North Korean must hang in their home—"with instructions to place one in their children's bedroom and the other in their living rooms beside but slightly below the one of his father, Lee Kun-hee."

Breen and his colleagues were out for a night of karaoke, he told me, when he got a call from the editors.

"Samsung is really upset about your column," an editor on the other end explained.

The company claimed it was filing three lawsuits against Breen, the paper, and its top editor, each alleging $1 million in damages, along with a criminal complaint that could lead to prison time. The newspaper agreed to retract the article and issue an apology on the front page. Samsung later dropped the lawsuit against the newspaper and the editor but pursued the lawsuit against Breen.

Summoned for five hours of questioning by prosecutors, Breen was defiant.

The prosecutor's assistant went through each point in the article.

"Did you research and check your claim that every employee of Samsung received the portraits?"

"The column is satire," he said. "The whole point is these claims are not true." South Korea, though, didn't have a tradition of public satire.

"If it's not true, it's not funny," the prosecutor barked.

"Under Korean law," Breen told me, "if somebody takes out a defamation suit against you, they don't have to prove that you've damaged them. You have to prove that you haven't."

"You have to somehow try and assuage their [the defendants'] feelings," he said.

Breen explained his conundrum to the UK's ambassador, sitting in the pub in the basement of the British embassy.

The ambassador called Samsung CEO Lee Yoon-woo and tried to get him to drop the case.

"His hands are tied" was the message the ambassador delivered back to Breen. It suggested to Breen that someone higher on the food chain was the true "plaintiff" bringing the suit forward. Korean activist groups appealed to a visiting free-speech official from the United Nations.

Korean nationalists, however, failed to see why this petulant foreign journalist was deserving of lenient treatment.

"Michael Breen Who 'Mocked' Korea, 'Did He Even Apologize?'" read one news headline.

Mike tried to remain low-key, hoping that the lawsuit would blow over and Samsung would drop it. But eventually he realized the strategy wasn't working.

After four months, Mike gave an interview to the *Los Angeles Times,* telling the newspaper: "The reason I'm being sued is because the beast roared." He also wrote an apology letter to Jay Lee, the Samsung heir, without admitting guilt.

On the same day that the article was published, a month later, Samsung dropped the civil suit. And when Breen attended his criminal hearing, Samsung's representative failed to show.

"As Samsung has withdrawn its civil suit and is not here at present, there is no real victim and therefore there's no need to make a ruling," the judge told Breen at the hearing.

In 2010 the lawyer and whistleblower Kim Yong-chul published a bestselling memoir about Samsung called *Thinking of Samsung,* detailing what he had witnessed at the company and in the Lee family. Many Korean newspapers refused to review it, and nearly all refused to carry advertisements for the book.

"One newspaper reported on its popularity," reported *The New York Times,* "but did not print its title or detail its allegations. It became a best seller thanks to strong word of mouth on blogs and Twitter."

The left-wing newspaper *Kyunghyang* later issued a front-page apology for rejecting a college professor's column praising the book and criticizing Samsung.

"But I wouldn't say I came out the winner," Yong-chul told me that day in his conference room.

"I was unemployable. After all that, I returned to Gwangju, my hometown, and I opened a bakery," he told me. "No one would hire me. Then I got a job as a government auditor."

IT WAS THE CHAIRMAN of Samsung, in fact, who emerged as the winner in the conflict.

"Mr. Lee's wisdom and experience are urgently needed," Samsung's communications chief told the newspaper the *Hankyoreh*. Only with the vision and guidance of the chairman, he claimed, could Samsung continue to rise among the ranks of global technology companies.

The executives at Samsung had been hard at work in Lee's absence. In 2009 they announced a bold new campaign called "Vision 2020." Some questioned its feasibility. Samsung planned to hit $400 billion in sales in the next decade, rise to fifth in the world in brand value, and invest $20 billion in five industries in which it didn't currently have a strong presence: biotech drugs, solar panels, e-vehicle batteries, medical devices, and LED lighting. Its goal was to discover and invest in new growth areas, as it was clear the older ones—LCD displays, cellphone batteries, flash memories—would not continue to grow forever.

The company made organizational changes as well. G.S. Choi was appointed CEO in December 2009, and Jay was elevated to chief operating officer. At forty-nine years of age, Jay was inching toward succession. His father had been forty-five when he became chairman.

In March 2010, the company declared a "management crisis." Almost two years after his resignation, Chairman Lee returned to his chairmanship to preside over the Samsung empire once again.

"We are now facing a real crisis." he announced. "We must start again. We have no time for hesitation. Let us move forward looking only ahead."

The "Vision 2020" areas of growth were not Samsung's only targets.

Increasingly, executives realized that they had an Apple problem. The iPhone ban in South Korea had been lifted in 2009, and Apple was poised to hurt Samsung badly.

A group of executives gathered to plot Samsung's next great campaign: to beat Apple.

18

Guardians of the Galaxy

AT 9:40 A.M. ON February 10, 2010, twenty-eight executives gathered in the Gold Conference Room on the tenth floor of the Samsung building in Gangnam. J.K. Shin, the relatively new head of the mobile communications business, took the floor. "[Our] quality isn't good," a company memo would quote him as saying, adding that the designers were under pressure in terms of the schedule of their products because the company put out so many different models.

J.K. Shin, a lifetime Samsung employee, was an engineer who had risen through the research and development department and been promoted to head of mobile communications two years after K.T. Lee departed. It was Shin who was now leading the charge to overtake Apple.

"He was a very aggressive leader," a former vice president told me. "You had to get up and say that you were going to accomplish something almost impossible. Otherwise you'd get thrown off the stage." If you pledged to reach a goal in a year, he'd respond, in dead seriousness, "That's great, but can we do it next quarter?"

Kim Titus, a former Samsung spokesman, said of him, "He was

semifamous for his kind of showmanship. He would pretend he was a magician. He would pull out a phone from his suit pocket. And he would pull out another one from his pants pocket. And he'd lay them all out, and try to get the customers [the carriers] excited."

Samsung and others were pursuing carrier-centric marketing, putting more of their efforts into winning over carriers rather than consumers. Samsung found itself ranked near the bottom of the mobile category in sales. Apple was on top. Samsung's executives weren't sure yet whether so-called smartphones—still a nebulous and ill-defined concept—should be their focus. Why go head-to-head with Apple, they wondered, when Samsung could simply supply the chips and displays for the iPhone and other models, making money by riding off Apple's success?

Many in Samsung's headquarters were keen on this risk-averse route.

"Samsung had a revolutionary new screen technology called Super AMOLED that it at first wanted to put in someone else's device," *Business Insider*'s Steve Kovach reported, "perhaps a phone built by a major wireless carrier like Verizon."

J.K. Shin had already tried to compete with the iPhone once in a panicky response, rushing out a Windows-based smartphone called the Omnia II in October 2009. It was an irredeemably ugly phone, even by the standards of the time. The phone dropped calls and auto-rebooted and had a clunky touch screen.

"Some customers burned the product on the streets or hammered it to bits in public displays of disaffection," Reuters reported.

Compared with the iPhone, said J.K. in a company memo in 2010, "the difference is truly that of Heaven and Earth." He added, "It's a crisis of design."

After the iPhone 4 came out, Samsung Electronics saw its $885 million in revenue from telecom from the first quarter of 2010 tumble to half that number in the following quarter. Samsung's scattershot mobile strategy needed a reboot. It needed a single premium smartphone brand that could stand up against the iPhone onslaught and carve out market share for Samsung.

The Samsung Men, it seemed, had forgotten the chairman's lessons of the past two decades. Too busy chasing Motorola, Sony, and Nokia, they'd developed tunnel vision as a result of their insular and reactive culture. The tech world, meanwhile, had shifted under their feet.

ON THE NINTH FLOOR of the Samsung headquarters, in the corporate design center, Samsung's head designer, Chang Dong-hoon, told me he had a "single direction . . . all the way from the top" to put out Samsung's first premium smartphone iteration, the Galaxy S, in June 2010. It happened a full three years after Apple's first iPhone, at an obscure and little-watched product launch in Singapore.

Why the name "Galaxy"? Samsung has never publicly told the story. But over coffee in Palo Alto, California, former Samsung senior vice president Ed Ho told me about a $95 bottle of wine enjoyed by its top executives, the Terlato family's "Galaxy" red blend. It inspired Samsung executives to later choose the name "Galaxy," which had to them a premium ring, for their phones.

Unfortunately, the first Galaxy smartphone didn't hit the mobile phone sweet spot. It came in four variations, confusing users: "Epic" was sold by Sprint, "Fascinate" by Verizon, "Vibrant" by T-Mobile, and "Captivate" by AT&T. Apple, meanwhile, had a single phone on the market. One. As a result, the name "iPhone" conjured up a cohesive image among consumers and craftsmen, creatives and hipsters. "Galaxy" was the second-moving wannabe with no cohesive brand.

Steve Jobs was livid when Samsung released its smartphone. As he told biographer Walter Isaacson, he wanted to launch "thermonuclear war" on Android, the operating system used in Samsung phones. Samsung was the Apple iPhone chip supplier that dared to compete directly against Apple by making a similar-looking smartphone, and with the Android operating system, which Jobs abhorred. Jobs was prepared to sue. Tim Cook, as Apple's supply chain expert, was wary of endangering the relationship with a supplier that Apple depended on.

When Jay Lee visited the Cupertino campus, Jobs and Cook expressed their concerns to him. Apple drafted a proposal to license some of its patents to Samsung for $30 per smartphone and $40 per tablet, with a 20 percent discount for cross-licensing Samsung's portfolio back to Apple. For 2010 that revenue would have come to $250 million.

In the end, Samsung's lawyers reversed the offer. Since Apple was copying Samsung's patents, they argued, Apple had to pay Samsung.

In April 2011, Apple filed multiple lawsuits, spanning dozens of countries, against Samsung for patent infringement. It demanded $2.5 billion in damages. Samsung quickly countersued for infringement of five patents relating to its wireless and data transmission technology.

The war was on.

WAS SAMSUNG, RESORTING TO "fast-following" in terms of new products, doomed to fail at executing Chairman Lee's vision of creativity and global-mindedness?

It was at that point that, years later, I met Todd Pendleton. Pendleton strode into Dallas's Stoneleigh Hotel full of energy, with a big smile on his face, wearing a chain necklace and bracelets and slightly spiked hair, holding the hands of toddler twins.

"I'm Todd!" he exclaimed, "and these are my little ones!" He sent his son and daughter, accompanied by his wife, up to their hotel room.

Todd was the chief marketing officer for Samsung Telecommunications America at the time, where he had become a legend. It took me two years to convince him to sit down and talk. Now living in Los Angeles, he would initially exchange emails with me, and then he'd go dark for months at a time.

"He likes to be mysterious and keep a low profile," a former employee told me. "Despite his energetic and optimistic personality."

I'd heard a lot about Pendleton from his colleagues.

A South Korean Samsung employee, speaking to me in violation of company media policy, told me, "I didn't like Todd's team. But man, they accomplished a lot. He was our guy in the Samsung-versus-Apple wars."

His tenure as CMO of Samsung Telecommunications from 2011 to 2015 was unusual. He was clearly no Samsung Man in the traditional sense. His stories about Samsung tended to be overwhelmingly positive, and in fact he was known for his incredible optimism. His employees sometimes remembered things differently.

STUDYING THE TESTIMONY FROM the *Apple v. Samsung* trials, and speaking with dozens of his colleagues and employees, I learned the story behind the Samsung war on Apple under Todd's watch.

"We need more creativity!" Dale Sohn, the CEO of Samsung Telecommunications America, the Texas mobile phone office, had exclaimed in a meeting in 2010, according to a senior manager who was present. Dale reported to mobile chief J.K. Shin. He had been tasked with turning things around in America, Samsung's toughest market, given the iPhone's huge popularity. "I want someone who's got tattoos all over his arms and earrings!"

Dale had been heading Samsung's mobile sales and marketing unit in Richardson, Texas, since 2006. He had been promoted to CEO of the unit shortly before Pete Skarzynski decided to leave the company in 2007.

Dale had his complaints about his American marketers.

"The approach was very traditional . . . selling products on an individual basis but not really telling a unified Samsung story," he said. Samsung was producing a dozen phones a second, but it had no brand, no identity that stuck out in the minds of consumers. The early sales and publicity momentum around the Galaxy and its brand was already spiraling downward. Dale knew that his employer was "in trouble," as he put it.

Dale, who joined Samsung in the early 1980s and rose to export manager, spent time taking wholesale orders in Texas, putting the client's logo on products, and sending them off to various vendors. Samsung, he knew, was invisible to the end user. And its executives had gotten too comfortable in this obsolete arrangement. The result?

"We were reading forecasts on our wholesale buyer, the carriers," he said, skewing Samsung's product offerings. The idea was to convince consumers to go into a Sprint shop and ask for a Samsung.

Samsung employees were surprised that Dale, a rough and tough traditional Samsung Man, was the one kick-starting Samsung's next big transformation.

The "Dale Sohn regime," as one person called it, brought mixed feelings.

"When you worked with Dale, you got used to being berated in public," explained former acting CMO Paul Golden. Said Pete Skarzynski: "He brought in a lot of Korean managers. . . . [The makeup of the office] changed quite substantially."

Office morale, Pete recalled, had fallen during his last year there. Dale yelled, chided, and told people they were fired but then called them to come back to the office. He called weekend meetings that felt more like exercises in patriotic solidarity. According to former CMO Bill Ogle, as punishment for failing to triple sales—his American team only doubled them—Dale ordered his executives to work Saturdays from Memorial Day to Labor Day.

He unveiled inspirational slogans and banners that were standard practice in South Korea but had the American employees scratching their heads.

"He put a slogan up in front of the room that said something like 'We only want young and energetic people in Samsung,'" recalled Skarzynski. "Dale, take it down," a human resources employee urged him, worried about a discrimination complaint.

But Dale was equally hard on himself. He devoured books on leadership and practiced a sort of personal asceticism, denying himself the typical luxuries of a Samsung CEO.

He was also unusually accessible. He occasionally dropped in to the cafeteria—an unthinkable act for a South Korean CEO—to ask about life in America. His office was "right beside everyone else's offices along the hallway where everybody walked," said a marketer. "If you wanted to talk to Dale, you could just go in there and talk to him." The Americans who took to Dale confided in him. And since, as a Korean, he had access to the opaque and all-powerful Samsung headquarters, they would depend on him over the course of the Apple wars to come.

WHEN DALE PUT OUT a call for a new chief marketing officer, a headhunter zeroed in on Todd Pendleton. Pendleton had been an unconventional marketer at Nike, an impresario and master brand builder. He had been offbeat and irreverent in the ads he crafted and sharp and to the point in the way he communicated.

When Todd received the call, he wasn't aware that Samsung even made smartphones. But he flew out to Las Vegas in February of 2011 to meet Dale at the Consumer Electronics Show (CES), where Samsung frequently wowed the industry with large booths that dominated the conference hall.

"I was very impressed," Pendleton said. "I think the quality of the products, the big screen, things that I hadn't seen before were very intriguing."

Apple didn't have a booth at CES. But outside the CES walls, it was a vastly more respected company among the public.

Coming off a fifteen-year career of excellence at Nike, Todd could have stayed on at Nike in comfort. But he was getting bored. He wanted to do something new.

"It was a dream job," he said of Nike. But at Samsung, he claimed, "there was no place to go but up." At a company run by engineers, his goal of creating "cultural moments," as he put it, was novel—"being in places that are culturally relevant to consumers here in the U.S." He would use a mixture of planning and improvisation, rather than a rigid hierarchical approach, to craft those cultural moments.

For two decades, Todd had learned the art of cultural branding in high-profile negotiations with basketball players and other sports stars. Samsung, in many ways, was butting up against the same challenges that had afflicted Nike and Reebok in the 1980s and 1990s: too much focus on engineering, and not enough focus on clever marketing that "knits the whole organization together," as Nike co-founder Phil Knight put it. Each year the big shoe makers put out catalogs full of ribbed ankle collars, textured fabrics, and CO_2-filled bladders of cushioning, which appealed to serious athletes and shoe nerds but were seen as gimmicks by consumers, fads that came and went. A relentless race to release new, innovative shoes in high volumes drew Nike into a price war.

"We were trying to create a brand . . . but also a culture," Knight wrote in his memoir *Shoe Dog*. "More than a product, we were trying to sell an idea—a spirit." That was the recruiting pitch to Rob Strasser, the lawyer who would join Nike—a struggling company—and then approach a relatively untested basketball player named Michael Jordan to convince him to sign a five-year contract worth $2.5 million.

The first Air Jordan shoes, released in 1984, made $130 million in revenue in their first year for Nike and three decades later still had eight times the sales of Nike's newer signature line, the LeBron James Collection. It was a masterstroke of symbiosis: Jordan made Nike, and Nike—through cultural branding—helped make Jordan.

By the time Todd joined Nike as a baby-faced advertising manager in 1996, just a few years out of Northeastern University in Boston, Nike dominated the cool factor in shoes and sports clothing. The swoosh and "Just Do It" were not yet annoying corporate sloganeering—they were entries into the pop-culture canon.

Todd spent the next fifteen years as one of the architects of some of Nike's iconic campaigns, helping guide the company into the post-Jordan world as His Airness prepared to retire in early 1999. In 2002 Todd, after his success working on an iconic Nike basketball commercial called "Freestyle," was promoted to become the company's first-ever basketball brand manager, giving him more power to sign stars. Nike was seeking a rookie promising enough to replace the Air Jordan gold mine. In 2003 Todd led the team that convinced LeBron James to sign a $90 million seven-year endorsement deal. It proved a fruitful relationship; he'd later take LeBron with him to Samsung for a high-profile endorsement.

In a surprise a month later, his team won against Reebok in a bid to sign Kobe Bryant in a four-year, $40 million deal, winning Pendleton further acclaim at Nike. The deal showed off his ability to act decisively and take huge risks in a novel strategy of splitting the Nike brand between two rising stars.

But in June 2011, Todd called it quits. He packed up his boxes at Nike's verdant Oregon campus—leaving behind the lake, running trail, soccer field, orange loaner bikes, and "Hall of Champions"—and settled into the staid Samsung office in Richardson.

Pendleton himself was far from staid and formal. He was as much a relationship builder as he was a brand builder, securing the trust and friendship of high-profile sports stars and signing deals on the backs of napkins. His sports marketing expertise was a promising marketing approach for Samsung. But Todd had never worked at a tech company before and didn't know the industry. As a tech specialist, the company reached out to a former BlackBerry digital marketer named Brian Wallace.

But, as Wallace said, "no one wanted to go to Samsung at that time." He'd had two interviews at Disney Studios and at Kraft, but they weren't ready to make a decision yet. So Wallace figured he might as well continue to brush up on his interview skills by talking to Samsung.

Wallace showed up at the staid, beige-brick corporate building in Richardson. The area surrounding the building included a hotel and corporate offices and a strip mall with manicured grass. Wallace described Samsung's building as an "old cubicle farm out of the eighties. Beige and brown, stained carpets, and stale air. Very depressing.

"I was escorted into the sparse boardroom adjacent to Dale Sohn's office. It's just a table, some prototype phone and literally on the table looking at me, I kid you not, they had a picture of Steve Jobs," he said. Sohn sat there for ten minutes thinking, *What the hell is going on?* "Then the door opens and Dale walks in. Doesn't say 'hi' or anything."

"Do you know who that is?" Dale asked, pointing to the photo.

"Yes."

"Do you think you could beat him?" Dale asked. He deliberately kept a photo of Jobs on his desk as a reminder of his mission.

"Yes, I think I can."

"What would you need? And how long would it take?" Because Apple was an important Samsung customer, the executives at headquarters were pushing for a cautious approach. They wanted to take down each competitor, from HTC to Motorola to BlackBerry to Apple, one by one over the next five years.

"No, no," Wallace said. "We don't need to do that." Approaching the whiteboard with a marker in hand, Wallace proposed his own version of Samsung's strategy.

"I'm a rather petite Canadian," he said. "So one of my biggest fears is going to prison." If you want to up your chances of survival, you don't punch the second-, third-, and fourth-biggest guys in the room. "You punch the biggest guy in the face, because that's your statement of intent. That's what we have to do with Apple."

Something lit up in Dale's face, Wallace recalled.

"Okay. What do you say?" he said. "Want to come here and beat Apple?"

Wallace appreciated the seriousness and outlandish goals of Samsung, as well as the energy of its American CEO. He was also happy to hear that Nike marketer Todd Pendleton was joining soon. So Wallace next called his wife.

"You know how we were going to move to either Chicago or California?" he said. "I think I'm gonna say yes to this job in Richardson, Texas."

His colleagues at BlackBerry couldn't believe Dale's decision.

"That's such a step down," said one, referring to Samsung. Little did they understand that it was their company that was facing collapse.

"OH MY GOD, THERE'S this jalopy of a marketing department here," Wallace said to Todd Pendleton on day one, when he joined Pendleton a couple weeks later, in June. The marketing department was, in Wallace's words, "dejected, beat up," and a "black hole" with "no respect in the organization. No respect amongst themselves." The average turnover time for a marketing executive at Samsung Mobile in the Texas office was seven and a half months.

The work of the marketers at Samsung was frustratingly subpar. Samsung didn't use people in its commercials—"just product and voiceover and talking about the product benefit," Todd Pendleton said. Rather than pitch consumers on why Samsung was great, marketing stories were framed around the telecom carriers—"telling a story around their network and why their network is great."

The South Korean headquarters, meanwhile, sent over goofy and culturally inappropriate commercials that incited rebellion among the Americans on staff.

"They wanted us to use butterflies," said former marketing vice president Clyde Roberson. He called the ads "Hello Kitty."

"One commercial showed a woman in the mountains, surrounded by a flower bed, spinning around and holding a phone that blossomed into flower. It was reminiscent of *The Sound of Music,*" said Bill Ogle, former CMO. The American team refused to run it.

The South Koreans' respect for rank, prestige, and luxury didn't always translate well into the American market. Proposed spots out of Seoul that were designed for the Asian market tended to showcase "trendy, Eurotrash, rich, white people . . . that have a certain look and feel," explained a Samsung marketer. The marketing team in Texas worried about the potential blowback in the United States. U.S. ads, ideally, would have "a sense of humor" and would be "hipsterish"—an art that Apple had mastered.

Pendleton and Wallace quickly got to work. "At Samsung, you had one day to get settled in," said another marketer on the team. They acted like good cop and bad cop, visionary and executor, with Todd bouncing ideas back and forth with his team and Brian doing the managing and yelling to make them happen.

"Todd's strengths are that he's an extremely good marketer in terms of creative," said a marketer who worked for him. "He wasn't great at managing people day to day in the corporate bureaucracy."

At office talks and public events, he'd show up with an earring, stubble, jeans, and an untucked button-down shirt inside a sport coat,

Todd Pendleton, the Samsung Telecommunications chief marketing officer,
who led the advertising war against Apple.

COURTESY OF GETTY IMAGES ENTERTAINMENT, PHOTOGRAPH BY NEILSON BARNARD. OCTOBER 24, 2012.

sometimes featuring checks and eccentric colors. Bright-red sneakers and other zany patterns made appearances in the office from his prized collection of six hundred pairs of Nike shoes. But he also brought a refreshing sense of creativity and insight to the team. At ad agency meetings, "he would point out things which, me watching [the commercial], I wouldn't really think about it," said the marketer. "It just was very clear that he was thinking creative on a different level." Sometimes, however, he got too creative for his own good.

"He was a total conspiracy theorist," said another team member, recalling his penchant for seeing 9/11 as an inside job. He could come off as aloof—a "ghost" who "didn't talk to anybody," joked one colleague—spending a lot of time at Los Angeles commercial shoots and not always accessible in Texas.

But the Nike veteran was also a tough perfectionist, and working under him could be hard on one's ego.

"He said no more than he said yes," Brian recalled. "He wouldn't compromise. He would keep the pressure on. The positive pressure always produced great work."

Sometimes he'd push everyone to think about the most outlandish things they could achieve in the world of advertising, no matter how silly or crazy. Some of the team's ideas—like a plan to remove the letter H from the Hollywood sign and float it down the Mississippi River—were exactly that.

"How much would it take to get that spot?" Brian once asked Pendleton as they stood in Times Square looking up at the glowing neon display screens.

"What would it take to get all of them?" Todd said. They went back to the office to figure it out. Later Samsung actually did do a momentary Times Square takeover of all display signage that was available, though Todd's team didn't oversee it.

And Brian Wallace's role on the team?

"At Samsung, there's always a guy who goes and yells at people. He was not really a pleasant guy to work for. Brian was actually cut from a very good Korean cloth," said a marketer on the team. But he added, "That was Brian's strength."

Todd's inaugural marketing team was "a start-up with the billion-dollar backing of a major corporation," said one colleague. The team was dysfunctional, in her eyes, but in a way that somehow worked.

"You could be partying with Jay-Z one day and then getting audited by the South Korean headquarters the next. You'd stay up forty-eight hours straight getting your ass kicked."

The two marketing executives brought aboard thirty-six marketers and treated the office as a black-box operation. "We had to be somewhat insular to be able to pull some of this stuff off," said a team member. They were worried about meddling from South Korea's bureaucracy. Dale provided air cover from headquarters, giving them an unusual degree of latitude and space to get their work done.

Then, just as they got started, Dale informed them that five years was too long a time period to overtake Apple. Dale shortened the time frame to two years, on orders from Samsung headquarters. In fact, the team completed their work in eighteen months.

19

Cult of Steve

ON DAY ONE AT Samsung's U.S. headquarters, Pendleton gathered about fifty people into a meeting. He approached the whiteboard and wrote: "Samsung = ?"

"Who are we?" he asked. "What do we stand for?" Then he went around the room and asked everyone to fill in their idea. "I got about 50 different answers," he said. For Todd Pendleton, it was alarming. "If we can't answer [that] as employees, consumers are not going to know who we are."

After a long back-and-forth conversation, the fifty marketers in the room converged on a Samsung strength.

"We were always going to have the best hardware, first," he said. On the whiteboard Todd wrote: "Samsung = relentless innovation."

The group convened each week at Dallas's Hotel Palomar, now renamed the Highland Dallas Hotel, a boutique hangout in the city's University district. Here they plotted their understanding of how people saw and understood Samsung's brand on a perceptual map, based on the market research they'd gathered.

On a chart of competitors in their space, with "style" for the vertical axis and "innovation" for the horizontal axis, they placed Apple and Sony in the upper-right quadrant, marking them as both stylish and innovative.

Samsung, on the other hand, still lacked brand power: It was raised only slightly on the style axis, while it was far to the left on the innovation axis. In other words, consumers saw Samsung as having little of either. "Less stylish, less innovative." "More functional." "Good quality and value." With Apple and Sony commanding and fiercely protecting that stylish and innovative space, could Samsung find an opening?

In focus groups and surveys, the marketers noticed, there was a growing divide between two camps: those who used Apple's iPhones and those who used smartphones from HTC, Samsung, and Nokia, which ran Google's quickly growing open-source operating system, Android.

"Android people consider themselves to be smarter than Apple people," a marketer under Todd concluded from his data. "They consider themselves to be smart shoppers; they consider themselves to make informed decisions, unlike Apple sheep, who just do what [Apple] says." Samsung's data crunchers concluded that Android users were looking for someone to say, "Hey, look, this Android phone is just as good as your iPhone."

In fact, the team had to split up focus groups that included both Apple and Android fans, as they'd get particularly raucous and unproductive. There was always at least one Apple fan in the room who scolded the Android fans, and vice versa, with Android users pointing out how much more flexible and customizable their operating system was. "There was this growing base of Android users who could become a tribe," Brian Wallace said, crunching a new trend in the social-media chatter. "But they needed a leader."

Samsung wanted to be that leader.

Pendleton showed his colleagues side-by-side hardware comparisons between the iPhone and the Galaxy phone in *The Wall Street Journal,* which showed Samsung leading in a number of areas. The problem was that Samsung, up to this point, was not attempting to tell a story. Apple was commanding the narrative: It had the cult of Steve Jobs, a massive following, and glowing media coverage, and it had unleashed a barrage of aggressive legal action arguing that Samsung was a copy-cat in terms of new products and innovation.

Could Samsung reverse the narrative? What if its Android phones were actually the smart person's alternative to the iPhone, and Steve Jobs's worshippers were the mindless followers? What if Apple's lawsuit vindicated Samsung? What if Apple had patented something as silly as a black rectangle with rounded edges out of desperation, attempting to bully its way into a monopoly?

The outcome of the lawsuits—showing that this or that square, icon, or color wasn't copied—wasn't the concern of Todd's team. More urgent was the big-picture narrative; that is what built emotional appeal for the customer. The court case was only one aspect of the Samsung war; final victory, they knew, would go to the company that told the best story to the public.

ON OCTOBER 6, 2011, *The New York Times* published an obituary on its front page: "Steven P. Jobs, 1955–2011: Redefined the Digital Age as the Visionary of Apple."

In some alternate universe, the loss of Apple's guiding hand could be seen as good news for Samsung. But not here on earth, in our current universe. Techies and passionate Apple users exalted Jobs to the saintly pantheon of Mahatma Gandhi, Martin Luther King, Jr., and Nelson Mandela.

"Outside the flagship Apple store on Fifth Avenue in Manhattan, people had left two bouquets of roses and some candles late Wednesday," the *Times* reported. "By 11 P.M., the crowds gathering outside the store were thickening."

"Here's to the crazy ones, the misfits, the rebels, the troublemakers, the round pegs in the square holes, the ones who see things differently," a recording of Steve Jobs called out at his memorial. The only Asian executive invited to the memorial service was Samsung heir Jay Y. Lee.

Michael Pennington, vice president for sales, shot off an email to Pendleton and Dale Sohn two days after the Apple co-founder's death.

"Unfortunately, Steve Job's [*sic*] passing has led to a huge wave of press coverage of Apple's and iPhone's 'superiority,' all created by the 'passionate, tireless, perfectionist. . . .' The point here is that there is an unintended benefit for Apple," he wrote to them. "What consumer wouldn't feel great about purchasing a device developed by such a person. . . . I know this is our best opportunity to attack iPhone."

"As you have shared previously, we are unable to battle them directly in our marketing," Pennington wrote in another email.

Every time the U.S. marketers mentioned Apple by name, "someone from headquarters would call and yell," as a team member put it. Samsung executives in Seoul didn't want to anger Apple, as it was such an important customer.

But in the face of the Apple lawsuit and Steve Jobs's death, Samsung's reluctance to act was increasingly hurting it in building its consumer brand in the United States. In a series of meetings, the American executives disagreed with the cautious, incremental approach of Samsung headquarters, who continued to want to focus on the company's other competitors—HTC, Motorola, BlackBerry—first. The American marketers wanted to move against Apple without Samsung headquarters' permission.

"If it worked, we could say, 'Oops, sorry, it worked,'" said one marketer. Momentum was building for a sudden and swift marketing counterattack.

Later on the day of the emails, Todd Pendleton responded to Michael Pennington with his own email.

"Hey Michael, we are going to execute what you are recommending in our holiday GSII campaign and go head to head with iPhone 4S," he wrote.

Their plan?

"We will demystify the perceived Apple advantage (ecosystem/services) by showing how consumers can easily switch to Android and have more personalization/more choice by being part of the Samsung ecosystem. More to come soon."

The American marketing team wanted to take advantage of Jobs's death to charge into battle against Apple, head-to-head. One marketer said that the team briefly held back to avoid looking petty, then charged forward.

Trucks carrying fresh apples started arriving at the Texas headquarters of Samsung. Bushel baskets were placed in the elevator banks and break rooms, so that wherever Samsung employees took a coffee break, they were reminded of their mission—to take a bite out of Apple.

"We had one objective—beat Apple. I'm not kidding you. That's it," Wallace said. "I really believe the guy who filled up the Coke machines at Samsung had to demonstrate that he was supporting the beat-Apple strategy. And it worked."

20

Coke Pepsi Redux

"IF THIS DOESN'T WORK, you're both fired," Dale Sohn informed Pendleton and Wallace.

In its heyday, Coca-Cola, like Apple, had mastered the art of pop-culture marketing. The curvy bottle with its red label and white lettering was an American consumer institution. It was a remarkable victory for what really amounted to a sugary, carbonated, caffeinated beverage.

A raft of other cola drinks entered and exited the crowded soft-drink market. Nearly all were eclipsed by Coke. Then Pepsi, the young person's cola, came along and made it a two-horse race.

Pepsi confronted the industry leader head-to-head with its "Pepsi Challenge" campaign in the 1970s. In a series of commercials shot at shopping malls and parks, random people were given blind taste tests sipping cups of Coke and Pepsi, the labels hidden.

The majority opted for Pepsi, the sweeter drink. It was a legendary but controversial stroke of marketing, breaking the adage that you avoided attacking a competitor head-on. What if more people chose Coke?

Coke, in response, became convinced, as a result of the Pepsi challenge, that Americans preferred a sweeter drink. So it changed its sacrosanct recipe for the first time in its history, creating a new and sweeter Coca-Cola called "New Coke." But in the process, it made the unforgivable error of abandoning a huge market of traditional Coke lovers. Sales collapsed, and people accused Coke of gutting a drink associated with America itself. Within three months Coke restored the original version, now called "Coke Classic."

Coke came back from the disaster over time, but from that point on, it was always competing in the presence of Pepsi.

By attacking Apple head-on, Samsung's marketers thought they could establish themselves as the challenger brand, turning the competition with Apple into a Coke-versus-Pepsi war for the smartphone world. But how do you attack Apple without looking petty, without giving it free advertising, without acting like the smaller dog in the pack who barks the loudest and then gets laughed at?

Wallace was pushing for the blunter Coke-versus-Pepsi strategy. But Pendleton and others wanted to tone down the approach.

"You never attack the people who are buying the products," he told Brian.

Samsung's approach? *The customers were Apple's victims.* "I think this is done with a wink and a smile," Pendleton told *Business Insider*.

"It was very confusing for Samsung, the Apple of Korea . . . not to be number one in the U.S.," Brian Wallace said. It was as if "Steve Jobs had gone to Paris and couldn't understand why Apple was losing market share in France."

The team turned to a consultant named Joe Crump, senior vice president for strategy and planning at Razorfish, one of the world's largest interactive agencies, to help them convey the depth of the brand problem in America to Samsung's senior executives. Crump had an idea to get that across: He'd send camera crews around Times Square, each carrying two duffel bags. The first bag, people on the street would be told, contained the next unreleased iPhone. The other had a Samsung phone.

"What would you give us for each?"

Here was the response to the question when they thought the bag contained the new, unreleased iPhone: "I'd give you my brand new BMW. . . . I'd give you ten thousand dollars. . . . I'd give you my sister."

And the response for the Galaxy: "I don't know. Five bucks?" One guy offered his half-eaten ice cream cone.

"The Samsung [response] was just blistering," Brian recalled. "We had to even take some of it out because it was just so harsh."

A visiting delegation of South Korean executives huddled in a conference room to watch the video of these Times Square interactions. They were aghast. Suddenly Pendleton had their ear. The research—the field testing—had been done for internal consumption only. It was designed by Pendleton to get the South Korean executives to grasp the size of the problem.

Step two was to ensure that the economics of the coming marketing war on Apple made sense. Samsung had built, under Pete Skarzynski, a carrier-driven model, jumping through hoops to ensure that Sprint and AT&T received their own customized Galaxy phones to sell, using Samsung's marketing money. If Todd pulled a maneuver akin to the Pepsi challenge too soon, swarms of customers might show up at AT&T stores—AT&T was the exclusive carrier for the iPhone at the time—only to have the advertising throughout the stores nudge them toward Apple.

The solution? To redirect Samsung's marketing budget. At the time, Samsung was putting about 70 percent of its U.S. smartphone budget in so-called marketing development funds (MDFs), which were cash piles allocated to the carriers for advertising and rebates. About 30 percent of the budget went to Samsung's own branding efforts. Pendleton's team convinced Dale Sohn to reverse the figures: to put 70 percent behind Samsung's own efforts and devote 30 percent to the carriers.

Once Samsung had the marketing budget to reach out directly to customers, Pendleton could initiate step three: hiring an ad agency. He annoyed Samsung headquarters by going around their established Madison Avenue and Seoul agencies and instead putting in a call to relative newcomer 72andSunny, a boutique advertising firm with offices in Los Angeles, New York, and Amsterdam that had a special zing for cultural marketing.

"I just need some help to get this thing really rocking," Todd told 72andSunny's Glenn Cole, John Boiler, and Matt Jarvis on the phone. "We've got the best phone. Nobody knows it."

Known for their edgy and rebellious approach to their craft, the trio didn't even describe 72andSunny as an ad agency in the traditional sense. Earlier that year, 72andSunny had been in trouble for a campaign called "Unhate" that it had crafted for United Colors of Benetton. In a push for what it called "global love," it featured fake images of President Barack Obama kissing the former Venezuelan dictator Hugo Chávez, a smooch between Israeli and Palestinian leaders, and an especially controversial lip-lock between the pope and an Egyptian Muslim religious leader.

The Vatican threatened legal action; its spokesman told *The Guardian* he was dismayed that "in the field of advertising, the most elemental rules of respect for others can be broken in order to attract attention by provocation."

Todd's team chose 72andSunny specifically for its edginess. On a conference call with 72andSunny, he laid out Samsung's goal, as handed down by Dale Sohn.

"I expect us to be number one in a couple years."

John Boiler called the goal "laughable." Samsung, after all, was a speck on the smartphone radar. But, Boiler recounted, one important ingredient to success was "high pressure."

"We had to get concepts to Todd in four days and we had to produce [the campaign] in two weeks," he said. "We got it, and we nailed it out of the gates because we were fighting mean and lean."

Pendleton needed to move quickly to promote the upcoming Galaxy S II under Samsung's demanding time line. He assigned Joanne Lovato (known to colleagues as Jo), a California native and UCLA graduate, to coordinate with 72andSunny and the other agencies.

They worked to inject more life into Samsung's engineering jargon describing new products. A few weeks before each product launch, the team would receive a thick technical tome consisting of hundreds of pages. They had to tear through it, decipher it into everyday language, and select the top six or seven features to promote to consumers. Then, without being allowed to see the new phone before the launch, they'd crunch the engineering-speak into a PowerPoint presentation called a "product positioning document," which they'd present to the agency.

Both Samsung's team and 72andSunny were always racing the clock. Product unveilings, called "Unpacked" events, were typically held in March or April. The engineering manual arrived in February. Three times a week, Jo held a teleconference with the agencies to ensure everything was on track.

Short on time and eager to create commercials immediately after Apple's iPhone product unveilings, Pendleton decided that the actors they hired for commercials would ad-lib their parts, rather than go through a longer, more structured creative process.

The creative executives at 72andSunny got to work and churned out their first approach for Pendleton, who was present at the shoots and the editing, eager to maintain his creative hand. In one early version of a commercial, two characters waiting in line outside an Apple store had a conversation about the features and quality of their Apple and Samsung phones, followed by a cut to another scene of two characters talking about their phones.

It was slow, boring, and dull. Samsung's bid to take on Apple, Todd's team feared, would be finished before it had even started.

"We don't have a campaign here, guys," Pendleton said.

With the holiday shopping season closing in, the only solution was to chop up and redo the film then and there. During a frantic all-nighter, someone in the room suggested that they turn the commercial into a single scene, rather than two separate, awkward, forced moments of chatter between disparate characters.

21

The Next Big Thing

THE NEW COMMERCIAL WAS completed the next afternoon.

It started, as before, with a line of apparent Apple lemmings waiting all night around a street corner for the release of the next big iThing—presumably an iPhone, though Apple was never mentioned by name.

"Guys, I'm so amped I can stay here for three weeks," says an apparent Apple idolater.

"Nine hours down and we're almost in the door," chants another.

"If it looks the same, how will people know I upgraded?"

One guy notices a woman on a sidewalk tapping away at some weird gadget that—*what?*—doesn't look like an iPhone.

"Whoa, what does she got there?"

Then another pedestrian hails a taxi on the sidewalk, holding the mystery device.

"Hey, bro, can we see your phone?" The mob of Apple fans snatch the device and pore over its hardware and features.

"It's a Samsung Galaxy," the pedestrian tells them.

"Check out the screen on this thing—it's huge."

"It's pretty massive," says another guy at the sidewalk.

"And it has 4G speed."

What is this thing?

"It's a Samsung," they repeat to each other. "Samsung?"

A priggish kid with a beard, sitting in solitude on his laptop and wearing headphones, raises his chin with an announcement.

"I could never get a Samsung. I'm creative."

"Dude, you're a barista."

"It's a Galaxy S II. This phone is amazing," says the Samsung guy, showing off his smartphone before getting into a taxi, bidding farewell to the crowd of Apple zombies.

The message? You don't need to wait in line. You don't need to follow the hype.

"The Next Big Thing Is Already Here," the commercial finishes.

"God damn!" Todd exclaimed after looking at it. "We've got a campaign!"

Pendleton's staff sent the commercial to South Korea for approval. Five days later, they'd still heard nothing back. At six o'clock on day five, Dale Sohn stood up, put on his jacket, and got ready to go home, before leaving a word of advice on the silence from Seoul.

"It means they've given you enough rope to hang yourself," Sohn said.

It was up to Todd's team to make the leap and take the risk. And if it failed, they'd have to answer for it.

They proceeded to leak the film to the popular tech and culture website *Mashable,* which unveiled it on November 22, 2011, before Samsung posted it "officially" on its Facebook page later that day. Pendleton was abandoning the marketing world's older, more vanilla strategy of going through print and TV news outlets, opting for the Web first, appealing to millennials. Then, on Thanksgiving weekend, the commercial debuted in minute-long spots during the NFL games.

The campaign was a phenomenal success, beyond anything the team had anticipated; Samsung had hit precisely the sweet spot, with viewers responding that they were tired of swallowing what they thought was Apple's unjustified pretentiousness. The commercial transformed Samsung Telecommunications America into one of the fastest-growing brands on Facebook, with more than 26 million fans in sixteen months.

"We are the fastest-growing brand globally on Twitter, with almost two million followers," Pendleton later recounted at a press conference.

"Get ready to take out your designer pitchforks, Macheads. Your hipness is under attack as we speak," joked CBS's Chenda Ngak.

During the third quarter of 2011, Samsung surged past Apple to the number one spot among phone manufacturers, based on shipments. No longer was the smartphone war a battle between Apple and a tangle of obscure Android me-too phones. Now it was a two-horse race. Apple was Coke. Samsung was Pepsi. Everyone else had fallen by the wayside.

But as Todd's team celebrated a deluge of press reports against Apple, the mood at the U.S. headquarters was quickly tempered.

Executives at Samsung's headquarters in Seoul, initially silent, suddenly took notice. Concerned emails from South Korea popped up with grimmer media reports attached. Samsung, declared Ben Bajarin, writing on *Tech.pinions,* appeared more interested in winning over iPhone users than in targeting first-time shoppers looking for a smartphone.

In very short order, Todd and Brian were told by the executives in Seoul that they were going to be fired, according to Brian.

"Todd and I were fired probably six times," said Brian. More than a dozen former colleagues concurred with this version of events. One remembered Pendleton telling the story differently, claiming that the Korean headquarters and American mobile office had a good relationship, though he acknowledged tensions at what he called the "working level."

"Don't quit, don't quit," Dale Sohn told team members who marched into his office, frustrated at being punished for a job well done.

"Todd went rogue. He spent too much money," a South Korean marketer said of Pendleton's team.

Frustrated that the U.S. team was acting on its own, the South Korean executives demanded Wednesday-night conference calls (Thursday morning in South Korea), commanding oversight of the American marketing team.

"Every time we broke a rule, we'd get yelled at by the headquarters," said a marketer on Pendleton's team. They didn't trust the American marketers. To them, Pendleton's instinct to make fun of Apple was nothing short of dangerous.

WHEN SAMSUNG'S WORLDWIDE MOBILE marketing chief, D.J. Lee, traveled to Seattle, Brian Wallace was assigned to give him a Power-Point presentation to show off the team's marketing successes. He didn't realize the cultural minefield he was entering—outdoing your boss would make him or her look bad. In his 1996 book *Trust: The Social Virtues and the Creation of Prosperity,* the political scientist Francis Fukuyama defined South Korea and Japan as "low-trust societies." I found that to be true. Korean businesspeople joked to me that they needed to document "every paperclip."

"I'm thinking we are gonna be heroes," Wallace told me. "[Lee] sits there quietly, listening to this presentation. I'm showing him the social media metrics on the projector screen. I thought I was killing it and he'd be so impressed."

D.J. sat with a stony expression, saying nothing. Then he pulled out a marker.

"He starts telling us a story about this marketing organization in Russia that they had to audit. It turned out . . . they were corrupting the data and buying the results," Brian said. That was all D.J. said in response to the presentation. Brian was dumbfounded.

"I think he's accusing me of being corrupt," he said, leaning over to a colleague.

The reality? How successful was the new campaign? As *Business Insider*'s Steve Kovach reported, "The US team was outperforming Samsung's headquarters in South Korea, and other international offices were itching to adopt 'The Next Big Thing' in their respective countries."

Nonetheless, the more successful Pendleton and his team were, the more complicated their relationship became with Samsung's headquarters, as a number of the team members recalled.

Todd Pendleton was unable to make the trip to South Korea for Samsung's global marketing meeting. Instead, Brian Wallace and five other American marketing executives were dispatched to the Suwon headquarters to listen to a series of inspirational speeches and, they thought, receive awards.

The Samsung auditorium was filled with several hundred marketers and executives. Wallace was convinced that the team was going to receive an award of some sort for the success of their marketing efforts in the United States. And indeed, the executives on the stage had a special announcement. They asked the American team to stand up.

"The [Korean] executives told the employees [in the auditorium] to clap for the US team as encouragement since they were the only group failing the company, even though it was clear to everyone the opposite was true," *Business Insider* reported.

It was a Samsung ritual that I'd heard recounted by countless Samsung executives: Just when you thought you would be recognized for your achievements, you were scolded. "The concept of limited good pervades Samsung," said a marketer. "If someone else is doing well, that means you're not doing as well."

Brian Wallace was furious. At that moment, he wanted nothing more than to quit.

"Stand up. Stand up," an executive beside him said, grabbing his arm and pulling him to his feet.

He realized at that moment that Samsung "is the most bizarro place I've ever seen," as he told me. Later, back in the United States, Brian raced into Dale's office and announced that he was quitting. Dale convinced him to stay for the moment, but Brian's mettle was wearing thin.

Fortunately, the heir apparent to the Samsung throne, Jay Lee, had taken notice of Pendleton's work and loved it. He flew out to meet his team for dinner. The U.S. team was protected as long as the ruling family was behind them—and as long as they could maintain their stellar performance.

22

Galaxy Trilogy

PENDLETON'S MARKETING TEAM WAS at work translating the next five-hundred-page manual of engineering mumbo jumbo for the follow-up phone from Samsung: the Galaxy S III, due out in May 2012.

Though they hadn't seen the phone, they sensed this one was different. Samsung had been following its path of incremental innovation, making small adjustments and adding new features to the first two models, and they were making a big difference.

During the eighteen-month development process for the new phone, Samsung's engineers were under a secret lockdown. They were allowed to carry three prototypes around the building in boxes to meet with their colleagues, but they were unable to show their colleagues a picture. They had to describe the phone as best they could in project meetings.

"To be honest, it was quite tiring and frustrating," said senior engineer Lee Byoung-sun.

The South Korean engineers homed in on the social media–like features of the phone. They were novel in terms of their ability to let people share multimedia instantly. The engineers decided to develop the Galaxy S III's ability to connect to a television and to automatically share content with friends when tapped against another phone—sharing files instantly. They also documented several hardware firsts or near firsts: two gigabytes of RAM, a superior camera, a bigger display, and one of the highest-definition smartphone screens available on the market, the Super AMOLED, designed in-house by Samsung. But the U.S. marketing team's strategy was to build a phone that would transcend the individual features, to create a brand that customers would remember.

"A lot of those features didn't catch on," according to one of Todd's marketers. Since many features were transient, coming and going with uncertain futures, brand building was especially important so that the Galaxy could stand above the features alone.

"You know, Todd," Chief Product Officer Kevin Packingham said in front of reporters at a press conference announcing the new phone, "our closest competitor [Apple] in this same time frame has announced a marginal software update. . . . Since then, Samsung has launched three massive products, and now we're introducing the Galaxy S III."

The Galaxy S III roared through the markets, selling nine million preordered units and forty million units in the first six months. *TechRadar* declared it the number one handset in the world in September 2012. At the Mobile World Congress, the Galaxy S III won the "Best Smart Phone" of the year award, beating out the iPhone. *CNET*'s Natasha Lomas called it "the Ferrari of the Android circuit"; *CNET* UK readers voted it the "best phone of 2012."

Samsung's profits soared that quarter, hitting $5.9 billion, a 79 percent increase from a year earlier. Pendleton's team celebrated as Samsung hit a landmark: It had overtaken Apple in volume (though not in profit margins). Samsung was crowned the world's largest maker of smartphones.

Apple responded with a victory lap of its own on August 24, 2012, as lawyers from Samsung and Apple and a mob of reporters were called into a Cupertino courtroom.

At 2:35 P.M., after plowing through seven hundred highly technical questions, the jury unanimously informed the court that Samsung had infringed on many of Apple's patents, and awarded the company $1.05

billion. Apple, meanwhile, the court decided, had infringed on none of Samsung's patents.

"We wanted to make sure the message we sent was not just a slap on the wrist," jury foreman Velvin Hogan told Reuters. "We wanted to make sure it was sufficiently high to be painful, but not unreasonable."

Judge Lucy Koh found herself something of a celebrity for snapping at high-powered Silicon Valley attorneys. She told them during the trial, "Unless you're smoking crack, you know these witnesses aren't going to be called when you have less than four hours." She later decided that the jury had miscalculated Samsung's damages and cut Apple's award down to $600 million.

The legal war between the two companies, however, was far from over. A second trial between Apple and Samsung was already in the works. Samsung took home some legal victories as well in the UK, Japan, and South Korea.

ON THE MORNING OF September 12, 2012, Apple CEO Tim Cook took the stage in Cupertino for his first product launch as successor to the late Steve Jobs.

"Today we're taking it to the next level," he said. After years of waiting to deliver something dramatically new, Apple was releasing the iPhone 5.

A few hundred miles south in Los Angeles, Pendleton and the team were camped out in a war room at a Wolfgang Puck restaurant in Los Angeles, huddling around tables with laptops and TV screens, monitoring social-media reactions to Cook's remarks on each new feature. Headphone jacks, new software, the size of the screen—the marketing team compiled the responses.

"As the data flowed in," reported *Fortune*'s Michal Lev-Ram, "writers from the company's advertising agency"—72andSunny—"who were also camped out in the restaurant turned war room, scrambled to craft a response."

As Cook stepped off the stage two hours later, Pendleton was already getting ready for the next barrage in the Apple-Samsung marketing wars. He shot his next commercial in a rapid-fire manner; it was quickly ad-libbed by the actors, with their talking points pulled from social-media chatter on Tim Cook's iPhone launch, intending to satirize each feature. One week later, on September 19, the commercial was aired. It tried to steal the show by beating Apple's release date for the iPhone 5, which took place two days later, on September 21.

In the commercial, the Apple herd is once again waiting in all-night lines stretching around street corners in Chicago and San Francisco and New York, jabbering about this or that rumored feature in the coming launch.

"The headphone jack is going to be on the *bottom!*"

"I heard the connector is all digital. What does that even mean?"

"I heard you have to have an adaptor to use the dock on the new one," one guy complains.

"Yeah, but they make the *coolest* adaptors."

The Samsung kid from the first commercial is waiting in line this time. "Welcome back," says an Apple fan. "Guess that Galaxy S III didn't work out."

"No, I love the GS III. It's extremely awesome. I'm just saving a spot in line for someone."

A pair of stodgy old folks show up. The Samsung kid, it turns out, is holding a place in line for his markedly uncool parents.

The Apple fans are deflated. The protagonist bids adieu, taking his Samsung phone with him.

"The Next Big Thing Is Already Here," reads the tagline.

ONE DAY IN SEPTEMBER, the U.S. division's South Korean dispatcher (its liaison with Samsung headquarters), a man named Ji, hurried into the office, frantic and fearful. A plane full of auditors from Samsung headquarters was on the way from Korea to go through the American team's books. They would land in a few hours.

"Help them and be open," Ji said. He told his team to give them "whatever they need."

For three weeks, the American marketing team was forced to go through their records, proving their successes.

"They were accused of bribing the media, falsifying sales, and a bunch of other damaging claims that hurt morale in the office," reported *Business Insider*'s Steve Kovach. "The same US-based office that helped turn Samsung into a brand as recognizable as Apple was suddenly being punished for its work."

This is our reward for success? Todd's people wondered.

While his employees were distracted by the audits, Pendleton managed to keep his head held high. For him, the audits weren't something to fear. Samsung's headquarters just wanted to learn what his team was up to. Employees claimed that it was his unrelenting optimism that was the magic behind his success.

The auditors found nothing out of the ordinary. And the team's feeling of being under attack by South Korean headquarters had the unintended consequence of strengthening the sense of unity among Pendleton's high-performing marketers, bringing them closer together as a tribe.

Whatever the executives at Samsung's headquarters in Seoul thought, the team's "Next Big Thing" commercial garnered seventy million online views. It was the most popular tech ad of 2012.

Todd was ecstatic when he opened *The Wall Street Journal* on January 28, 2013, to an uncanny story.

"Has Apple Lost Its Cool to Samsung?" the headline read.

"Samsung, the market leader in smartphones, on Friday said its fourth-quarter profit surged 76% to a record high on the strength of smartphone sales, including its Galaxy S line. The latest version is considered comparable by many shoppers in both design and technical features," wrote tech reporters Ian Sherr and Evan Ramstad.

"Apple, meanwhile, reignited concerns about demand for its iPhone 5 after reporting flat earnings for the holiday quarter, sending its stock down 14% in the past two days. The stock has also dropped 37% since hitting an all-time high on Sept. 19, just two days before the iPhone 5 launched in stores."

The *Wall Street Journal* piece pointed to Pendleton's "marketing onslaught" that had allowed Samsung to close the "coolness gap with Apple Inc." The article riveted the tech industry.

But the Super Bowl was fast approaching (six days away), and Pendleton wasn't finished attacking Apple's position in the marketplace.

He had a new $15 million ad-libbed commercial set to air during the Super Bowl, made with 72andSunny, featuring comic banter among Paul Rudd and Seth Rogen and *Breaking Bad*'s Bob Odenkirk, plotting their own fictional ad spot for the Super Bowl.

"We actually can't say 'Super Bo—'?" asked Seth Rogen.

"No! It's trademarked," snapped back Bob Odenkirk. "You get sued!"

"Can I say 'San Francisco'?"

"Sure."

"But I can't say 'the forty-nine—'?"

"Eee!"

"Can we say 'the San Francisco fifty minus ones'?"

"Love it."

The commercial was more a brand-building exercise than an effort to sell more phones, poking fun at Apple's lawsuit. Samsung was riding a powerful wave of positive publicity.

Forbes described the ad as "a barrage of not-so-subtle jabs at competitor Apple Inc."

Phil Schiller, Apple's senior vice president for marketing, was livid at Samsung's marketing campaign. He shot an email to Apple's ad agency, TBWA\Chiat\Day, with a link to the *Wall Street Journal* article.

"We have a lot of work to do to turn this around," he wrote. He emailed Apple CEO Tim Cook that Apple "may need to start a search for a new agency."

"We feel it too and it hurts," wrote an advertising executive at TBWA back to Schiller. "We understand that this moment is pretty close to 1997 in terms of the need for advertising to help pull Apple through this moment."

Schiller blew up at the response.

"To come back and suggest that Apple needs to think dramatically different about how we are running our company is a shocking response," he wrote.

"In 1997 Apple had no products to market. We had a company making so little money that we were 6 months from out of business. . . . Not the world's most successful tech company. Not the company that everybody wants to copy and compete with."

Samsung's Super Bowl ad was "pretty good," Schiller went on, "and I can't help but think these guys [TBWA] are feeling it, like an athlete who can't miss because they are in the zone while we struggle to nail a compelling brief on iPhone."

In response to Samsung's new marketing savvy around new releases of its Galaxy phones, Apple seemed to follow Samsung's lead in terms of new products, expanding Apple's limited lines to include bigger screens, additional sizes of iPhones, and more variety for a wider swath of customers.

"Consumers want what we don't have," read a gloomy Apple presentation slide in February 2013, lamenting that growth was slowing, that people were crying for bigger screen sizes. It was an idea that Samsung had pioneered.

Advertising at Apple, too, needed a reboot. As Pendleton was lampooning Apple, the company had already been releasing commercials around the "Genius Bar," featuring a feisty, squeaky-voiced fix-it geek running around telling people about basic features on their Apple products on an airplane, at someone's front door, and on a city street. The spots were called "embarrassing" by *The Verge* and so derided by loyal Apple followers that Apple pulled them.

In 2013 Apple raised its U.S. mobile phone advertising budget from $333 million the previous year to $351 million. (Samsung had a $401 million budget in 2012.) It brought its advertising work in-house and forced its traditional outside agency to compete with Apple's in-house talent.

As Apple struggled to get its mojo back, Samsung was already moving forward with its next assault, this time in software.

23

The Ecosystem

"THERE IS NO OBSTACLE that cannot be overcome. We must work harder, push harder, and go faster!"

At the product planning meetings for the Galaxy, Kang Tae-jin (T.J.), senior vice president for software and content, was sitting through a tongue-lashing from his superiors. The engineers had been building blazingly fast hardware for the Galaxy phone, overtaking Apple in many areas. But their software—the animus that breathed life into each device, that allowed for communication with others in an ecosystem—was buggy and clunky.

"Readers Hub, Video Hub, Music Hub," T.J. recounted to me in his corner office at the new company he now worked at, sipping tea, thinking about the Galaxy's original (and disastrous) Kindle and iTunes wannabes. "You had to sign up separately for these individual services run by completely different people, different companies." They had contracted with Samsung "so the sales and marketing offices could check off their boxes, with little thought to quality. It was a terrible experience."

"We had a Kindle, we had an iTunes, we had a Netflix. There was no exchange of information, so all payments were different. If you registered a credit card to purchase music, you had to do another registration to purchase an e-book."

T.J. spoke with a certain resignation in his voice. A well-known software entrepreneur in Silicon Valley, he'd built a name for himself as the founder of ThinkFree in the late 1990s. ThinkFree was a free Web-based word processor back when such software, dominated by Microsoft Word, was strictly a hard-drive affair—before the days of Google Docs and cloud computing.

In the end, ThinkFree was too small to overtake Microsoft. But Samsung, a vastly larger company, had infinitely more resources and potential.

"I felt that at Samsung maybe I could realize that dream"—to build software that everyone uses.

T.J. had been recruited by Samsung three years earlier, in March 2010, enticed by an old friend, a former director at IBM Korea named Ho Soo Lee.

After T.J. joined, Samsung's Chairman Lee II, in failing health, was promulgating his vision to find new growth areas for Samsung. And software was one of them.

Kang Tae-jin (T.J.), former senior vice president for software and content,
brought to Samsung to build a software ecosystem, an area where it struggled to beat Apple.

COURTESY OF T.J. KANG

"They came to the realization that in order to succeed in the pre-mium mobile phone market, they needed to create an ecosystem," T.J. said.

Over lunch, his friend Ho Soo Lee told T.J. about a new office he was heading at Samsung, called the Media Solutions Center (MSC), a five-hundred-person group housed in a sprawling old VCR factory. Their mandate was to build a software ecosystem linking every Sam-sung device—"to match what Apple was doing," T.J. recalled.

"I very much wanted T.J.," Ho Soo later recounted to me in his of-fice at his new company, near the top of a skyscraper southeast of Seoul. "If anyone could get this done, it was him."

"Why don't you come work for Samsung?" Ho Soo asked him.

When T.J. started at Samsung, however, he found himself in a disas-ter zone when it came to software.

"Samsung had eight or nine operating systems running our de-vices," he said. Software developers were struggling to design an in-house operating system called Bada ("ocean" in Korean). It was released in February 2010, just a month before T.J.'s arrival, and the company quickly realized it was a failure. Basic applications like Skype couldn't be used because of restrictions on VoIP (Voice over Internet Protocol) software, and the GPS system was poor.

Samsung's executives realized they had a chicken-and-egg problem. Few people were willing to use a junky operating system without ap-plications, and few developers are willing to make applications for their software without users. The solution was twofold. They needed to build a community of developers, some in-house, some out of house, and from there they could build a community of users. But how should they begin?

T.J. sat in a presentation with G.S. Choi, the CEO, around a confer-ence table of executives as they discussed their next move.

"You must have some idea about this, right?" G.S. Choi asked the fresh-faced software developer. "What's the best way forward?"

T.J. looked around at the conference table of executives. "Acquisi-tions," he told them.

It was a bold proposal. The blundered acquisition of the PC maker AST Research in 1995 had ripped through the psyches of Samsung's leadership. It made them distrustful of acquiring other companies. The more globally minded G.S. Choi was keen on acquisitions, but mobile chief J.K. Shin and his engineers opposed them. Shin thought that acquisitions were too risky, according to numerous employees familiar with the company's strategy.

T.J. made the case that if Samsung wanted faster results, it would have to forgo building new capabilities in-house, as Samsung traditionally did. It would have to explore the much-fabled promised land on the other side of the Pacific: Silicon Valley.

"They didn't want to wait seven, eight years before we started having some traction," he told me. "Meanwhile, Samsung was making money left and right." With the success of its smartphones, Samsung's cash pile would soon grow to $28.5 billion, money it could put to use in acquiring software companies.

Chairman Lee II and his son Jay Lee, Samsung's crown prince, came out in support of acquisitions, and as a result the strategy won out. Over the coming half decade, Samsung snatched up start-ups as if panning for gold—you'd wade through a lot of junk, but the rare nuggets would hopefully make up for it. Samsung set aside $1.1 billion in venture capital for start-ups and began poaching software engineers from Apple and its ilk. Work would eventually begin on a $300 million research facility in San Jose, which broke ground in 2013. If Samsung wanted to hire software talent, Silicon Valley was the place to find it.

"There was a rumor that Palm was being shopped around," said T.J. He saw Palm as a good buy for an operating system for Galaxy phones. Just as he flew out to begin talks, Hewlett-Packard beat him to it with a $1.2 billion acquisition. Then HP approached T.J. about splitting ownership with Samsung.

T.J. was prepared to negotiate with HP, when its CEO Mark Hurd became "embroiled in this sex scandal," in T.J.'s words, resigning when news broke about a tryst with a female contractor. "There was a fight for Hurd's position," he said. "The guy who led the conversation [with Samsung about Palm] got distracted."

In Cannes at the music technology conference Midem, T.J. stumbled on the chance to bring his dream to life. He'd heard from an assistant about a pair of California software entrepreneurs, Daren Tsui

mSpot co-founder Daren Tsui. After his software company was acquired by Samsung,
he became vice president for content and services, running Samsung's
streaming music service, Milk Music.

COURTESY OF DAREN TSUI

and Ed Ho. One of them, Tsui, was giving a talk about "digital locker"
technology.

Tsui and Ho became early movers in an industry shift, releasing in
2004 a service called mSpot Music, a tool for audiophiles to upload
their music libraries to a "locker" on a server and access it anywhere
with a connection. Consumers no longer had to carry around an iPod
with a music library. Suddenly music could be streamed on demand on
their mobile phones.

"Until then, you stored your music library on your PC, and from
there you could download it to your iPod with a cable," T.J. said.

Before the days of Spotify, mSpot worked across all different devices, from television to speakers and phones and tablets. It was the
ecosystem that Samsung sought. T.J. set out to meet the pair and urged
his boss, G.S. Choi, to look into an acquisition of the company.

DAREN TSUI AND ED Ho were software visionaries. They could have
become billionaires, featured on the cover of *Fortune*—save for their
premature timing.

"It's not just the idea you have. Timing is everything—that's the lesson I learned," Ed Ho told me over coffee near his home in Palo Alto, California.

Daren was the deal maker, running around and impressing partners and signing deals, while Ed was the programmer at the computer, the one who attended to the product's coding.

Daren kicked off his career in 1994, that allowed publications to bring their classified advertising and articles to the Internet. In late 1997, Pantheon was acquired by Elon Musk and bundled into his first company, an online city guide called Zip2. It offered "door-to-door directions and Yellow Pages and so forth," Daren told me.

It was at Zip2 that Daren and Ed met, developing the door-to-door technology that Nokia used on an early mobile phone. They went on to co-found a mobile advertising company called SkyGo that was also acquired. "The problem we had is that the idea was perfect, but the timing was too early," Daren said.

Why let themselves be acquired by Samsung?

"This would be a great chance to become a formidable company," Daren told me, "to put out software used by millions of people with Galaxy phones."

But what started out as talks to preload mSpot on Samsung phones turned into a full-blown acquisition. And Daren was uneasy about the way Samsung approached it.

"We really thought that these guys were playing with us," Daren told me, "that they were just stringing us along to get the information and then possibly build it themselves."

Samsung's due diligence was intense. Streams of accountants and engineers and lawyers filtered through mSpot's nondescript, warehouse-like space in a California suburb, driving the pair crazy.

"There's a team of people that sit around a U-shaped table," said Ed. "You're sitting in the middle and there's a conclave of people just typing the whole time. They ask a question and then someone's typing. They actually have a bunch of people off-site whose whole job was to take whatever was typed and create a daily report at the home office so whoever there was managing the process could then ask follow-up questions the next day."

The only breaks were for lunch and dinner. The Samsung executives repeatedly brought up their disastrous acquisition of AST Research nearly two decades earlier, explaining that they were going ahead with the acquisition only because Jay Lee was forcing them to.

As the pair sat through meeting after meeting, they could see that Samsung's managers didn't get the point of software or building an ecosystem. They treated software as an afterthought to the Galaxy's hardware. One team didn't understand coding basics. Another didn't get the point of building a user community. "Why do you do it this way?" they would ask.

"The acquisition negotiations just kept dragging on because they needed J.K. Shin's final sign-off," Daren said. But Shin, the crusty old Samsung mobile chief, remained skeptical. He couldn't oppose Jay outright, but he could stall the acquisition process. Samsung employees would report back to South Korea, and then additional bureaucrats would show up. Each would pass the buck to someone higher, a bewildering process of consensus building. No one had the power to say yes. Everyone, it seemed, had the power to say no.

After nearly a year of dragged-out due diligence and negotiations, Tsui and Ho were thinking of pulling out of the deal.

"We were overwhelmed," Daren said.

Then things reached a decisive moment.

"I kept pushing and pushing," T.J. said. He finally got approval. G.S. Choi forced the issue at a meeting in Barcelona, going around the conference table and getting yes votes from all the executives. The voting concluded with J.K. Shin, who finally conceded to his boss. The deal went through in May 2012, reportedly for between $50 million and $60 million.

Daren Tsui and Ed Ho now reported to T.J. at the Media Solutions Center.

"Our goal was to be able to launch the mSpot locker at the time on as many devices as possible across the world," Tsui said. But right away, Samsung's speedy hardware culture didn't jibe with the mSpot way of doing things.

The Galaxy marketing team in Korea demanded that mSpot's services be made available in 170 countries as soon as possible. They wanted to conflate the process of developing, releasing, and patching software programs with the manufacture of hardware. With hardware, you designed a set order of parts, cleared out your inventory, and sent it off. Job done. In software, the release of the program itself was hardly the end of its development.

Crafting a software ecosystem, as T.J. pointed out to them, was softer, was trickier, and took time and patience. To make his point to the hardware engineers, he pointed to Apple's experience.

"After the launch of the original iTunes service," he said, "it took Apple three or four years to enter the next country, which was Canada. And it took them seven years to make that service available in twenty countries."

It would take years to navigate a tangle of copyright laws in each country, he pointed out, and to negotiate with music labels and publishers all over the world. Then there was the censorship.

"We decided not to launch a video service in Saudi Arabia," T.J. said, over concerns about bans on showing a woman's bare ankle, among other problems. He told the marketing executives, "Let's start with the U.S. and Korea and see how it does."

In a compromise, Samsung rushed out mSpot's new service, called Music Hub, to seven countries—a much more manageable number than 150 but still far more than what the software developers thought was reasonable. "There were a lot of bugs inside of the code because, you know, we were forced to operate fast and there were so many features," Daren Tsui said.

Samsung's local sales offices complained about pre-installed bloatware that annoyed customers. Daren, Ed, and T.J. encountered lackluster online reviews.

Nor were they the only ones struggling against Samsung's corporate culture.

"Samsung worked with Silicon Valley in the same way they worked in Korea," said Sumi Lim, a former business development manager.

One such story was rampant in Samsung's offices. Android founder Andy Rubin offered to sell his operating system to Samsung in late 2004, as he told journalist Fred Vogelstein in the book *Dogfight*.

"You and what army are going to go and create this? You have six people. Are you high?" was basically what the Android founders were told. Rubin found himself butting up against the Korean company's preference for working with large corporations.

"They laughed me out of the boardroom. This happened two weeks before Google acquired us."

Google paid an estimated $50 million for Android. The operating system would become the backbone of Google's products on just about every non-Apple smartphone. Google vice president David Lawee later called it his company's "best deal ever."

What a missed opportunity for Samsung, employees thought. Samsung now depended on Google to power the software on its phones.

In February 2013, seeing little progress in developing its own operating system, Bada, Samsung quietly shut it down and transferred its efforts to a new operating system, Tizen, co-designed with the struggling microprocessing company Intel. According to T.J., the two realized they needed to reboot some of their software projects and decided to team up. The plans for Tizen were strategic. In addition to attempting to build an ecosystem, Samsung's new operating system was a hedge against Android. If relations with Google deteriorated, Google could use Android as leverage to win competitive concessions, hampering Samsung's access to Google's prized open-source operating system. Tizen was Samsung's plan B.

In October 2013, Samsung put on its first developers' conference at San Francisco's St. Francis Hotel. Charging $299 a ticket, Samsung barely avoided disaster.

"We arrive the morning of this workshop and the attendees start downloading this stuff and it's not working," recalled an exasperated Hod Greeley, a Samsung developer-relations director who helped set up the event. "This is stuff that we've spent hours and hours and hours prepping for and testing, and it's ruined by this update."

Samsung's developers, it turned out, were writing new software code without clearly communicating things to those outside their department, as they raced to push out the apps, hardware style.

"Being in hardware," he said, "they don't do good [software] testing."

Greeley's struggles didn't end there. As we sat together in a San Francisco coffee shop, Greeley, a tall and soft-spoken software engineer who got his PhD from MIT, told the story of Samsung's bumbling bid to acquire Waze, the popular GPS app that gathers user-reported data to identify traffic patterns and plot routes around delays and congestion.

Waze CEO Noam Bardin was candid in a meeting with Samsung, frustrated over his dealings with the Korean company, as well as big telecommunications companies that wanted him to pour time and resources into developing exclusive features only for them, "believing they needed some special feature to make their platform unique," said a Waze executive. Samsung, Greeley recalled, treated the prospect of a software partnership as permission to assert its authority, asking developers to make software of questionable value—say, for using Samsung's beloved S Pen stylus—but without recognizing the time, resources, and potential impact on the reputation of the developers if the project fell through.

"It's a new feature that they have to test and maintain, and it affects their reputation if it's crappy," Greeley said. "Samsung didn't realize that there's no such thing as a free lunch."

In 2013 Google acquired Waze for $1.15 billion.

No amount of pleading, meanwhile, could convince WhatsApp's co-founders to partner with Samsung. CEO Jan Koum and co-founder Brian Acton "sat there and looked at the devices and said, 'Thank you, but we don't want to deal with you,'" recalled Greeley, who was present at the meeting. Koum and Acton didn't want to change the app's sacrosanct user interface, which seemed incompatible with Samsung's little-used operating system, Tizen. Later, WhatsApp agreed to develop the software for Tizen, but Tizen users reported bugs and software errors with the app. In October 2014, Facebook acquired WhatsApp for $19 billion.

By that time, Samsung had already released its own alternative, ChatON, in September 2011. Just over three years later, it was discontinued.

The conundrum for Samsung? "A ten-person startup should be struggling; lots of people should be knocking on Samsung's door to be working with them," said Greeley. "But with Samsung, the attitude of start-ups was 'Thank you very much, go away.'"

As Samsung was sullying its reputation in Silicon Valley, it was at least piquing the interest of Hollywood. Daren and T.J. had a strong network in the music and film industries and reached out to old friends to get better licensing deals for Samsung's streaming services.

At the Universal Music headquarters in Santa Monica, California—a sleek sandstone-colored building surrounded by palm trees—T.J. walked past the iconic Universal globe outside the entrance to attend a one-on-one meeting with Rob Wells, president of the global digital business. Frustrated with Apple in the year after Steve Jobs's death, Wells told T.J. that Jobs was "ruthless," "cutthroat," and "a bully." But a former Universal Music Group executive also insisted to me that Jobs was "the only partner that comprehended the value of music to his business.

"We already knew back then that Spotify was going to blunt Apple's edge internationally," the former Universal Music Group executive told me. "We were looking for a global device manufacturer that could step up and do what Apple had done with a download store and a massive marketing push."

The mSpot team responded by playing off Hollywood's concerns over Apple's dominance. The company wielded too much influence, T.J. claimed, over the people who actually owned the films and albums.

"Right now you're beholden to iTunes," Daren said in negotiations. "They're the only game in town. You need to counterbalance that. You need other successes in the industry so that you . . . have more control of pricing and all that. And Samsung really is the only credible company that can do that for you at a global scale."

If Hollywood helped nurture Samsung as a media company—say, by offering it favorable licensing deals—it'd be a chess move that would play to Hollywood's favor.

Samsung was already getting a good deal of buzz in music circles.

"We did everything properly and we licensed content, so they loved us," Daren said. Samsung's team reported back stellar news: Universal and Sony were eager to license music for Galaxy's Music Hub, and the terms for Samsung were favorable.

T.J. and Daren explained to Samsung's hardware executives how to license music for streaming services. Sony Pictures and Universal doled out licenses for their copyrighted films and songs with minimum guarantees each year. For instance, the smartphone manufacturer would

have to pay a minimum of $10 million to a music label in a hypotheti-
cal agreement.

The catch? Whether Samsung was actually successful in selling $10
million worth of songs fell on Samsung. The fee was guaranteed. The
big four labels "usually require this minimum guarantee," said T.J.
"They want you to commit to a certain number of licenses that you
have to promise them. You will be serving so many millions of users.
And you have to pay accordingly up front." The risk, in other words,
fell almost entirely on Samsung. There were no credits or refunds. But
the deals offered by Universal and Sony were still very good.

Samsung's hardware executives, led by J.K. Shin, on the other hand,
"just couldn't believe that we had made such a stupid deal," T.J. told
me. The lifelong Samsung executives in hardware and finance believed
selling music should be like selling semiconductors: The order came
in, you manufactured a certain number of chips under your client's
parameters, and you shipped them off. You didn't have to bet on the
success or failure of the product. But Samsung was wading into new
waters with software developers. Samsung's hardware people believed
that if customers weren't using their app to download music, Samsung
should get some refund from the labels.

T.J. and Daren were disappointed. They were being asked to institute radical change in how such deals were structured. But their hands
were tied under the short-term hardware dictates of racing to bring
out the next smartphone. Chairman Lee's long-term strategic vision
was missing from the company's discussions.

T.J. and Daren flew back to Los Angeles and attempted to renegoti-
ate terms—terms that the music executives thought were already settled. They managed to cut some licensing deals in half. T.J. and Daren
conveyed the terms of their new deals back up Samsung's hierarchy,
only to see them rejected once again as too risky. The process continued until the labels and movie studios finally abandoned Samsung out
of frustration.

Daren called it the Samsung tax—take the deal a Hollywood studio
had with everyone else and increase the studio's costs by 50 percent;
that's the burden of working with Samsung. The company clearly
didn't get it.

"I ended up burning most of my music industry relationships because of Samsung," Daren said.

By the end, T.J. and Daren were pushing for stratospheric and completely unrealistic streaming rights. Within a year, Sony executives, once excited about Samsung's content foray, were incensed. Daren had to sit through unpleasant negotiations with Sony Music's digital media president, Dennis Kooker, who clearly was not happy.

"You guys are clowns," Daren recalled another executive basically saying. "We don't want to deal with you."

Talks with Universal also came to a dead end. A former Universal Music Group executive told me, "They [Samsung] blew it."

But what if Samsung didn't need Hollywood labels? T.J. and Daren asked themselves. What if the company could use its weight as a smartphone manufacturer to bypass grumpy gatekeepers like Sony and Universal? What if it could use its technology to woo the artists themselves, who were frustrated with the big four record companies?

24

White Glove

IN RICHARDSON, TEXAS, TODD Pendleton was already plotting Samsung's new music strategy in a black-box operation called White Glove. Rising star Jo Lovato, senior director for integrated brand marketing, was appointed to head it.

The Samsung executives were frustrated at the omnipresence of iPhones in the hands of celebrities like Ellen DeGeneres, who appeared with their iPhones for free, without being paid to sponsor Apple's devices. They seemed simply to get lost in their love affair with Apple devices. It was incredible free word-of-mouth advertising, an unbeatable way of forging a cultural movement.

White Glove was Samsung's response. It was a form of guerrilla warfare.

"We would give products to different people," said one person on the team. "It may be actors, it may be musicians, obviously a lot of NBA players," since Todd Pendleton had such a close relationship with the NBA. Seeking to convert consumers into Galaxy users, the White Glove team would show up at a gala party, an event of some sort, and help celebrities transfer everything from their old device, usually an iPhone, to a Galaxy.

How did they do that? When a member of the team noticed a celebrity complaining on Twitter about his iPhone, he'd get the White Glove. Need an influencer with media access? White Glove her. Know of a filmmaker or music producer who might use a Samsung phone on the film set or in the recording studio? White Glove him.

Samsung executives in the White Glove program emphasized the quality, build, and customizability of Samsung phones to hardworking, high-pressure, and demanding VIPs. They suggested that it was a more flexible alternative to the rigid functions of an iPhone, a device produced for the masses.

"They're gonna give you a phone," said Dana Brunetti, president of the film studio he co-founded that was behind *The Social Network* and *House of Cards*. That's how Samsung advanced its cause—through low-key negotiations with celebrities, producers, and hosts. In the scheme of things, it was an inexpensive way to build word of mouth.

"To some extent it was pay to play. Or we would 'host' the events with already-friendly partners," said a marketer on the team.

Samsung hosted a celebrity dinner party at the estate of Cameron and Tyler Winklevoss—the twins who sued Mark Zuckerberg claiming he took their ideas to create Facebook. On a swanky terrace overlooking an infinity pool and the Hollywood Hills, a Samsung representative was waiting for Dana Brunetti, Galaxy S III ready to go, customized with a background picture of the producer's name.

Yeah, I'm going back to the iPhone as soon as dinner's over, Brunetti thought. He politely sat through the tutorial before joining the party. But after playing around with the Galaxy for a month, something clicked. And he did exactly what Samsung hoped. He switched to the Galaxy permanently.

Pendleton and Lovato enlisted Mitch Kanner, chief of an off-radar brand-building firm called 2 Degrees Ventures, whom *Fast Company* called one of Hollywood's "quietest influencers," to deploy methods and reach out to contacts that were a closely held secret. Yet he got actual results from celebrities, chopping out the middlemen involved in official sponsorships.

"Anybody with a checkbook can pick up the phone and buy [a celebrity sponsorship]," Kanner told *Fast Company*. "But if it's going to work, it has to be authentic."

———

THE TEAM QUICKLY CAME to see the importance of locking into American pop culture and not appearing inappropriate or cheesy.

The launch of Samsung's next flagship product, the Galaxy S4, kicked off in March 2013 with a chatty master of ceremonies who introduced Samsung mobile chief J.K. Shin, followed by the usual presentations. Except this time something was off in terms of the subsequent commercials. And not in a good way.

The commercials were an awkward and jarringly sexist attempt to imitate a series of Broadway musical skits. In one, a bevy of housewives gathers for a cocktail party, with one squealing over her wedding ring and another phoning her husband, who is golfing with the men. The housewives are coiffed like June Cleaver in *Leave It to Beaver*.

"Why would you need that?" a voice asks in the background when the commercial introduces a feature called "air gesture," which allows users to browse without actually touching the screen.

"My nails are wet!" . . . "Sunscreen!" . . . "I really don't want to put down this drink." The ad was widely derided as sexist and tone-deaf.

In shock at the cultural insensitivity of the South Korean headquarters, two of Pendleton's marketers told me that they had hoped the company was beyond that.

"In the middle of a red-hot conversation about women in technology, the resurgence of equal-pay discussion, and Sheryl Sandberg reigniting the very concept of feminism in America," wrote *CNET*'s Molly Wood, "Samsung delivered a Galaxy S4 launch event that served up more '50s-era stereotypes about women than I can count, and packaged them all as campy Broadway caricatures of the most, yes, offensive variety.

"To be fair, *everyone* in Samsung's bizarre, hour-long parade of awkward exchanges, forced laughs, and hammy skits was a stereotype. The kid was lispy, tow-headed, and tap-dancing. The little girl did ballet, of course. Will Chase, the emcee-as-actor, was orange and desperate for fame; his in-skit 'agent' was clueless, abrasive, self-absorbed and vaguely Jewish. The backpacking guys were horny, the Chinese actor in his 60s was an 'old guy.' So, it shouldn't come as a surprise that the women would *also* be a little, let's say, underdeveloped, as characters."

I broke out in laughter as I watched the debacle from my apartment in Seoul. I knew this was a typical advertising campaign in South Korea; it was hilarious seeing it play out in an American setting. I'd seen these kinds of cultural expressions at K-pop concerts, on television, even at political speeches and rallies in Korea. I attended a product launch for which Samsung had hired young models to stand around its new OLED televisions.

It was a typical culture clash between South Korea and the West. The housewives were demonstrating a Korean ideal called *aegyo*—a slang word that means to radiate virginity and innocence, to warble in a high-pitched, childlike voice, to prance around and dress as if they were in a cutesy Hello Kitty cartoon.

Pendleton's office was relieved when, despite the horrific product-launch skits, sales of the Galaxy S4 in the opening months turned out to be strong. In fact, two months after the faux-glamour event, in June 2013, Samsung surpassed Apple as the biggest smartphone maker in the United States.

Clearly something was working. And now the company had a chance to put its silly launch behind it.

THROUGH HIS FRIEND AND sponsorship partner LeBron James, Todd Pendleton linked up with the rapper Jay-Z. The rapper's representatives first approached Samsung after hearing about Pendleton's deals with NBA stars.

With the help of Mitch Kanner in the White Glove program, someone who ate pizza with Jay-Z in his mansion, Samsung began its campaign to convince the rapper to enter a massive sponsorship with the company.

The thing was, Jay-Z had to like his Samsung phone if he was going to be of use. Luckily, Jay-Z quickly became enamored with the Galaxy for all sorts of reasons. One of them was business.

"The soul of a hustler, I really ran the streets / A CEO's mind, that marketing plan was me," he once rapped. This from the ex–drug dealer who once quipped, "I'm a business, man."

Jay-Z was an entrepreneur with his own label, Roc Nation, who got his start on the streets of New York, struggling in his midtwenties to sell CDs out of his car, unable to find a label willing to take a bet on him. His early career, defined by failure, had an upside: He wasn't beholden to a sluggish music industry that struggled to foresee trends and that, in the age of social media disruption, was quickly approaching its sunset.

"First of all, we're in a dying business, everybody sees that," he told *Fast Company*. "So what am I supposed to do, just sit here and wait to get to zero before I do something?"

Samsung was Jay-Z's answer to the music industry's decline, and Todd Pendleton saw an opportunity. They began lining up a deal to tie the release of Samsung's next smartphone to a stellar event: the release of Jay-Z's next album, *Magna Carta . . . Holy Grail*.

During a blazingly fast thirty days of talks that Pendleton negotiated himself shortly after the botched Galaxy S4 launch, Jay-Z rhymed and rapped as he talked business, a sight that dazzled Samsung's side of the table. Samsung agreed to pay close to $30 million to Roc Nation. The first million owners of the Galaxy S III, S4, or Note II who registered—using a downloadable software app by an independent contractor—would get a free download of Jay-Z's next album. Jay-Z got an up-front payment that underwrote the making of the album, a set number of downloads, and a guarantee of profitability. Samsung got a major endorsement for its brand.

Jay-Z was set to become the face of Samsung in America. But the purpose of having Jay-Z was not only to sell more Galaxy phones. It was to turn Galaxys into little music machines.

THE CAMERA PEEKED AT Jay-Z at work from behind the sliver of an open door to his apartment, a shot meant to create a sense of mystery, to invite the viewer into the rapper's private place of creation.

"How do you navigate your way through this whole thing, through success, through failure, through all this, and remain yourself?" asked Jay-Z, at work on a music recording in this makeshift studio. The Samsung Galaxy Tab and a Galaxy smartphone rested on a desk as two of the tools of his craft.

Viewers watching the NBA Finals during halftime on June 16 saw the rapper in a commercial fashioned as a documentary, with nothing scripted. The piano riff behind him pulsated.

"We need to write the new rules," he said.

Sixteen hours later, Jay-Z took to Twitter with a flippant message for the music establishment:

> If 1 Million records gets SOLD and billboard
> doesn't report it, did it happen? Ha. #newrules
> #magnacartaholygrail Platinum!!! VII IV XIII.

Jay-Z believed that by selling one million albums to Samsung, he was at the edge of rewriting the rules of the industry, an industry with which he had tumultuous ties. He believed that the incredible volume of the Samsung deal would place his album at the top of the *Billboard* chart.

Business Insider declared that the deal would "change music forever."

"Samsung made Jay-Z's new album profitable before it was even released, scoring a major win for both brand and artist with the announcement of their deal," reported *Billboard*'s Andrew Hampp. CNN dedicated six minutes of every hour to the story for a day. It was a marker that the winds were shifting in the music ecosystem.

1 million, 2 million, 3 million, 20 million
Oh, I'm so good at math
Might crash ya Internet.

Neither Jay-Z nor Samsung realized that when he rapped these lyrics on the upcoming album's song "SomewhereinAmerica," he was speaking prophecy.

Pendleton and his team were out partying with Jay-Z and his wife, Beyoncé, at midnight on July 4, when the album went live for download on Samsung devices—three days before its release to everyone else. On the app's main screen on the Galaxy phones, a clock ticked down to midnight and displayed an image of the album art—two sculptures out of Greek antiquity.

At the stroke of midnight, fans around the world raced their fingers across their Galaxy screens to start the download.

Then the app froze.

"Jay!" Pendleton called over to Jay-Z at the launch party. "We broke the Internet!"

The picture of the album cover art sat there without downloading. Seconds passed. Then minutes. Frustrated Jay-Z fans frantically rebooted their phones again and again to no avail.

"#SamsungFail" became an explosive hashtag on Twitter. Yet Samsung, true to its style, remained quiet. "#JayZ's sponsors at #Samsung prove themselves not only intrusive, but technically inept . . . #SamsungFail," Korean American journalist Hannah Bae later tweeted.

The server had been overwhelmed, destroying much of the marketing value of the campaign. Pirated files went up online, and some American radio stations had already begun streaming the new album—hurting the exclusivity for which Samsung had paid millions and had promised to owners of Galaxy phones.

"Magna Carta Holy Fail?" joked a writer at the website Contently.

Finally, around 2:00 A.M., two hours after the promised release, the downloads kick-started.

Pendleton, as usual, managed to see the bright side of the disaster: Demand for the album was so overwhelming that the servers couldn't keep up.

Still, the debacle exposed the risks of high-profile celebrity deals without the capabilities to back them up.

"There's a reason why Twitter rang out with cries of #SamsungFail, and not #JayZfail, after all," wrote Contently's Henry T. Casey.

Nearly a week later, Jay-Z appeared on a radio talk show in New York and called the experience "disheartening" and "not cool" but played up the incredible demand for the album, without blaming his sponsor.

"It was twenty million hits for the app and it broke," he said. "No one is expecting it, there's no way in the world for you to calculate twenty million hits. It's not even a number you can fathom. You cannot prepare a service for that."

Jay-Z didn't get everything he wanted. *Billboard* never did change its rules to account for the Samsung–Jay-Z alliance. But the Recording Industry Association of America (RIAA) was forced to change the way it measured sales. And after the album went on sale through traditional channels three days later, the album hit the *Billboard* number one spot instantly, selling 528,000 copies in the first week.

25

Milk

"I TOLD YOU THAT wasn't going to work," mobile chief J.K. Shin told T.J. whenever the two crossed paths, a dig at the buggy and flawed software that was Music Hub, still struggling to catch on.

Samsung had established a task force to reshuffle its software leadership and had repackaged Music Hub under the new name "Samsung Music." But once again it failed. "If you continue to crank out mediocre services, it's not gonna help the sale of the device," T.J. said. "They will just think of Samsung as losers, and the moment that they buy a Samsung device, many people will turn off the software and try to delete it."

T.J., Daren, and Ed were determined to give it one last shot. Like Todd's marketing team, they would maintain a veil of secrecy. They would cut out headquarters, with its bureaucracy and stuffy hardware engineers.

T.J. traveled back to San Jose, where he stumbled on an underutilized team called the Samsung UXCA Mobile Lab, founded in 2010 with centers in Mountain View and San Jose. It had been created to design user interfaces for Samsung products using Silicon Valley flair and expertise.

"We had a terrible reputation in the American job market," said director and co-founder Han Kuk-Hyun. This was Samsung's chance to change that. The lab poached Bay Area talent from Apple and Google and elsewhere with the promise that while their current employers were throwing out their user experience (UX) ideas, Samsung was a fast-growing newcomer that would actually put them to use. The center had a reputation for coming up with creative ideas that the company's rank and file would have otherwise suffocated.

When Samsung beat Apple to market with the release of its first smartwatch, the Galaxy Gear, in September 2013, "there was no mass production for perfectly round-shaped displays yet," said Han. The designers at UXCA thought up an unusual user interface that could accommodate a square watch instead.

When T.J. showed up, he found a circular dial intended to be used as an alarm clock. But since it wasn't used in the Galaxy, the designers were searching for a new purpose. "We created this prototype where moving the hand along the circumference of the circle would play different music that's stored on the device," said lead designer Neil Everette.

It was an aha moment. T.J. connected the UXCA lab with Daren and Ed.

"This is perfect for the radio interface [for the Galaxy]," he told them. "You should only report to me and don't let anyone in HQ know, because once HQ people realize that there's a project going on, then they'll want a report on this, and then hundreds of hours will be spent creating those reports."

Daren met with Ed to get the project moving.

"So look, this is our last chance," Daren told everyone. "We gotta turn this thing around, because if we don't, they will think that the mSpot acquisition was a failure."

The Music Hub team and the UXCA hit the ground running. They carved the dial into the Music Hub software. It took many long hours over weeks at the office; the teams knew that if they got too caught up in "feature-itis" and checklists, they'd ruin the project. It needed to be simple and usable by anybody. Daren deployed what he called the "mom test." He gave it to his mom to see if she could figure out how to use it to spin the dials and change the channels. And she could.

"Okay, great. This is something that everybody can understand. It's reminiscent of the old way of changing radio stations, but it's digital," he said.

It helped that there was a leadership reshuffle at Samsung Telecommunications America, with a Korean American marketer named Gregory Lee replacing Dale, the previous CEO. An audiophile and jazz connoisseur, Gregory was ecstatic about bringing the Galaxy into the realm of music and video.

T.J. was in South Korea when Gregory came by his office. He'd heard about the project and was interested in backing it in an exclusive launch on the American market. T.J. showed him the confidential program. He asked, "How soon would you want to launch this?"

"Can you launch in March or April?"

An internal steering committee headed by Todd Pendleton convened to think of a name for the new music-streaming service. Relying on the advice of a large outside branding agency in New York, they studied the demographics of the target audience, millennials.

Daren didn't like the practice of what he called "mimicking" Apple's naming convention of using its recognizable *i* in its devices and software applications, for names like "iTunes" and "iPhoto." They were memorable and easy to say.

In the same vein, Samsung had been using an *S*, likely for "Samsung," at the beginning of each new software app's name. This led to some clunky and silly names, like "SMusic," "Svoice," and "SHealth," that didn't stick in the minds of consumers. And the services struggled to gain traction.

"We couldn't run away from that fast enough," Daren told me. "We thought a name that's easy to remember and evoked positive thoughts would be ideal."

The name the committee came up with was "Milk."

"It was a bit of a rebellious move from HQ," Daren said.

The name "Milk" was controversial even within Samsung's committee. But the committee went forward nonetheless, recommending Milk to CEO Gregory Lee, who signed off on it.

In the new year, Chairman Lee, despite his deteriorating health, managed to press out more scriptures and wisdom, telling his employees, "Let's throw away the hardware-oriented system and culture."

T.J. delivered Milk Music to Jay Lee, Chairman Lee's heir, who tested it and loved it. Milk Music, he declared to a roomful of top executives, had his support. With the blessing of the crown prince, news spread rapidly within Samsung, and executives hounded T.J. for demonstrations.

Finally the software entrepreneur was summoned to the executive offices by mobile chief J.K. Shin, one of his biggest skeptics.

As he waited in the lobby, he could hear screaming in the conference room behind the receptionist. Fifteen awkward minutes later, she ushered T.J. into the conference room.

J.K. was pacing the floor, furious. He walked up to the thermostat in the room and flipped it open.

"Horrible, horrible user interface," he said. "How does one expect to use a thermostat like this?"

J.K. was deliberately using the thermostat as an allegory for Samsung's software failures. T.J. recalled him saying that he didn't "trust the software people in Samsung to create anything that could be used by any human."

T.J. was prepared for a tirade from the mobile chief. But J.K. calmed down and watched the demonstration of Milk with interest.

"Why can't others do the same thing?" J.K. asked. "Good job," he told T.J., who left the conference room feeling vindicated.

26

The Selfie That Shook the World

WAS ELLEN DEGENERES A White Glove candidate or not? The marketers at Samsung disagreed.

A dedicated iPhone fan, she'd be hard to convert, some thought. Internal profiling had shown that iPhone fans typically didn't want to invest the time to learn Android.

But the comedian was likable and had influence. And the timing was right. Ellen was going to host the 2014 Oscars. Approaching her was risky, they decided, but worth it.

Samsung was in its fifth year as an Oscar sponsor, going back to 2009. In 2014 Samsung had spent an estimated $20 million on ads, giving Pendleton's marketers five and a half minutes of airtime.

Backstage, just off the main Oscars stage, in an iconic cocktail and coffee lounge called the Greenroom, stars were greeted by a giant Oscar statuette and a wall mosaic of eighty-six Samsung televisions, smartphones, and tablets. On the other side of the lounge, celebrities could take a selfie with the Twitter Mirror, a mount consisting of a Galaxy Note and a rococo frame, and then send it out from the Academy's Twitter account.

Two Samsung marketers were dispatched to try to obtain the ultimate product placement: a photo of a Galaxy in the hands of the Oscar host.

"These things are like a lightning strike—you can't plan for something like that to happen," said a marketer in Pendleton's office, "but what you do is to put yourself in the position to be able to take advantage of things like this when they do happen."

"Todd provided the leadership," he said. Todd sent Jo Lovato to negotiate the amount of time Samsung product placements would appear on air with ABC. As head of the White Glove program, Lovato already headed up Samsung's Hollywood talent outreach.

Pendleton also dispatched Amber Mayo, director of media and partnerships, to Los Angeles's Dolby Theatre to work with Ellen DeGeneres during rehearsals in person. Mayo instructed her assistants not to tell anyone else—even her colleagues at Samsung—until the Friday before the Oscars.

"Amber drove the brilliance of our media integration," said another marketer. Mayo knew that no matter how much money Samsung spent, even a multimillion-dollar sponsorship had its limits. They'd heard stories about corporate marketers showing up at Oscars rehearsals over the years, needling scriptwriters to insert their products. It hardly ever worked. Airtime was notoriously tight. Each award was planned down to the second—and yet the stars frequently ad-libbed their presentations. Few marketers were sure exactly what their sponsorship would earn them, even if they had formal guarantees of a product placement. Corporate sponsors came and went, but only the cleverest were written into the annals of Oscar history.

Samsung had to find a way around that.

"You have a loose idea, and there's a certain amount of time and a certain point [where it has to fit] in the show," said another person on the team. Another problem? "We didn't want any rogue shots of iPhones being posted by anyone saying it was from a Samsung phone." That would be a disaster. To get around that, Samsung marketers had to train the backstage staffers on the Galaxy as well, since they might tweet something on behalf of Ellen or the Oscars.

Samsung wasn't sure how Ellen would choose to use its Galaxy phone, until DeGeneres announced her intent to take a selfie, part of a lineup of stunts and slapstick to lighten up the ceremony.

The selfie, the script declared, would happen after a commercial break, when she would sit down next to Meryl Streep. "NOT IN TELE-PROMPTER," read the script, emphasizing its quasi-planned nature. Hopefully, she'd bring in Angelina Jolie and Julia Roberts for a second shot, according to the script.

"Everyone say, 'Hashtag Oscars 4 Ellen,'" she'd shout, followed by the snap.

The comedian brought her idea to ABC, which delivered the news to Samsung for approval. Pendleton's team was ecstatic. Amber's team worked with DeGeneres at the rehearsal to introduce the Galaxy Note III. Since Ellen was an iPhone user, her muscle memory wasn't attuned to the Galaxy, and she needed training from the White Glove team.

After a good deal of troubleshooting, Mayo called Pendleton two days before the event with good news: They'd secured the use of a Samsung phone in the ceremony itself—not only in the commercials. Samsung would provide Ellen with two Galaxy Notes. They'd alternate as needed.

The social media was the market's democratizing force. Oscars viewers who had once passively watched from their couches could, after Ellen's tweet, pick up their smartphones and join the online conversation. It was an amplified form of the field's most powerful tool, word of mouth.

The catch? The spontaneity meant fewer guarantees, and Mayo wasn't sure how it would play out. The bad news: This was either going to work well or flop, maybe even get Amber fired. She'd done all she could. Now all she needed was luck.

A DAY BEFORE THE Oscars, Ellen didn't look like her sprightly self as she finalized matters backstage. She sat in a conference room with the writers, her eyes baggy, her face slightly haggard, as they reviewed jokes and gags for the ceremony. Usually her tone was carefree and jovial, but when they arrived at the moment in the script for the selfie, she got quiet.

"I really hope that everyone jumps in this picture," she said. "I keep thinking, I mean, that it will be amazing if everyone really gets in."

At a table scattered with pencils, papers, to-go coffee cups, and 7-Eleven Super Big Gulps, the team gazed wearily at the script and its yellow highlights, scratching chins and scribbling notes, their stares somber and intense, running their hands through their hair.

There could be no mistake tomorrow, with so many people watching. Amid the grind of planning, DeGeneres occasionally jolted the room awake with a joke or two.

"And he is from Somalia? Sommelier! Who's the wine captain now?" she shouted, a planned dig at Best Supporting Actor nominee Barkhad Abdi, who was in *Captain Phillips*. His character, a Somali pirate, had seized Tom Hanks's ship and asked, "Who's the captain now?" The room lightened and chuckled. Things were looking good for the Oscars selfie. Ellen could improvise. She could put people at ease.

It was Oscars Sunday. Todd Pendleton, riding on Samsung's massive sponsorship, showed up with invitation in hand at the Dolby Theatre in a tuxedo and gray checkered bow tie with his wife, stopping to pose for the photographers roaming the entrance, making his way along the red carpet where the glitterati gathered. He entered the auditorium and took his seat.

Some 43 million people had tuned in to the Oscars, the most-watched entertainment telecast in a decade.

"We know that the most important thing in life is love and friendship and family," Ellen said in her opening monologue, "and if people don't have those things, well, then, they usually get into show business."

She mingled among the audience, ordering impromptu delivery from Big Mama's & Papa's Pizzeria, bringing onto the show the delivery-man, Edgar, whom she later called her "new best friend," and roaming the aisles and serving up greasy slices to tuxedo-wearing celebrities. Cameras were conveniently not looking when Harrison Ford splattered tomato sauce on his white dress shirt.

"Happy" singer Pharrell Williams thankfully hadn't forgotten his bulging bowler hat—deep enough to hold an equally deep tip—which

Ellen took and passed around for tips. She eventually handed the pizza man a $1,000 gratuity.

"Harvey!" Ellen shouted across the theater for producer Harvey Weinstein, looking for his tip. "No pressure, only a billion people watching!"

After a commercial break, she came out onstage as Glinda the Good Witch from *The Wizard of Oz* in a fluffy pink dress: "You have the power . . . ," she said in the character's motherly alto, waving her wand, before dropping the act midsentence. "Oh, never mind."

"For those not at the event, we had a conference room in a hotel where everyone was camped out, as well as conference rooms in the various offices," said a marketer on Pendleton's team. They monitored social-media reactions and took note of the positive reactions to Samsung's advertising.

Then, after another commercial break, the camera opened on Ellen, standing in the audience.

"Meryl, here's my idea, okay?" Ellen DeGeneres dawdled down the aisle, smartphone in hand, shooting zingers at the stars before she arrived at her target. "So you were nominated a record-breaking eighteen times, right?" Meryl Streep glanced up with a look of slight bewilderment.

"So I thought we would try to break another record right now with the most retweets of a photo," Ellen announced as she held up the phone. "So right now, I'm gonna take a picture of us. And then we can see if we can break the record for the most retweets."

She dipped down slightly and extended the phone between herself and the legendary actress. But Meryl wasn't satisfied with a two-person shot. She turned around to the woman behind her.

"Get her in," Meryl insisted.

Seated behind her, with her orange-dyed hair and wearing a slim black dress, was four-time Oscar nominee Julia Roberts. This year she was up for Best Supporting Actress for *August: Osage County*. Roberts raised her hands and made bunny ears, saying, "I can just do this!"

Bunny ears weren't in the script. But Ellen played along.

"No, lean in," said Ellen. "Channing, if you can get in also," calling over to Channing Tatum, the former male stripper who had broken out nine years earlier in *Coach Carter* and had become a regular Oscars presenter.

"Bradley, would you come? I want you in it. Jennifer, I want you in also!"

Down the row and next to each other, Jennifer Lawrence and Bradley Cooper, nominees for Best Supporting Actress and Best Supporting Actor in *American Hustle,* got out of their seats and gathered around Ellen. The impromptu scrum of stars was getting messy.

Cooper, who towered over five-foot-six Ellen, reached for her phone.

"I'll take it," he said. "I'll take it."

Cooper, unbeknownst to the audience, was a White Glove member.

"No, I'm doing it. Brad, get in here!" said Ellen, keeping her Samsung Galaxy away from Cooper.

She called for another Brad—Brangelina's Brad, Brad Pitt, who produced that year's Best Picture front-runner *12 Years a Slave*. Kevin Spacey crept up from behind, smiled, and raised a thumb. "Angie! Lupita!" Lupita Nyong'o, the Kenyan Mexican nominee for Best Supporting Actress for *12 Years a Slave,* scurried to the back of the group, her bright-blue dress parting the sea of black tuxedos. She was followed by her younger brother Peter, a college freshman who was hardly a celebrity, but hey, why not?

Long-haired and red-bow-tied Jared Leto—another nominee for Best Supporting Actor, for his role as an HIV-positive transgender woman in *Dallas Buyers Club*—bolted in from somewhere across the auditorium and stuck his face in front.

The rowdy thirteen celebrities were talking over one another, making faces, one raising a fist, another sticking out a tongue and inciting minor chaos. Ellen stuck out her smartphone and struggled to get a decent frame. Then Jennifer Lawrence chimed in.

"Somebody should just drop a boob if we want it to be retweeted this many times!"

"Meryl . . ." Ellen turned to Meryl Streep for help, unable to fit everyone in the frame.

"Drop a boob, somebody drop a boob," repeated Channing Tatum.

"Meryl. Meryl . . ."

"Want me to do it?" asked Meryl.

"Here, I'll do it," said Bradley Cooper.

"No, Meryl. Can you take it, Meryl?"

"I'll drop a boob. It's fine, I swear," announced Tatum.

"What does that mean, drop a boob?" asked Julia Roberts.

"Meryl, can you take it? I can't get everybody in here." Meryl reached out for Ellen's phone but failed to grab it.

"All right," Tatum interrupted. "I got a boob. I got a boob."

"Is my arm further?" Jennifer Lawrence said as she lunged at the Galaxy Note and missed it.

Bradley Cooper, kneeling at the front of the group, took charge in the chaos. "My arm's probably better," he said.

He reached up and nabbed the phone out of Ellen's hands. "Here we go," Cooper said, in firm control of the phone.

"Hey, that's good! Look at us," said Ellen, looking at the frame with a nod of approval. The crowd went silent as the celebrities squeezed into the picture, except for Jared Leto, who only managed to get the left half of his face in.

"That's it. Ready? Ready?" said Cooper.

Snap.

"Nice!" he shouted to applause and cheers.

"We did it!" Ellen shouted, getting pecked on the cheek by Meryl.

"Oh, I've never tweeted before!" the sixty-four-year-old exclaimed.

Ellen wanted one more. She glided across the aisle to Chiwetel Ejiofor, a Best Actor nominee for *12 Years a Slave,* and snapped a second selfie with him in his seat, turning around in time to catch the two photo bombers behind them, Brad Pitt and Benedict Cumberbatch.

"God, you're a photo hog," she barked at Pitt.

Then Ellen made her move and set the image to a tweet blasted out on her personal account.

> If only Bradley's arm was longer. Best photo
> ever. #oscars.

This for an image that was blurry, poorly lit, and slightly off-kilter, devoid of the perfectionism of the fading epoch of print media.

Which made this moment all the more significant. Eleven of the greatest legends of Hollywood—among them the winners of ten Oscars who together had $9 billion in earnings—had spontaneously gathered together into a single shot on the theater floor, an act that would otherwise have required the butting egos of throngs of talent agents and high-powered lawyers and quite possibly months of planning.

But all were happy, all were natural. And the moment was historic. Within minutes the photo had thousands of retweets. Then tens of thousands. Hundreds of thousands. Soon, millions.

Keyboard paparazzi chattered about Kevin Spacey's goofy gaping mouth, about Ellen's warm smile, and about whether the moment was planned, or maybe not, or maybe it secretly was. Then the Twitterians noticed the peculiar logo on Ellen's phone. Afterward audiences noticed that when she went backstage she was tweeting from her usual iPhone.

But this was a Samsung.

Todd Pendleton was watching from the audience when he turned to a colleague and remarked, "This is going to be one of the biggest product-placement moments ever." Since his arrival at Samsung, it had been three years in the making. It was spontaneous, inventive, and improvisational. And it had just the sliver of planning needed to make a splash.

Then Ellen emerged from backstage.

"We just crashed Twitter," she announced. "We got an email from Twitter and we crashed and broke Twitter. We have made history!"

It was the first time the site had crashed in Twitter's eight-year history. Users who loaded the home page were directed to an error message: "Something is technically wrong. Thanks for noticing—we're going to fix it up and have things back to normal soon."

"See, Meryl, what we did? You and I? It's amazing. We really just made history. It's fantastic."

Ellen went back to the Greenroom, passed the Samsung mosaic of screens and phones, and approached Jennifer Lawrence, who was conversing with Glenn Close, cocktail glass in hand.

"Did you hear we broke Twitter?" Ellen asked.

"Well, I'm shocked!"

"We made history. We made history."

"Someone should have flashed a boob. That's the only way we should have shut down the Internet."

Ellen held up her Galaxy Note.

"We have time!"

Already drowning out talk about whether *12 Years a Slave* was better than *Dallas Buyers Club,* or who had the gala's most stunning dress, was this little simple selfie.

Within an hour, the selfie had more than one million retweets.

"IT'S NOT LIKE YOU can just sell products," said the British comedian Russell Brand the next morning, holding up the UK's biggest newspapers, many of the front pages dominated by the Oscars selfie. "You need to sell the entire context for products. You need to sell the concept of glamour. . . . All of it creates a frequency of consciousness that's constantly spellbinding you in your state where a Galaxy phone seems like a good idea."

By that day, the selfie had more than three million retweets.

It was as if Sotheby's had discovered a new Andy Warhol piece and put it up for auction.

"It may seem odd that a poorly framed, slightly blurry, group portrait taken on a smartphone should have generated more buzz than the Oscars themselves," opined *The Telegraph*'s Harry Wallop. *Time* later included the image in its one hundred most influential photos ever.

"Hey, @TheEllenShow!" Twitter employee Lauren Mitcheom tweeted at *The Ellen DeGeneres Show.* "We painted a picture of you at Twitter HQ. Come take a #selfie with us!" Twitter had commissioned a painting of the Oscar selfie and hung it in the lobby.

"I thought it was a pretty cheap stunt myself," President Barack Obama, the previous record holder for most retweets, teased Ellen on a telecast from the White House. About a year and a half earlier, upon his re-election victory in 2012, Obama had had his staff tweet a romantic shot of him and Michelle Obama in a romantic embrace.

"Four more years," read the tagline; the image captured the world record at the time by surpassing 500,000 retweets.

How much was it worth for its copyright owners, Ellen DeGeneres and Bradley Cooper, who ended up snapping the photo? "The earned media—all the buzz which has been done around the Oscar [selfie]—represents roughly a value between $800 million to $1 billion," Maurice Lévy, chairman of the global PR agency working with Samsung, Publicis Groupe, estimated, "because it has been all over the world."

The number was contentious. In the traditional world of commercials, a thirty-second nugget of airtime cost about $1.8 million. So why would a selfie fetch an infinitely higher value?

"It does seem high, but if you think about the value of customers and social shares, then potentially that valuation is correct," Jonah Berger, an associate professor of marketing at the University of Pennsylvania's Wharton School of Business, told *NBC News*.

In an age when anyone could skip commercials on video devices, studies were suggesting that a word-of-mouth or a social-media ad was worth perhaps ten times more than a paid one. NBC pointed out that Samsung would have had to pay a "king's ransom" to get all those stars into the same advertisement. And while the Samsung logo wasn't visible in the selfie, the company still got nine hundred Twitter mentions a minute during the Oscars.

"Samsung is telling better stories and just plain out-innovating its arch-rival in Cupertino," *Advertising Age* wrote. *Adweek* determined through advertising metrics that Samsung had the most effective Oscars campaign of any of the Oscar sponsors, beating Pepsi, Google, Netflix, and numerous others.

This was Samsung's moment. That quarter, Samsung would ship 85 million smartphone units—a combined total greater than its four biggest competitors—Apple, LG, Lenovo, and Huawei.

"If you look at that picture—the faces of all those people—they truly are joyous," Pendleton later said. "It brought joy to a lot of people."

27

Return to Tradition

TODD PENDLETON'S TIGHT-KNIT TEAM was tired. They had fought, sacrificed, and succeeded in achieving so many impossible goals, transforming Samsung into the only smartphone maker mentioned in the same breath as Apple. But they felt they were repeatedly being scolded by executives at Samsung headquarters for a job well done.

"I left Samsung destroyed," said Brian Wallace, who was among the first to get out of the company, just over a year before the Oscar selfie. "Oh, my god, dude, I gained thirty pounds. I hadn't seen my wife and family. And my marriage was under stress. I'd been traveling three weeks in a month, working seventy-two hours a week. Just crazy. I was drinking too much 'cause it was so stressful."

"We called that weight gain the Samsung fifteen," another employee said.

The pressure from headquarters was rising to bring Pendleton's insular team of Texas cowboys back into the orbit of their Korean bosses. Internally, Samsung dubbed the new campaign "One Samsung."

MILK MUSIC WAS GETTING buzz at the South by Southwest festival held in Austin that spring. And Samsung brought back Jay-Z to perform a concert with Kanye West, in an exclusive to Galaxy owners.

The Milk user interface—T.J.'s and Daren's round dial that users could rotate to select tunes—was launched at the SXSW festival with its own demonstration booth, where users could try out high-fidelity headphones and toggle between streaming stations on an imposing translucent screen in front of them.

"Within a matter of hours, there were more than seventy thousand downloads, which was crazy," said Daren. "And then it just took off from there."

Finally, the Galaxy was closing in on that elusive third force in the smartphone ecosystem: its iTunes, its software. Apple had been slow to innovate, and Samsung was catching up.

That's when Google stepped in, putting up resistance to Samsung's forays into software. The gentlemen's understanding between the two companies was that Samsung would stick with hardware, Google with software, and everyone would be happy.

"Why are you guys doing this?" politely asked Jamie Rosenberg, Google's vice president of digital content, clearly not pleased with

Samsung's Milk Music streaming service launches at the
South by Southwest Festival in Austin, Texas, in March 2014.

PHOTO BY NADIA CHAUDHURY. COPYRIGHT BIZBASH.

Milk Music and its movie counterpart, Milk Video. "We should just all work together, right?"

Sitting around a conference table, the Google executives laid out what they knew about Milk. Daren was unsettled.

"They were tracking us," Daren said. "They knew more about Milk than I thought they would."

Daren later added, "Rosenberg suggested that we give up Milk and adopt Google Music as a default [app]. In return, he would customize a version of Google Music just for Samsung."

Daren resisted, and Samsung pushed on with its software crusade. The South Koreans wanted to rework Google's open-source Android operating system into its own variant, a strategic move that would distinguish Samsung from the muddy mess of hardware manufacturers converging on an increasingly homogeneous and accepted standard for smartphone design.

In January 2014, at the Consumer Electronics Show in Las Vegas, J.K. Shin had unveiled Samsung's new design flourish for the Galaxy Note, called the Magazine UX. Samsung's interface designers were using in-house software called TouchWiz to modify their operating system.

It was "a revamped interface that resembled the table of contents in a magazine, allowing users to click directly into videos and articles," reported *Bloomberg Businessweek*'s Brad Stone. "Manufacturers often apply their own design flourishes to Android, and this one looked good—not groundbreaking, maybe—but slick, intuitive, and completely unobjectionable."

Behind the scenes, Google executives were furious; the Magazine UX was an affront to their own software. It hid the Google Play app store and forced Android users to use something other than Google's design.

It was a dangerous prospect: If Galaxy fans started learning a new user interface, they might have trouble going back to Google's original Android variant. Google's designs could fade into obsolescence.

Google's Android chief, Sundar Pichai, was dispatched to deal with Samsung. The son of a factory manager, Pichai had grown up in India and moved to the United States after college, getting master's degrees from Stanford and the Wharton School.

Pichai was diplomatic and a good negotiator, the type of emissary you would want handling sensitive disputes with companies you depend on. But he also knew the importance of expressing Google's strategic frustration, of turning up the dial.

"Pichai set up a series of meetings with J.K. Shin, chief executive officer of Samsung Mobile Communications, at the Wynn hotel on the Vegas strip, at Google's offices in Mountain View, Calif., and again in February at the Mobile World Congress convention in Barcelona," *Bloomberg Businessweek*'s Brad Stone reported. "Pichai says they held 'frank conversations' about the companies' intertwined fates."

"I view Tizen as a choice which people can have," Pichai told Stone. "We need to make sure Android is the better choice."

Google was known for the idealistic spirit embodied in its slogan "Don't be evil." But this was a new kind of fight.

Pichai told Shin that Google was willing to "walk away" from its Samsung partnership. It was a bold statement; nearly three years earlier, Google had acquired Motorola for $12.5 billion, putting it into direct competition with Samsung and its smartphone hardware.

Android was feeling less and less like open-source software and more like hardened Google territory. The company fought back by pumping up the terms of the Android licensing agreements in its favor. Within three years, Silicon Valley news service *The Information* reported, it had upped the number of Google-made preloaded applications— a requirement for hardware manufacturers to use Android—from nine to about twenty for one manufacturer.

Then there was the return to the Google search app, which one contract stipulated was required to be "set as the default search provider for all Web search access points on the Device." And Google's search widget now had to be placed on the default home screen, along with an icon for the Google Play app store.

Behind the scenes, some Samsung executives thought the company that had once heralded the age of "Don't be evil" was becoming a bully. One app maker felt Google's move to bundle its software was "reminiscent of the monopolistic heyday of Microsoft," *Recode* reported. Samsung was being forced to rethink their software efforts, which were now in conflict with their previously indispensable partner.

D.J. Koh, a star who would eventually rise to CEO, later likened Samsung's relationship with Google to a marriage in an interview with *Bloomberg Businessweek*. He said you need three, not two, marriage rings. "An engagement ring, a wedding ring, and another ring is always necessary: suffering."

But Pichai's people knew that carrots had to follow the sticks. The meetings in Mountain View continued; T.J. recalls a Google executive yelling at one of them. Samsung's leaders, he said, were getting "cold feet" from the prospect of conflict with Google. It was time for a grand bargain.

On January 27, 2014, "Google and Samsung signed a wide-ranging global patent deal which will last a decade," Gordon Kelly wrote in *Forbes*. "Buried within it was an agreement that Samsung would tone down TouchWiz, refocus on core Android apps over its own customizations and cancel more radical customizations such as its 'Magazine UX' interface." Two days later, in January 2014, Google publicly announced the sale of Motorola Mobility to Lenovo. Google would no longer be overseeing the manufacture of smartphone hardware, removing it from direct competition with Samsung, at least for now.

Google kept most of Motorola's trove of twenty thousand patents, however. The patents served as a buffer between Samsung and Apple. Google had the patents to hold off lawsuits from Apple, while staking out a hilltop position from which to keep Samsung at bay in software.

Jay Lee, meanwhile, was skeptical about continuing to develop in-house software at Samsung.

"G.S. Choi and Jay Y. Lee did not always see eye to eye, especially on the software side," said T.J. "It's not that Jay Y. didn't understand the importance of software. He saw what Apple did . . . but he just didn't believe that Samsung had the DNA to have that capability grown organically."

Adding to his growing doubts, the profit margins on smartphone margins had peaked. Samsung executives had a gnawing fear of the difficulty of continuing to pour money into software acquisitions in the face of declining profits.

Arguments broke out between the crown prince and his regent. Jay believed that Samsung was a hardware maker at its core, and that it was time to go back to hardware. G.S. Choi responded that it took many years to build a software ecosystem. "We have to try," he'd tell Jay. "We'll never get there if we don't try."

While the South Korean media painted Jay Lee as a cultural reformer, inside Samsung the reality wasn't so straightforward. His father, the fire-and-brimstone chairman of years past, had promulgated a vision of software. But Jay ended up questioning the very core of that idea by moving away from software. Jay's rising leadership, ironically, was in many ways a return to a hardware tradition at the company. He was cautious, careful, and reticent. He wasn't promulgating a vision for a bold new future like Apple's. Rather, he was starting to look more like his prudent grandfather, the Samsung patriarch B.C. Lee, who'd died twenty-seven years earlier.

On the other hand, others thought, perhaps Jay Lee made a wise move.

"The story of 1993 needs to be forgotten," Tameo Fukuda, Chairman Lee II's Japanese adviser and author of the company's legendary design report, told South Korea's *Chosun* newspaper, referring to Samsung's Frankfurt Declaration, when Chairman Lee II stood before his executives at the Kempinski Hotel Gravenbruch and commanded them to "change everything except your wife and children."

"In 1993," Fukuda said, "there were fewer employees and the business was smaller, so innovation was relatively easier. But now, on a larger scale, things have become far more difficult. [Samsung] needs to be more careful than they were in 1993. Unless they think seriously about what to do from this point onwards, there will be no future at Samsung."

"I cannot protect you guys anymore," G.S. told his software executives in a meeting shortly afterward. The crown prince was losing faith.

In May 2014, Chairman Lee suffered a heart attack in his home and underwent CPR. He was admitted to the Samsung Medical Center and put into a medically induced coma. Two weeks later, he opened his eyes and awoke from the coma, according to Samsung reports. But then the chairman had a stroke, *Bloomberg Businessweek* reported. The chairman's death, many feared, was imminent.

28

Vulture Man

JAY LEE HAD A conundrum. He was inching closer to the throne, but challenges continued to stand in his way. He and his two sisters were facing a combined inheritance tax that could reach $6 billion, a stratospheric amount. If he was unable to pay it, he could be forced to sell some of his Samsung stakes. It could block his effort to become Samsung's next chairman.

"To put that in perspective," Reuters reported, "the United States expects to collect just $16 billion this year in estate and gift taxes—a levy that has long been a political bone of contention and which many rich Americans go to great lengths to minimize."

For almost twenty years, starting with the scandalous share sale in 1996, his father Chairman Lee's lieutenants had been preparing for moments like this. Some had been criminally convicted because of their efforts, a bewildering labyrinth of buyouts and sales from one Samsung affiliate to another with the aim of raising the needed cash for the inheritance tax—among the highest in the world at 50 percent in the highest tax bracket—and passing control of Samsung to Jay Lee.

They stayed mere steps ahead of the South Korean government,

which had moved to sew up its financial regulations shortly after the
failed share sale twenty years earlier. By early 2015, chatter was flying
in coffee shops and investment funds that another such financial chess
move was imminent.

"Is it possible?" a minority shareholder asked me at a coffee shop in
Yeouido, Seoul's financial district. He was nervous that he might lose
money if Samsung ignored its shareholders to elevate its crown prince.

"With Samsung, anything's possible," I said.

"Is the chairman dead?" he asked me.

It was the question of the day. Rumors were spreading throughout
South Korea that Samsung was attempting to suppress news of his
death. The company vehemently denied the rumors. When renewed
rumors spread in April 2015, Samsung's stock price, ironically, went up,
signifying that shareholders wanted an end to the uncertainty or
thought Jay Lee, his successor, would be a better leader than his father.

But if the chairman died now, Jay Lee was in a difficult spot. The
royal succession, which rested on the delicate success of complex share
sales and mergers, was far from complete—and far from guaranteed.

Under South Korean law, his father's shares would be divided at a
ratio of 1.5 for his mother to 1 split between him and his two sisters. If
she didn't see eye to eye with him, she could attempt to oust her son.

I had no idea what the truth was about the chairman's health. My
company sources told me that no one, apart from the ruling Lee family
and its lieutenant, Vice Chairman G.S. Choi, was permitted access to
the chairman's hospital suite.

On May 26, 2015, it was reported that Samsung C&T, the construction and trading arm of the company, would be acquired by Cheil Industries, the de facto holding company of the Samsung empire.

"Under the merger agreement, subject to approval from the two
companies' shareholders, Cheil Industries will acquire Samsung C&T
by offering 0.35 new shares for every Samsung C&T share," Cheil Industries said in a press release.

"That seems . . . odd?" observed Bloomberg View's Matthew Levine
on the battle to come. "Why would you sell your company for less
than nothing?"

Combing through the financial details of the merger, Levine wrote that this was "an all-stock deal in which C&T shareholders will get a total of about 55 million new Cheil shares, at a ratio of 0.35 Cheil shares for each C&T share. At the time, Cheil shares traded at 163,500 won, meaning that the deal was worth about 8.9 trillion won. (At the time, C&T's public equity was worth a bit more than they are now—about 12.5 trillion won, or 9.5 trillion after tax.) So C&T shareholders were getting paid considerably less in the merger than the easily calculable value of the public stocks that C&T owns, and C&T's operating business was being valued at considerably less than zero."

Why leave Samsung C&T shareholders feeling bilked in an obvious way that was sure to stir up terrible PR?

Samsung argued that this was an attempt to consolidate business units, but it was ultimately about the primacy of the ruling family. Jay Lee owned a 23 percent stake in Cheil Industries, the company acquiring Samsung C&T, which in turn owned a 4 percent stake in Samsung Electronics, the group's crown jewel. The merger would solidify and simplify Jay Lee's control of Samsung Electronics through this shareholding web, starting with Cheil on the top.

That, in effect, would raise his stake in Samsung's empire. Not only would he rise closer to the throne as chairman of the Samsung empire, but it would help him pay his inheritance tax.

In fifty-two days shareholders were slated to convene for the vote. Manhattan investors were livid.

"The attempt by the boards of Samsung C&T and Cheil Industries to force through this unlawful takeover proposal," hedge fund Elliott Management announced in its opening salvo, "represents an effort to divert, without any compensation, more than 58% (equivalent to approximately KRW7.85 trillion) of Samsung C&T's net assets out of the hands of the shareholders of Samsung C&T and into the hands of Cheil Industries' shareholders."

The hedge fund was determined to "fight against charlatans who refuse to play by the market's rules," as its founder, Paul Singer, once said.

Elliott Management controlled $27 billion in assets and was a sizable shareholder in Samsung C&T. It was not an institution to be trifled with. It was known for its hard-fighting style and culture of paranoia, and its founder, Paul Singer, was an ex-attorney and New York investor, a hardened financial fighter unafraid to confront the world's most powerful companies and even some governments. *Bloomberg* called Singer "The World's Most Feared Investor."

Elliott needed to convince Samsung shareholders to go up against their nation's economic pillar. And he had influential backers.

"Vote AGAINST the transaction," warned Institutional Shareholder Services, the proxy advisory firm with enormous credibility in shareholder circles.

On June 3 Elliott bought a 2.17 percent stake in Samsung C&T that allowed it to up its voting rights, raising its total stake to 7.12 percent and making it the third-largest shareholder. It signaled that the financial war was on.

There were forty-four days until the vote.

Jay Lee was having breakfast with JP Morgan CEO Jamie Dimon when he learned about the fierce opposition Paul Singer was putting up. Samsung executives realized that winning approval of the merger from shareholders would not be easy. Jay Lee needed a 55.7 percent shareholder vote in favor of the merger.

Paul Elliott Singer, the founder of the activist hedge fund Elliott Management, is known as "The World's Most Feared Investor."

On June 9 Elliott filed for a court injunction against holding an upcoming shareholders' meeting to vote on the merger, arguing it was
unlawful. The next day, the construction company enlisted the help of
an ally in a shareholding maneuver that would be impossible in the
United States. It tapped into its bank of treasury shares, a key tool for
maintaining family control, and sold all of them for $608 million to
South Korean chemical and auto parts manufacturer KCC Corporation, instantly turning the previously obscure manufacturing firm into
Samsung C&T's fourth-largest shareholder.

"The move is necessary," the company said in a statement, "to protect the company and shareholders from overseas hedge funds that
pursue profit-taking, and to improve its liquidity and financial health."

The use of treasury shares was a trick of South Korean trade. It is a
type of share that *chaebol* ruling families give to allies in exchange for
supporting controversial shareholder decisions. Once a company sells
its treasury shares, their voting power is activated. It is a way of creating shareholding power out of what seems like nothing.

"That was dirty," an Elliott employee told me. Samsung C&T's
stock fell, effectively sabotaged by the company's own managers.

The sale was "deeply alarming," Elliott said in a statement, explaining that it "intentionally diluted the voting rights and the value attrib-
utable to Samsung C&T shareholders." The next day, Elliott filed for
another injunction to stop KCC from voting.

On July 1 a South Korean court shot down the hedge fund's petition and declared that the merger vote must go on—since the sale
didn't constitute an illegal transfer of wealth to a ruling family, and
Samsung had followed the law in calculating the merger ratio through
a predetermined ratio based on recent stock prices.

On July 3 Elliott appealed. It was fourteen days before the scheduled vote.

The gears of the Republic of Samsung were spinning at top speed.
Samsung ran ads appealing to the emotions of shareholders and the nation in more than a hundred newspapers, as well as on eight broadcast-
ers, six cable channels, and two Internet portals—all on a single day.

"Elliott is trying to defeat the merger. We implore our Samsung
C&T shareholders," announced a TV spot.

Next, Samsung executives pulled up the company's shareholder list and sent out five thousand employees with walnut cakes, watermelons, and other fruits to shareholders' homes across Korea, serenading them on the merits of the merger. "We beg of you, meet us in person," a Samsung employee pleaded over an apartment intercom to a retiree holding about 0.004 percent of Samsung C&T stock.

"That's shady," a representative of a minority shareholder said to me in a coffee shop. "Very shady."

In the meantime, Jang Choong-Ki, a powerful Samsung president from the Tower, the Future Strategy Office, received updates on Elliott's actions via text message from South Korea's spy agency. Such practices are not unusual in South Korea, where government and corporations work hand in hand for the national interest.

"Korea-related issues [at Elliott] are handled out of Hong Kong and said to be on strict lockdown," wrote the strategy chief of the National Intelligence Service (NIS), South Korea's equivalent of the CIA, to a Samsung executive. "Will be meeting with a source on Monday to find out more and will find out further details from a friend."

Later the agent reported that the "lawyer actually handling this case is Jang Dae-gun. Twenty-second graduating class of the institute," likely referring to his graduating class at South Korea's judicial training institute.

The Samsung president and the spy contacted each other at least 150 times by text message and phone, according to a leak obtained two years later by South Korean investigative magazine *SisaIN*. The magazine learned from messages it obtained that the government spy agency had been watching Elliott since 2013.

It was now nine days before the vote.

"'Jewish ISS [Institutional Shareholder Services]' blatantly supported 'Jewish Elliott,'" read a headline in the *Munhwa Ilbo* newspaper.

"Jews are known to wield enormous power on Wall Street and in global financial circles," wrote a columnist at the business tabloid *MediaPen*, adding that it is a "well-known fact that the US government is swayed by Jewish capital.

"Jewish money," the website reported, "has long been known to be ruthless and merciless." The website called Paul Singer the "greedy, ruthless head of a notorious hedge fund."

"Elliott is led by a Jew, Paul E. Singer," the business tabloid website *Money Today* wrote, "and ISS [Institutional Shareholder Services, the advisory firm that told shareholders not to vote for the merger] is an affiliate of Morgan Stanley Capital International (MSCI), whose key shareholders are Jewish.

"According to a source in the finance industry, Jews have a robust network demonstrating influence in a number of domains."

The Anti-Defamation League and other rights groups called on the South Korean government to condemn anti-Semitism. The organization pointed to its survey that found that 59 percent of South Koreans believed "Jews have too much power in the business world."

Seven days before the vote.

South Korea is home to the third-largest pension fund in the world, the National Pension Service (NPS), tasked with overseeing the retirement savings of the South Korean people. The fund had about $450 billion in assets. The retirement body also happened to have the single largest stake of Samsung C&T stock, at almost 10 percent. It also had 5 percent ownership in the other Samsung company acquiring Samsung C&T, Cheil. South Korea, with its aging population, depended on the integrity of the pension service and its myriad investments in companies like Samsung. It needed good returns to sustain its retirement payouts. Was it willing to risk taking the hit for Samsung's ruling family?

On July 10, in a closely followed saga, the National Pension Service's investment committee convened in secret to cast its vote on the merger, which would then be delivered at the shareholder meeting a week later. "Pension Fund Could Be Samsung Kingmaker," read a headline in *The Wall Street Journal*.

To the consternation of media and shareholders, the pension fund passed on the option of consulting an outside panel of experts and academics. Consulting with an outside committee was the usual practice in weighing controversial shareholding votes. "The NPS is shooting themselves in the leg and trashing their reputation," Kim Woo-chan, a former NPS committee member, told me about the decision.

The NPS's internal estimates showed that it stood to lose $120 million should the merger go through. During many hours of deliberations, the pension service's executives debated the merits of the merger.

Its decision makers acknowledged the impending losses for the retirement service as a result of Samsung's actions.

"Our stake in Cheil Industries can't cover all the losses from C&T," said Lee Soo-cheol, head of investment strategy, according to minutes leaked to me later. "There needs to be enough synergy to make up the difference."

The head of research went so far as to propose an investment in another Samsung company to offset the impending losses.

"In order to counterbalance the disadvantageous merger ratio," he said, according to the minutes, "there needs to be approximately a 6 percent (2 trillion won) increase in enterprise value. . . . We may be optimistic about increased enterprise value through the profit created by new growth businesses like Samsung BioLogics." But the pension service never ended up investing in Samsung BioLogics, one of Samsung's hot new businesses.

After hours of deliberations, the panel cast its vote in favor of the merger: Eight voted in favor of the merger and four against it. The exact reasons for the decision were a mystery even to those inside the pension service. It made little sense, when the pension service knew it was going to lose money on the merger.

As per standard procedure, the NPS informed the press that it was not disclosing the results of its vote until well after the shareholder vote at Samsung headquarters the following week. It was a nail-biting decision. The pension service could easily swing the vote for or against Samsung.

"Construction? It's not looking good as an industry," said Kim Seon-jeong, a former Samsung financial executive who became head investment officer of the pension service from 2008 to 2010. "It wasn't looking good when I was at the NPS, and it's not looking good now either."

Four days before the vote.

Samsung C&T put up a website called "Vulture Man" to argue its side of the merger question. Samsung's website slide show depicted a vulture, presumably a caricature of Paul Singer, whose sadistic practices consisted of plotting and preying on the poor and the disenfranchised around the world. In one cartoon, he descended on the Congo, where he watched a soldier point his assault rifle at a gaunt, crying child before running off with a profit.

Soon Vulture Man was splashed all over English-speaking media.

"The depiction of Jews as untrustworthy animals with huge beaks has an unfortunate history that hasn't ended well for anyone involved," commented the *Observer* on the Samsung slide show. "This 1937 depiction on the cover of *Der Sturmer* bears an uncomfortable resemblance to the image viewable today on Samsung's official website."

When the cartoons hit the English-speaking media world, Samsung pulled its advertising from the website that called "Jewish money . . . ruthless and merciless," took down the Vulture Man website, and unequivocally condemned anti-Semitism.

"I think it's a shame this element of anti-Semitism crept into what is a business dispute," Paul Singer said at a business panel in New York. "I don't think the Korean people are anti-Semitic."

It was three days until the vote.

At 3:00 P.M., a rowdy mob of shareholders burst into the lobby of the National Pension Service on rumors that it had voted to support the merger—though the accusation was not yet proven. They delivered an angry petition.

"If this is for the 'national good,' then why are all these shares being used to aid Samsung's ruling family?" the petition read. "This constitutes professional negligence. For NPS to go along with this without objective evidence to back them up constitutes abuse of authority. They will answer to this. If there was outside pressure involved in the decision, make a conscientious confession now and reverse the decision before it's too late."

"The NPS," the petition argued sarcastically, "should be renamed the Samsung Pension Fund so that the country understands why they are paying the pension."

It was now two days before the vote.

"How dare they attack Samsung while our Chairman is bedridden and ill!" declared a small shareholder, according to one story circulating in the coffee shops in the city's financial district. According to reports, the investor marched into the lobby of the Samsung C&T headquarters, offering to entrust the company with his shares.

The gossip mill was abuzz with chatter and innuendo as the vote deciding the future of the Republic of Samsung approached. It was in one coffee shop that I first met the Korean shareholding activist who went by the pen name "Jim Rogers" but whose real name was Kang Dong-oh.

Jim Rogers was leading a group of minority shareholders frustrated with Samsung's treatment of them. Until recently, Samsung and major corporations like it, he claimed, had called these small stock traders "ants" who could easily be squashed. Now, the shareholders told me, they were binding together in "solidarity." Yet almost none of them wanted to be named, for fear of angering Samsung.

"I've studied the pension service," Jim Rogers told me. "I don't think the merger will go through. The math doesn't add up. They're going to reject it." Rogers pored over the scenarios and the numbers at the table. Samsung was likely to lose, he concluded, unless the pension fund could be convinced to act against its own interests.

It was the day before the vote. In a last-minute hearing, the Seoul High Court turned down Elliott's appeal to stop Samsung's shareholder vote.

On the day of the vote, July 17, 2015, I spent the early morning at the conservative protest tents at Seoul's Gwanghwamun Square.

"The future of our nation depends on Samsung. Please understand!" explained a military veteran.

"Samsung bought this nation," argued a protester across the street. "Samsung bought our government. We must stop this merger today!"

Around 9:30 A.M. I headed down to Seoul's Yangjae district, where the vote was starting a half hour late at the Samsung C&T building.

The 553 stakeholders present were packed into a room built for a hundred people. Investors were crammed in along the aisles and in the back. Outbursts of emotion and frustration were evident everywhere. One shareholder stormed the stage and shouted into the microphone, "I strongly oppose the merger!" Securities guards quickly ushered him away.

Outside, I hoped to catch a glimpse of Samsung's ruling family. But as usual, they didn't show up to address the shareholders, the "ants."

Samsung C&T CEO Choi Chi-hun took the floor and, interrupted by jeers, presided over the tense meeting. He insisted it was not possible to change the merger terms.

"If the merger ratio was set at 0.6 or 0.7," proclaimed an investor, "the situation would not have turned out like this." He was feeling ripped off at the ratio of 0.3, which translated into far less money.

Over the course of the four-hour meeting, shareholders later told me, one could feel the resentment, the bitterness, the anger, as well as the loyalty to Samsung, and to South Korea, permeating the room. You either loved Samsung or you hated it. It was the type of meeting that could have easily blown up into a brawl. South Korea is famous for its parliamentary fistfights and physical scuffles in corporate shareholder meetings.

From newsrooms to coffee shops, from corporate offices to breakfast restaurants, South Koreans watched tensely as the vote began. A team of employees went around the room, holding up large plastic boxes for each stakeholder, including representatives from the National Pension Service, to insert paper ballots. Elliott's and Samsung's lawyers watched closely.

The ballots were collected and the vote counted. After a few knuckle-biting minutes, a Samsung representative stood at the front and announced the results.

69.53 percent of the shareholders voted in favor of the merger.

Samsung managed to get 14 percent more than the number of votes it needed. And many of those votes, many concluded, probably came from the National Pension Service, which had yet to reveal its vote.

"The approval is huge for us," CEO Choi Chi-hun told the room in a sentimental and emotionally charged speech. Samsung, he said, would remember the goodwill of its supporters.

Executives later admitted to the media that the turnout *against* the company was unnerving.

"Elliott is disappointed," Paul Singer said in a statement, "that the takeover appears to have been approved against the wishes of so many independent shareholders, and reserves all options at its disposal."

Elliott's lawyers noticed that the chairman, rumored to be incapacitated in his hospital suite, had yet managed to "vote" in favor of the merger.

"Elliott has questions about the validity of proxies of Chairman Lee," the law firm said. "The Chairman failed to appear due to health problems. If Lee provides proxies, then please clarify the timing. This is about whether or not Chairman Lee made clear his view on this issue."

The Washington Post had reported earlier that the chairman had been "unable to speak since May [2014]." Samsung C&T's CEO rebuffed the challenge from Elliott's lawyers, claiming the chairman had voted through legal representatives.

As the company prepared to execute the merger that fall, the Samsung C&T stock tumbled. By May 2016, shares in the newly merged company had lost 40 percent of their value. By November of 2016, more than a year after the vote, the National Pension Service had suffered $500 million in losses.

On paper, Jay Lee remained vice chairman, since in Korea's Confucian tradition he couldn't become chairman while his father was still alive. That would be an affront to the supreme leader. But as the victor, Jay raised his shareholding value and moved a major step closer to the throne. The nosy New York interlopers were held at bay, for now. But the battle for the Republic of Samsung was not over.

"My observation on Samsung, is that Samsung is pretty unique and a very excellent dynasty," said the former chief of the National Pension Service, a crusty and passionate pro-*chaebol* conservative named Choi Kwang, who told me the story of the merger vote over dinner almost two years later.

We sat in a private dining room off the historic Seoul Station in 2017, as a protest took place outside while we dined on spicy chilled noodles, pumpkin, and salad.

It had taken me months to get the interview. Choi Kwang kept away from the media. He insisted to me that as an administrator he hadn't meddled in the pension service's voting decision, nor had he been a part of the vote in any way. But as the pension service's top leader, he had his own opinion of Samsung and its merger, which he was overwhelmingly in favor of.

"I thought the merger would be unanimously approved. . . . I was surprised that there were some people who didn't approve, who weren't in favor of the merger.

"Right now, I think we lost a lot of money," he told me. But he said that in "six or ten years," the merger would be in the retirement body's interests.

"What would be the benefit of the merger, say, ten years down the road?" I asked.

"Nobody knows."

29

My Kingdom for a Horse

"THE PRESIDENT WOULD LIKE to see you."

It was September 15, 2014, ten months before Samsung pushed through its merger. Jay Lee was in Daegu at a gathering of executives with South Korea's president. The president's aide pulled him aside and quietly informed him of the president's wishes, Jay would later testify in court.

Jay was ushered into a room where the president, Madame Park Geun-hye, awaited him.

It was a meeting of scions. Both Jay and President Park came from families that had built South Korea into an economic power. President Park was the daughter of South Korea's former nation-building dictator, Park Chung-hee, the man whose ideas became the driving engines of South Korea's growth. When she was a teenager, an assassin's bullet intended for her father instead hit and killed her mother in an auditorium. Her father was shot dead by his intelligence chief in 1979.

When asked why she never married, President Park would answer, "I am married to the nation and its citizens."

"How is Chairman Lee Kun-hee's health?" she asked Lee, about his bedridden father. After the pleasantries, she got to the point.

"I ask that Samsung take charge of the operation of the Korea Equestrian Foundation. In the run up to the Olympics, buy the competitors good horses and help with their field training." In South Korea, corporations traditionally helped the government and vice versa—even if it meant buying the South Korean equestrian team horses for the Olympics, or whatever else the state felt was needed.

Jay acquiesced, and Samsung initiated its sponsorship of the South Korean equestrian team. He next met the president at her official residence, the Blue House.

But something seemed off at the meeting. The president's aide scribbled a note, according to evidence later revealed in Jay Lee's trial.

"Samsung management succession situation → use as opportunity. Ascertain what Samsung needs in regards to the management succession situation," he wrote. "Provide assistance where possible, and find ways to induce Samsung to contribute more to the national economy. Significant governmental influence can be exercised in solving Samsung's more immediate tasks."

I THINK I KNOW what people mean when they say the president shoots lasers out of her eyes when she's angry," Jay told his aides in July 2015, eight days after the merger. Scolded by the president for not supporting the equestrian team enough, Samsung had to take care of the matter.

The smell of a scandal was in the air.

A little more than a month after the merger, a team of Samsung executives from the Tower traveled to the InterContinental Hotel in Frankfurt. Awaiting them in the lobby was a small group of employees from a company called Core Sports, a firm that had, curiously, been registered in Germany just the previous day.

The Tower executives signed an $18.6 million contract with Core Sports to fund the training of South Korean horseback riders for the upcoming Asian Games and World Equestrian Games.

Core Sports was an odd company—it employed only one person who had a background in equestrian sports. It was not the type of company that Samsung, a global electronics manufacturer, would have normally been keen to shower with cash.

Few people knew the enigmatic woman, Choi Soon-sil, who ran the company. The daughter of a charismatic spiritual leader, Soon-sil had befriended President Park Geun-hye as a teenager, when the dictator's daughter was reeling from the assassination of her parents.

"Rumors are rife that the late pastor had complete control over Park's body and soul during her formative years and that his children accumulated enormous wealth as a result," the U.S. embassy had cabled Washington.

If there was power behind the president, perhaps it was here. Samsung bought an $830,000 racehorse named Vitana V for Choi Soon-sil's daughter, Chung Yoo-ra. She would train with her coach and her majestic new prize in the German town of Biblis. She'd already taken home the gold medal in the 2014 Asian Games, when Jay Lee and President Park were first in talks about a Samsung sponsorship. But no formal deal had been inked at the time. The young rider hoped to be on a path to greater glory now.

Her career, however, was cut short.

BACK IN SEOUL, JAY Lee was attempting to institute modest reforms in the Samsung empire. The Galaxy phone line wasn't recovering from slippery sales, and the Samsung-versus-Apple wars had evolved into a ground war in an increasingly tough smartphone market.

Jay Lee sold off Samsung's corporate jets and got rid of its sluggish chemical and weapons companies, raking in almost $6 billion that could be put to better use in businesses like smartphones. He sold the Samsung Life Insurance building, a symbol of the company's history beloved by his grandfather, founder B.C. Lee, for $496 million.

Then, on March 24, 2016, Samsung Electronics issued a potentially groundbreaking announcement. In an auditorium filled with over six hundred employees, senior executives signed a document promising major changes in the way the company was run. They pledged to do away with their authoritarian, top-down hierarchy. Going forward, they intended to transform Samsung's militaristic culture into that of a flatter, more agile start-up, under the initiative "Start-up Samsung."

Changes were afoot. The months ahead would be trying for Samsung executives. The Samsung Summer Festival, a weekend of mass games, was canceled. The company cut down on the number of job titles, reports, and meetings and encouraged employees to speak up more. Excited young managers were encouraged to act more like executives in Silicon Valley.

"People felt liberated," a Samsung marketer in Suwon told me. "They started immediately changing their office clothes into shorts and T-shirts and sandals. Huge change. Freer spirit."

To me, it sounded too good to be true. And before long, I was proven right.

"Nothing changed," the marketer told me. "In autumn they made an announcement [about going] back to the old clothing guidelines."

Jay Lee was trying to reboot the company culture at the very time Samsung was making shadowy equestrian deals on behalf of the president. The company was increasingly trapped between tradition and the need to modernize.

The older Samsung Men, who'd given their careers to Samsung and the chairman, did not welcome the company's new ways. Some doubted, in private, the capabilities and vision of Jay Lee.

"It looks like any other Western company in some sense," Nam S. Lee, a former aide to Chairman Lee, groaned to me in a meeting at a Starbucks.

"For the older generation," explained Ho Soo Lee, a former Samsung executive vice president, from his new executive suite at another *chaebol*, "the culture is sacred."

The breakneck growth of Samsung during the Samsung-versus-Apple wars had resulted in enormous success in some areas but a step back in others. Todd Pendleton's "Next Big Thing" marketing campaign had successfully overturned the narrative that Apple was the lone innovator in the smartphone space and that everyone else was out to copy it. But Samsung's old-school, pressure-cooker demands had already chased out much of Pendleton's team in the United States.

Pendleton's spirits remained high. But he also wanted new challenges. Disappointed that Samsung was reluctant to move into software and retail by opening its own Apple-like stores on a vast scale, he departed Samsung in April 2015.

——

AT THE SAME TIME, the public perception of Apple was changing. In December 2016, the Supreme Court heard *Apple v. Samsung,* which was now dragging on in its fifth year, and came close to rewriting the way patent damages were calculated. It handed the final decision on the drawn-out court case back down to a lower court.

Samsung's argument resonated with a growing chorus of Apple critics: American patent law, drawn up more than a century ago, is not attuned to the complexities of modern technology. Samsung cited patents for a spoon handle in 1871, a saddle in 1893, and a rug in 1894.

"A patented design may be the essential feature of a spoon or a rug," Samsung's brief argued. "But the same is not true of smart phones, which contain countless other features that give them remarkable functionality wholly unrelated to their design." Remove a component or a piece of software under a patent dispute, and the entire phone stops working. Apple had taken advantage of this pitfall, arguing it owned what were in fact generic software features, and the black rectangle shape of a smartphone, in order to shut out competition.

"Isn't it simply unworthy of such a great company," asked the popular patent blogger Florian Mueller, "to engage in behavior that increasingly resembles the conduct of patent trolls who seek to extract undue leverage from weak and dubious patents?"

Yet that success came in the wake of a significant marketing pullback. Todd Pendleton's successors at Samsung struggled to line up anything nearly as creative or effective as their previous campaigns.

Casey Neistat, on his popular YouTube vlog after the Oscars in 2016, revealed that Samsung was "gonna have me skateboard down the aisle, past Leonardo [DiCaprio], taper down the aisle and jump on the stage holding their new 360 camera in my hand during the broadcast."

"This has to happen," Samsung's top people insisted to Oscars counterparts, according to Casey.

"Okay, let's figure this out," the Oscars staffers responded. "Is it a four-second skateboard ride or a six-second skateboard ride?"

Two hours before the broadcast, the scene was cut entirely.

Then there was the decline and closure of Milk Music in September

2016, effectively the end of Samsung's biggest software inroad against Google.

"I'm just disappointed, because we had all the right ingredients to be able to execute on that and we didn't," Daren told me unhappily. He and Ed Ho left the company, their beloved start-up gutted.

And Samsung's struggling operating system, Tizen? An Israeli security researcher weighed in at Kaspersky Lab's Security Analyst Summit: "It may be the worst code I've ever seen. Everything you can do wrong here, they do it." The operating system was a long way from catching on.

It was amid these difficulties that the Galaxy Note 7 phones began their spectacular self-destruction, leading to talk-show jokes, the grounding of an airplane, a botched recall, followed by a second recall, and the cancellation of the product line.

Samsung's hard-earned brand—the castle Chairman Lee built—was tottering.

ON OCTOBER 24, 2016, a team of investigative journalists working for the news channel JTBC went on television with a scoop that would change South Korean politics forever.

It all started with a tablet.

"The JTBC reporting team obtained and analyzed Choi Soon-sil's computer files," said the news anchor on JTBC, referring to the adviser of South Korean president Madame Park Geun-hye.

"We were able to confirm the fact that Choi received the president's speeches. But the dates Choi received these speeches in the form of forty-four files were all before the president had made the speeches."

It was South Korea's Watergate. The journalists had obtained evidence that the president's adviser had been a sort of puppet master, editing her speeches and cabinet briefings and getting access to her private appointment information. She had the president's itinerary and even private chat messages.

Yet Choi was a private citizen with no government clearance. She had no permission for special access to state secrets. South Korean media began calling her the "daughter of Korea's Rasputin," a reference to the Russian mystic who influenced Russia's royal family in the years leading up to their imprisonment and execution during World War I.

Within weeks, President Park's approval rating collapsed to 4 percent. Swarms of protesters cried foul over what they called a "shadow government" run by a "shaman adviser." The largest protests in the nation's democratic history took place in the wake of the reports, at one point reaching more than 424,000 people, according to police estimates.

I watched the protests from the eighteenth floor of the Seoul Foreign Correspondents' Club off Gwanghwamun Square. Below me, luminescent dots lit up like fireflies as protesters held candles and smartphone screens in solidarity.

"You have been surrounded, Park Geun-hye! Surrender!" protesters cried over loudspeakers outside the Blue House.

"Step down, Park Geun-hye!"

Later that day, I spoke with a South Korean lawyer involved in politics. "It's unbelievable," he said. "Korea is vomiting out the legacy of military rule." Authoritarian rule had ended in 1987, he said, but its legacy remained in the ties between people like President Park and Jay Lee.

Ten days after the story broke, prosecutors detained Choi after her return from Germany. The tentacle-like reach of the corruption and cronyism was becoming clearer from the evidence that was seized, ensnaring virtually the entire Republic of Samsung government itself.

"It is hard to forgive myself and sleep at night with feelings of sorrow," President Park Geun-hye announced, her voice trembling.

But it was too late for apologies. An investigation spread across the highest levels of government and business. On November 8 prosecutors raided the Samsung Electronics building in Gangnam, leaving the building with boxes stuffed with documents.

Two weeks later, prosecutors descended on the National Pension Service in another raid, ransacking its offices and leaving with more stacks of papers and documents, followed by the government body connected to it, the Ministry of Health and Welfare.

"We raided the NPS and the health ministry to see if the NPS voted in favor of Samsung's merger last year according to a normal process," declared a special prosecutor.

"We got phone calls from people complaining that the NPS was wasting people's money," a pension-service staffer later told me.

The head of the pension service's investment committee, Hong Wan-sun, and the closely linked minister of health and welfare, Moon Hyung-pyo, were quickly put on trial, found guilty, and imprisoned seven months later, sentenced to two and a half years for political interference.

"The fact that a Health Ministry official used pressure to damage the independence of the state pension fund is highly reproachable," the judges said in their ruling.

In South Korea the consensus was that this would result in the sacking of a few Samsung executives.

"Jay Lee?" a business analyst passing through Seoul asked me. "They're not gonna arrest him. He's way too powerful."

Jay Lee's father, after all, had been in similar straits almost ten years earlier. He was tried, convicted—and pardoned.

But as the weeks went on, the prosecutors learned how enmeshed Samsung was in this web of corruption. Jay needed his kingdom. Buy a horse, it appeared to investigators, and get a merger. As the protests and media coverage raged, the government knew it had to respond. Jay, the "reformer," was suddenly persona non grata, a target in an investigation by South Korean prosecutors and, even more, the subject of an unheard-of grilling in Parliament.

A parliamentary hearing isn't the same as an indictment or a grand jury investigation. It carries no criminal charges. It was merely a request by Parliament, in the face of the intense public anger, for Jay and other business leaders to testify, to which they acquiesced.

"Are you indeed an accomplice?" a lawmaker asked Jay Lee, who sat, visibly nervous and meek, in the parliamentary hearing on December 5. It was an unprecedented event and fully televised, bringing the heretofore untouchable business leader and South Korean lawmakers together in one room to challenge Samsung's corporate behavior.

"Do you promise to cut ties between big business and politics?"

Jay Lee waffled, dodging the question.

"Promise clearly," the livid politician commanded. "Apologize to the people!"

As the grilling went on, Jay paused and hesitated, his denials issued in rapid-fire bursts, his eyes darting around the room.

"I don't know," he said. "I don't remember. I don't know." He claimed that he learned about the scheme after it happened.

"Jay Y. Lee seems like he has memory problems," a viewer texted to the show on her phone. "Maybe someone more capable should be in his position?"

The hearing went on for hours, captivating the nation. It was the subject of gossip everywhere I went that day. When the interrogators took a recess, it seemed as if Jay Lee had escaped the total annihilation of his image. But prosecutors were not done with him.

The investigation into the scandal went on for more than a month. According to prosecutors, Jay bribed Choi Soon-sil, the president's confidante, with $38 million to finance her Olympic horse-racing project.

On January 17 a cavalcade of black cars pulled up outside a Seoul courthouse. Jay Lee exited, summoned for another round of questioning before a judge on accusations of bribery, embezzlement, and perjury. The final charge stemmed from what was claimed to be false testimony in the National Assembly. Prosecutors had requested a warrant for his arrest. The judge agreed to listen to the case and decide whether to grant the prosecutors their warrant. The Samsung vice chairman passed through throngs of journalists as he entered the court, a vacant look on his face.

"Do you still feel that you are the victim of the president's coercion?" one reporter shouted.

"The retirement savings of the South Korean people were used for your management succession! Don't you feel any moral responsibility?"

Lee was interrogated for twenty-two hours, remaining at the courthouse all night. Once again he denied any wrongdoing.

Early on the morning of January 17, 2017, the judge rejected the prosecutor's arrest warrant, citing a lack of evidence. But prosecutors filed a second arrest warrant almost a month later, worried that Samsung was destroying evidence. If they were successful, Jay Lee could be held in a jail cell while he awaited trial.

"You've been summoned for a warrant hearing for the second time. Please tell us your feelings before going in," asked a reporter as Lee entered the courthouse on February 16, 2017.

"On the cross-shareholdings," another shouted, "are there any illicit favors you requested?"

After Jay was questioned for another fifteen hours, two black Hyundai cars pulled up at the detention center where he would be held until a decision was made. Jay exited one of the vehicles, his face emotionless, greeted by throngs of protesters and reporters holding up Samsung phones, ironically, to record the event and shouting for a comment.

After yet another seven hours of questioning, prosecutors and defense lawyers presented their case to the judge. He reviewed the evidence alone and announced his decision at 5:30 A.M. And this time he ruled against Lee.

"We acknowledge the cause and necessity of the arrest," the judge ruled.

Jay Lee was taken into custody on the same three charges: bribery, embezzlement, and perjury. He was put in a jail cell for the duration of his trial.

Samsung heir Jay Y. Lee had been rising to the role of chairman through a series of troubled shareholding mechanisms and acquisitions among Samsung companies. Here, he's accosted by reporters as he arrives at the Seoul prosecutor's office to answer a criminal probe for bribery relating to one acquisition. January 12, 2017.

COURTESY OF THE ASSOCIATED PRESS, PHOTO BY SEONG-JOON CHO

The scion was held in solitary confinement, in a cell with a mattress on the floor, no shower, and nothing but an LG television to keep him company. The reason for solitary confinement?

"There are concerns about destroying evidence," a prison official told Reuters.

Three other Samsung executives, including Lee family lieutenant and company legend G.S. Choi, were indicted for bribery. They stepped down from the company. All three were now defendants alongside Jay Lee.

Samsung instituted emergency measures. It ordered the Future Strategy Office disbanded, sending many of the executives and staffers from the Tower packing. Suddenly the empire was like an octopus without a head.

"This is a new experience," Samsung Electronics CEO D.J. Koh told *Bloomberg Businessweek*. "We must make our own decisions."

I'D BEEN STUDYING THIS company for years and saw few real changes in the country famously known as the Republic of Samsung.

The winners in this Korean drama of dalliances were easily sorted from the losers, it seemed to me. The victors had played a Machiavellian game of power, embracing methods more akin to *Game of Thrones* than to a Silicon Valley start-up.

Many assumed Jay would likely get a presidential pardon—as his father had twice before.

The chairman's cousin Miky Lee (who'd tried to secure a deal for Samsung with DreamWorks back in 1995, before securing the Dream-Works deal for herself) left for California, citing health concerns. She, too, had been targeted by President Park Geun-hye, who had put her on a blacklist out of anger that the films produced by Miky's CJ Group were critical of her.

But the scandal continued to generate political fallout. Almost a month after Jay Lee's arrest, a South Korean court upheld President Park Geun-hye's impeachment and stripped her of immunity from arrest.

Before dawn on March 30, a convoy of black cars carrying the former president pulled up at the detention center, where she entered police custody.

And there they waited, the former president of South Korea and the heir to the Samsung empire, with a trial date looming, a moment of reckoning that might determine whether or not the Republic of Samsung would live on in the same unfettered way.

"WE LOOK AROUND, AND our Samsung spirit exists no more," a former Samsung vice president lamented to me over coffee. "Our empire is not an empire. We are becoming like any corporation."

He reminded me of a dazzling tale from twenty-two years earlier, when Chairman Lee II, furious at Samsung's faulty phones, had organized his famous bonfire and instigated a campaign of rejuvenation among his employees. The same was not true in the burning embers of the Galaxy Note 7 fires. The mood in the company wavered between dejection and self-censorship. There were no grand spectacles, no eight-hour speeches, no phalanxes of Samsung executives and employees ready to fight for their company and country.

"We are becoming like you. We are becoming like Americans, too short-term and cold and logical," he said, a sadness in his voice.

Amid the saga of Jay Lee, Samsung was still trying to solve the riddle of the phones that burst into flames. Some 200,000 Galaxy Note 7s and 30,000 batteries had been deposited in a laboratory for testing, where some seven hundred engineers attempted to uncover the cause of the smoking phones once and for all.

The hundreds of thousands of disabled Galaxy Note 7s were lined up on racks, a magnificent mosaic of the empire's flawed creation, where robotic arms entered the aisles and retrieved the devices as requested. Grinding through all-night shifts and grueling meetings, the engineers tested, re-tested, and discarded each hypothesis in this lab, attempting to re-create the cause of the fires.

The flames were not the result of the software, the engineers concluded. Nor did the circuitry cause the fires. The in-house manufacturing process did not reveal anything unusual. Quality assurance was absolved of culpability.

Four months after their investigation began, they found their answer.

On the morning of January 23, roughly three weeks before Jay's arrest, CEO D.J. Koh took the stage at the Samsung Electronics building in Seoul, giving an hour-long presentation to the press and the public on the company's findings.

"We are taking responsibility for our failure to ultimately identify and verify the issues arising out of the battery design and manufacturing process prior to the launch of the Note 7," D.J. said in a presentation.

He offered a detailed presentation of the technical problems that led to the fires. Samsung, to the amazement of some in the audience, stuck with a similar line from before the botched recall: that the battery hardware was the problem. D.J. didn't talk about the breakdowns that led to the company's failure to identify the problems.

Samsung SDI's (called "Supplier A" in the presentation) batteries were the victims of a design flaw that caused the battery to short-circuit. Pointing to CT scans and diagrams on his presentation slides, D.J. said that the supplier had created a pouch—the term for the battery's outside casing—that didn't have enough space to allow the battery to expand and contract when going through normal charge and discharge cycles.

The result was that the positive and negative electrodes touched, short-circuiting the batteries and causing them to combust.

Hong Kong–based Amperex ("Supplier B"), he said, manufactured functional batteries in the earlier batch of Note 7s. The problems began when Samsung turned the supplier into its sole battery provider, introducing errors in its next batch of ten million batteries for Samsung's post-recall phones. Samsung's testers discovered protrusions left over from the welding process, which caused more short circuits and fires.

Together engineers designed what Samsung called an eight-step battery-testing plan, a series of much tougher tests on the batteries before they were shipped to the market.

"The lessons we have learned are now deeply reflected in our processes and our culture," D.J. said.

The case, as far as Samsung was concerned, was closed.

Some of the journalists and analysts watching the press conference were taken aback.

"The rather poor way they handled the first recall suggests that they have trouble accepting problems until they become quite big and they have no choice but to face them," Willy C. Shih, a professor at Harvard Business School, told *The New York Times*. "This time, it will really call into question how they communicate problems, whether management is open to hearing things from the front line."

After sitting through an interview with D.J. Koh in which he placed stacks of photos on the table about the testing that had been done, and made his case, *The Wall Street Journal*'s tech columnists Geoffrey Fowler and Joanna Stern gave Samsung's battery fix a C grade.

Two separate sets of bad batteries from two separate companies? "That's like a meteor striking your house—twice," the pair wrote.

Park Chul Wan, the battery sleuth who first told me, "It's not the batteries," and predicted the problems weren't solved, told me that the presentation, while true, left out points he wanted to see answered.

"Samsung claimed that the problem lay in the battery manufacturing," he told me. "If so, this means that the issues can be solved by tightening up the manufacturing process."

Samsung claimed it did upgrade its manufacturing process. Chul Wan showed me the upgraded design of the just-released Galaxy S8 as a counterpoint.

"Samsung's proposed solution and their adjustments to the product," he said, "resemble the solution that you would come up with to address a complex of factors, not just a simple battery failure.

"The packaging is completely different and the fastening mechanism in the battery body has been changed," he pointed out. "It's been assembled in a way that will no longer cause one part of the battery to be subject to a large force, as in the Note 7."

Whatever the reasons for the upgraded designs—perhaps a cautionary decision to address any number of other potential factors—the tech world had moved on.

CNET's Jessica Dolcourt called the new Galaxy S8, released in April 2017, "the most beautiful phone ever." *TechRadar*'s Gareth Beavis declared it a "brilliant phone."

Still, internally, one former Samsung employee told me that sales of the Galaxy S8 weren't as strong as they'd hoped—tainted, perhaps, by the Note 7. Nonetheless, I was inundated with calls from *Bloomberg* and CNN.

"Samsung's profits just hit record highs," a reporter said to me on the phone. "They're overtaking Apple as the most profitable tech company in the world. But why? Isn't Jay Lee in jail? Their Note 7 just caught fire six months ago."

It was part of the genius of Samsung's business model. Samsung was rocketing upward into record profits each quarter, pushing the South Korean stock market to all-time highs. With profits of $12.1 billion in the second quarter of 2017, Samsung had overtaken Apple as the most profitable tech company in the world.

But its success was about far more than smartphones.

"This is the house that Jay Lee's father and grandfather built. The investments they made twenty, thirty, forty years ago had the long term in mind," I explained, referring to Samsung's LCDs, NAND flash memories, and cellphone batteries, many of which were used in the iPhone. "Samsung is picking the fruits of the harvest. When the semiconductor business falls, Samsung can profit from smartphones. When the smartphone business declines, Samsung can make a profit from displays. After that, maybe semiconductors will get profitable again. It's the cyclical nature of Samsung's businesses.

"Jay Lee," I went on, "is not that relevant to the day-to-day business. The business can run itself."

Consumers are fickle. The love of a beautiful product in their hands and a compelling brand are ultimately what matter. Even after an embarrassing recall, consumers move on quickly and forget easily.

Samsung, with its incredible catalog of products, was poised to remain a major player in technology. All of which posed an awkward question for the company. If the empire was posting record profits while its king-in-waiting sat in jail, then what was the point in having a king-in-waiting?

Samsung's public relations department was noticeably silent about the matter.

"It's a kind of tragedy," said the CEO of Samsung's semiconductors division, Oh-hyun (O.H.) Kwon, at a Q&A session at the Economic

Club in Washington, DC, on October 19, 2017, shortly after he announced his resignation. "The business itself is going well today. That means for the short term, we have not [suffered] a big impact. But the long-term . . . We need some advice, some of the chairman's group's advice."

Jay was supposed to lay down the grand vision for the next decade or more. And now he was sitting in a jail cell.

ON AUGUST 2, 2017, Jay Lee appeared in court to stand trial; crowds of well-wishers and opponents gathered outside the courthouse.

"What did Jay Y. Lee do wrong? He was only trying to make our nation greater by making Samsung greater!"

"If you don't like Samsung, just go to North Korea!"

"Free Jay Y. Lee!"

"Punish Jay Y. Lee!"

Jay Lee entered the courtroom as throngs of demonstrators held up signs, chanting and shouting. G.S. Choi and three other former Samsung executives from the Tower joined him as defendants in the courtroom.

The windowless fifth-floor courtroom at the Seoul District Court was packed with lawyers, reporters, and courtroom recorders. Jay's defense attorneys were wiping their foreheads with handkerchiefs in the heat. The question they had to answer: Was Jay Lee a well-meaning patriot bullied by South Korea's president into making payments? Or was he a callous and corrupt businessman offering bribes to gain the favor of the highest office in the land?

On August 2 Jay took the stand. He'd never before spoken publicly at such length and depth. The prosecutors went after him hard, seeking an unprecedented sentence of twelve years.

"In retrospect I had a lot of shortcomings and didn't take care of the things that needed to be taken care of, and that's all my fault," he told the court. "It was my responsibility. I have no excuses.

"I've been anxious, and I've struggled under heavy pressure to follow their [his father's and grandfather's] footsteps and not lead Samsung down a wrong path."

But he continued to insist that he was far from the mastermind of the bribes but instead a victim. President Park Geun-hye had pressured him to make the payments. Jay insisted he didn't know the roles of key players in his own company, or for that matter the functions of the Tower and the shareholder mechanisms of the family succession.

"I do receive a summary of daily news, but my work is mostly electronics and IT," he said.

"Mastermind or Naïf?" asked *The New York Times*. Either way, he didn't look competent.

But Jay Lee tripped up while talking about his bedridden father. "Back when Chairman Lee was alive . . . ," he began and then abruptly stopped.

Laughter broke out in the courtroom, according to a Yonhap News Agency reporter.

"Back when Chairman Lee was healthy . . . ," Jay said, correcting himself. The conspiracy theorists, convinced that Chairman Lee had died some time ago, had a field day.

G.S. Choi stood up for his boss until the very end in a hearing that went until 2:00 A.M. Loyalty, after all, is the true mark of a Samsung executive. "Jay Y. Lee is not the group's final decision-making authority," G.S. testified. He claimed that he personally signed off on the donations to the equestrian foundation, not Jay. "If you were to hold Samsung responsible, please blame me. I am aging and lost judgment. . . . Others just trusted me and followed my judgment," he told the court.

SAMSUNG UNVEILED THE GALAXY Note 8, its next iteration of its smartphone, two days before the verdict was rendered.

"None of us will ever forget what happened last year," CEO D.J. Koh said after taking the stage on August 23 at New York's Park Avenue Armory, revealing the new device to the world. "But I know I will never forget how many millions of dedicated Note loyalists stayed with us."

At $930, the new Galaxy Note 8 would prove to be a hard sell. It needed to be flawless—reviewers and customers would be quick to make comparisons to the faulty Note 7s. But early indicators suggested Samsung had been successful.

"A year ago, I wrote that the Samsung Galaxy Note 7 was the best big phone ever made," wrote *The Verge*'s Dan Seifert. "In fact, in all of the important areas, the Note 8 is a better device than even the Note 7 was."

TWO DAYS LATER, ON August 25, Jay Lee and the four former Samsung executives showed up at the courthouse to hear the verdict. Protesters swarmed in the area outside the courthouse in a raucous scene.

"Jay Y. Lee, not guilty verdict!" shouted flag-waving Samsung supporters along the street leading to the courthouse. Most were senior citizens.

"Prosecutors falsified evidence!" read their pickets. "Sickening lies, spun forcefully!"

"Please wave your flags as hard as you can, everybody!" shouted a college-aged protest leader. "There's a news team to our left!"

Protest trucks blasted military songs from the 1960s over loudspeakers, driven by men bedecked in Cold War–era military garb. The police looked bored and slightly amused.

Inside, the judge prepared to read the verdict. In the courtroom, the media, colleagues, lawyers, and family waited anxiously.

Jay Y. Lee, the judge announced, was not believed to have sought "direct favors" regarding the "specific agenda" of the merger. But the judge believed that he shared in an "implicit awareness and consent" in the exchange of "illicit favors."

"Is he going to walk?" a journalist asked me.

"Looks like he's innocent," said another. "No smoking gun."

But the judge was not finished. He read his verdict over a nail-biting twenty minutes. Then he got to the point.

"The defendant, as a de facto head of Samsung Group . . . facilitated each crime, making his role in the crimes and his influence on them considerably large," the judge declared. "The defendant also provided false witness at the parliamentary hearing to hide his criminal acts."

Twitter and Facebook were alight; South Koreans posted guesses of "Guilty!" or "Not guilty!"

My email in-box was instantly flooded with queries from business analysts and reporters. I got a frantic phone call from CNN.

"What does this all mean?!"

Finally, the judge reached his definitive conclusion.

He declared Jay guilty of bribery, embezzlement, and perjury and sentenced him to five years in prison.

The courtroom gasped.

The ruling was unprecedented. It meant that Jay was not eligible to have his sentence commuted, as his father had. A sentence of three years or more prevented judges from suspending a prison sentence.

Jay left the courthouse in a daze, saying nothing, ignoring the phalanx of reporters and protesters. His police escorts led him to the bus that would take him to prison.

It was a verdict "I cannot accept as a lawyer," Jay Lee's attorney told reporters, his face red and flushed. His lawyers appealed immediately, continuing to deny that Jay had committed any wrongdoing. Jay retained his title of vice chairman and his position on the board of directors, despite being a convicted criminal—something that would be impossible in a publicly traded American corporation. And he remained in prison as his second trial got under way.

Eighteen days later, on September 12, 2017, Apple CEO Tim Cook took the stage to introduce the iPhone X—with a price starting at $999, the most expensive iPhone ever. Though the iPhone X received praise in some circles, Apple was no longer the lone innovator in smartphone technology. In fact, in many ways, in a reversal of roles, it had become more of a fast follower to Samsung in hardware.

"The iPhone X is basically Samsung's Note 8 plus animojis," wrote ZDNet.

"iPhone X Features: A Leap Forward for Apple but Samsung Is Still Ahead," ran a headline in *The Independent*.

"That pisses me off," a former member of Todd Pendleton's team said about the iPhone X. "The iPhone X isn't a great phone by itself. Samsung makes the displays and a lot of the components. Apple still depends on Samsung! I don't know what all this Apple frenzy is about!"

Samsung was poised to earn $110 for every $1,000 iPhone X that Apple sold, according to an estimate in *The Wall Street Journal*. That would add up to $4 billion in revenues for Samsung.

"These are two of the largest companies on the planet, deeply tied at the hip and directly competitive," David Yoffie, a professor at Harvard Business School, told the newspaper. "That makes this stand out compared with almost any relationship you can think of."

I THOUGHT THE STORY of the Republic of Samsung was finished, in terms of this book. I was meeting old friends to plan a quick trip in Cambodia, a country I once lived in and love. Shortly after leaving Cambodia for Thailand, I opened my Facebook to frantic messages from human-rights groups, journalists, and diplomats, who attached a frightening yet hilariously untrue news report about me from Cambodia's government-aligned media. A photograph of me and the daughter of Cambodia's opposition leader at a private dinner appeared all over primetime television.

"Recently," the Cambodian reports said, "Geoffrey Cain had a conspiracy with the opposition party in South Korea to topple [President] Park Geun-Hye by using the inciting newspaper articles to criticize her, and also by employing social media as a means to spread out the information.

"Geoffrey Cain was also a spy behind an opposition party in another country," the report said, not naming this mysterious country. It accused me of traveling to Cambodia to stir up similar shenanigans against the government there.

Of course, I wasn't a spy. The whole affair was ludicrous. But it was the kind of strange episode foreign correspondents sometimes have to deal with. And I could no longer go back to Cambodia. The Cambodian government had just arrested an Australian filmmaker on trumped-up spying charges and sentenced him to ten years in a Cambodian prison. (He was later pardoned by the king.) My dinner friend's father, the opposition leader, was also arrested, for treason, and put in a maximum-security prison after the fake news came out about our spy ring.

I retreated to Seoul, where I met a spokesman from the office of the new South Korean president at the Seoul Foreign Correspondents' Club. He thanked me, jokingly, for overthrowing the previous president and installing his administration.

"Does Samsung have business in Cambodia? Do you think Samsung was behind those reports?" a good friend who is a scholar of Korean history asked me.

I laughed.

I knew that the fall of the Lee dynasty would not impact the continued growth of the company. Samsung had lost its heir, for now. But South Korea was still the Republic of Samsung, and South Korea would not survive without Samsung, people everywhere confided in me. Its namesake republic would live on.

Epilogue

"[THE] PREVIOUS 10 YEARS, it was an era of the smartphone," Samsung CEO D.J. Koh told *The Independent*. "From this year, maybe a new era is opening because of the emergence of the internet of things, 5G, AI, and all these technologies mingling together. The new era is in front of us."

For almost a decade, Samsung's engineers toiled away at their secret weapon for this new era, a smartphone that folds open, like a wallet, to reveal a massive display, ideal for movies and artwork—and then folds shut and fits in your pocket. Samsung went public with a prototype in May 2011. Then there were rumors of a release. Then leaks. Then patents and company announcements.

"Samsung Files Patent for a Bizarre Folding Smartphone," read an *Engadget* headline back in November 2016.

For years, the industry buzz around the peculiar phone came in waves. But the foldable phone didn't materialize in the marketplace. "It is difficult to talk about the date of the launch now. . . . There are still durability issues that we need to address," D.J. Koh told the *The Korea Herald* in January 2018.

The smartphone wars were getting staid. With price tags close to $1,000, Samsung and others were having a harder time selling new smartphones that pretty much only offered marginal updates from the previous model. The pressure was growing to release a novel and groundbreaking device.

AS INNOVATION WAVERED, SAMSUNG also had to cope with the fallout of scandal on the home front and the tumult of South Korean political winds. After ousting President Park Geun-hye through impeachment, South Koreans had elected a left-wing president, Moon Jae-in, who pledged to clean up corporate corruption. President Moon appointed a prominent economics professor at Hansung University in Seoul, Kim Sang-jo, to head South Korea's corporate watchdog, the Korea Fair Trade Commission (KFTC).

The KFTC was Korea's trustbuster, a regulatory agency that targeted monopolies, opened investigations, and broke up cross-shareholdings and cartels. It had the power to fine companies and order corrective measures.

Kim was both respected and notorious, known to Koreans as the *chaebol* sniper for his fearless targeting of Samsung and other powerful firms as a corporate governance activist. In 2004, he heckled Samsung at a shareholder meeting for not doing anything when Chairman Lee II was accused of making illegal political donations—and rankled Samsung so much that its security guards dragged him out of the meeting.

"They were born as if they were princes in a kingdom," Kim told the *Nikkei Asian Review* about the next generation of business leaders, which included Jay Lee. They "have lost the aggressive entrepreneurship that was shown by the generations of their founding grandfathers and fathers."

Whether the government had the willpower to dig in against the *chaebol* or would dole out minor regulatory slaps, was the big test. "While there's no change in our belief that cross-shareholding is a serious problem, we have to weigh benefits and administrative costs of any such reform," Kim also told reporters, tempering expectations. "We have limited capital to push for policy changes and it is important to set priorities."

In February 2018, Jay had already served one year of his five-year sentence when he, his former aide G.S. Choi, and another former Samsung executive, Chang Choong-ki, entered the Seoul High Court for the verdict of their appeal, as they proclaimed their innocence of offering millions of dollars and a race horse as a bribe for political favors and sought to get their sentences overturned.

The appeal trial was a relatively quiet affair. The panel of judges maintained that Jay Lee bribed President Park by supporting the equestrian career of the daughter of the former president's friend. But Lee's involvement was "passive compliance to political power," one judge wrote. The court reduced the amount of the bribes Jay Lee was charged with having offered from $6.4 million to $3.3 million.

"[Former President] Park threatened Samsung Electronics executives," one judge said. "The defendant provided a bribe, knowing it was bribery to support [the friend's daughter], but was unable to refuse."

The prosecutors had won a partial victory with the verdict, since Jay Lee wasn't completely off the hook. But the sentencing itself was a victory for Samsung. Because Jay was cut some slack in terms of the bribery charges, the court reduced his five-year prison sentence to two and a half years. In addition, the judges commuted his sentence while upholding his conviction for bribery and embezzlement. The other executives were also given reduced, commuted sentences.

Jay looked stunned, *Bloomberg* reported, by the verdict. Despite confirming his conviction for bribery, the court had decided the heir was free to go, with the stipulation that he would be on probation for four years. He blushed and exited the courtroom, and went to the hospital to see his father.

"The past year has been a really valuable time of looking back on myself," Lee told reporters, his voice shaky at times.

FOUR MONTHS LATER, IN June 2018, after seven years since the start of the Apple-Samsung lawsuits, both sides decided to call it quits. After a jury ordered Samsung to pay Apple $539 million in damages, Samsung and Apple agreed to an undisclosed settlement.

On paper, it looked like a victory for Apple. The courts had ruled repeatedly that Samsung copied Apple's patents and that Samsung was required to pay damages to Apple. But Apple's victory was pyrrhic. The patent war had degenerated into a seemingly never-ending succession of court battles and appeals over how much Samsung owed Apple in damages. Finally, with the settlement, the appeals were finished.

"The smartphone patent wars are finally over," declared *The New York Times*.

"And if I had to characterize it, it didn't really accomplish anything," Brian J. Love, a Santa Clara University law professor, told *The New York Times*. "Close to a decade of litigation, hundreds of millions of dollars spent on lawyers, and at the end of the day, no products went off the market."

SOUTH KOREA'S POLITICAL LEADERS' reforms were tested again when, in July 2018, Elliott Management, the New York–based hedge fund, filed for arbitration against the South Korean government. It sought $718 million in damages for political interference in the 2015 Samsung merger—the merger that hurt shareholders and led to the arrest of Jay Lee and the downfall of President Park Geun-hye.

"As Elliott's claims are groundless, we view that we don't have a liability to pay compensation for any damages," a justice ministry spokesperson told the *Financial Times*.

"It is baffling that the Republic [of Korea] is currently calling claims of fraud and discrimination 'groundless' while actively continuing an investigation that shows further proof of the very fraud they are calling 'groundless,'" an Elliott representative wrote to me.

The government needed Samsung to build manufacturing plants and to invest for the South Korean economy, and Samsung appeared to be seeking government support in hard times. In September 2018, President Moon traveled to North Korea for a diplomatic summit with the dictator Kim Jong Un. Accompanying the South Korean president was none other than Jay Lee—despite his two-and-a-half-year suspended prison sentence, and four-year probation—alongside almost a dozen other business leaders.

As per South Korean business tradition, Jay Lee's criminal conviction seemed to be of little importance to political leaders, as long as he and his company helped the national interest.

"Along with becoming a regular fixture at government events, Lee has also appeared at more than 10 public events both in the country and overseas over the past year," the *Hankyoreh* newspaper later wrote.

"Most people in the business community think that Lee has gotten his mojo back."

MEANWHILE, FINANCIAL INVESTIGATORS PROBED irregularities at Samsung BioLogics—a biopharmaceutical company billed as Samsung's next growth engine—that dated all the way back to 2015.

Investigators believed Samsung BioLogics had changed its bookkeeping methods to inflate the value of its stake in Samsung Bioepis, another pharmaceutical company and a partner. As a result, its valuation of Samsung Bioepis soared by eighteen times.

Why would it have committed a possible accounting fraud?

"Some analysts said the accounting shift was made to facilitate the controversial merger in 2015 between two Samsung group units— Samsung C&T and Cheil Industries," the *Financial Times* reported, referring to the troubled merger that ignited the ongoing scandals and arrests at Samsung.

"At that time, Samsung group had to buoy the value of BioLogics to justify the merger ratio," Park Ju-geun, head of corporate analysis group CEO Score, speculated to the newspaper. He was referring to the controversial share value ratio between the two merging companies— Cheil Industries and Samsung C&T—that Elliott argued undervalued Samsung C&T and hurt its shareholders. Samsung Electronics and Samsung C&T, two important companies for exerting family control over Samsung's cross-shareholding structure, owned 75 percent in Samsung BioLogics. The inflated value of Samsung BioLogics could have influenced the undervaluing of Samsung C&T shares, regulators believed.

The investigation sparked a dramatic share sell-off. In a single day, Samsung BioLogics lost $6 billion in market value, an enormous amount for a $30 billion company.

On November 14, 2018, the KFTC suspended trading for Samsung BioLogics stock. "We concluded that the company violated accounting standards intentionally in 2015," the KFTC wrote in a statement. Samsung denied the accusations of shady bookkeeping and then filed an administrative suit against the KFTC. Samsung claimed it wanted to show the court that it used a legitimate accounting standard.

Samsung BioLogics avoided being delisted, and trading resumed almost a month later. But regulators were not finished with their investigation. From April to June, eight executives were arrested, accused of destroying evidence and manipulating accounting data. The court twice rejected an arrest warrant for one of the suspects, Samsung BioLogics CEO Kim Tae-han, claiming there was "room for dispute" over his role.

Investigators raided two plants owned by Samsung BioLogics and found a cache of about twenty computers and notebooks, along with a computer server, under the floorboards, which they believed were related to the case.

"They [Samsung employees] deleted all computer files and emails that contained keywords like 'VIP,' 'JY' and 'vice chairman,'" Yonhap reported. Prosecutors believed those phrases referred to Vice Chairman Jay Y. Lee.

"We deeply regret the unsavory circumstances resulting from disgraceful acts such as the destruction of evidence," Samsung said in a statement.

"We feel gravely responsible for the arrest of our officials and the subsequent difficulties in management."

The investigation is ongoing. And it's one saga among many. From the time Samsung's leaders were first targeted in early 2017 and through September 2019, "prosecutors raided Samsung offices with warrants 27 times, and 31 executives were indicted," on myriad accusations, from the accounting scandal to allegations that Samsung wrongly dissolved labor unions, the *Korea JoongAng Daily* reported. In December 2019, Samsung's chairman of the board, Lee Sang-hoon, was sentenced to eighteen months in prison for union sabotage.

ON FEBRUARY 20, 2019, Samsung senior vice president Justin Denison emerged on stage at the Bill Graham Civic Auditorium in San Francisco. He had a much-anticipated and potentially groundbreaking announcement—ten years after the release of the first Galaxy phone.

The long-awaited and long-hyped foldable phone, unveiled as the

"Galaxy Fold," was ready. It would cost $1,980. "The Galaxy Fold is a device unlike any that's come before it," Denison proclaimed.

Almost two months later, the media got their phones to review.

"The screen on my Galaxy Fold review unit is completely broken and unusable just two days in," *Bloomberg* technology reviewer Mark Gurman tweeted. "Hard to know if this is widespread or not."

"My Galaxy Fold screen broke after just a day," wrote *The Verge's* Dieter Bohn.

"Now [the Galaxy Fold] is turning into a bit of an embarrassment, evoking memories of another botched launch: the 'exploding' Galaxy Note 7 smartphone," BBC reported.

Samsung attributed some of the malfunctions to the product reviewers peeling off a protective plastic layer adhered to the display, mistaking it for a screen protector that seemed natural to peel off. But CNBC's Steve Kovach posted a video of his display flickering. He said he did "nothing" to the phone. "Just unfolded it!"

With advance notice from media reviewers, Samsung acted quickly before releasing the Galaxy Fold into the hands of the public. On April 22, four days before the release, Samsung announced it was postponing its launch of the phone indefinitely.

"It was embarrassing. I pushed it [the Galaxy Fold] through before it was ready," CEO D.J. Koh admitted to *The Independent*.

Stephanie Choi, head of global marketing strategy, issued a statement that older Samsung Men would have once seen as insubordination, worthy of banishment. She attributed part of the failure to Chairman Lee II's New Management Initiative—that fateful campaign, kicked off in Frankfurt in 1993, when the chairman sought to inject his vision into Samsung, telling his executives to "change everything."

"We make what can't be made, and do what can't be done," she told *The Independent*. "This [Galaxy Fold issue] is unfortunately sometimes part of this process."

One week later, Samsung's profits plummeted. The memory semiconductor market—the bedrock of Samsung's empire—was volatile and crowded.

Samsung announced a new expansion called "Semiconductor Vision 2030." It pledged an investment of $115 billion in another promising field of semiconductors—non-memory chips—that power rising technologies like autonomous vehicles, medical robots, and devices that depend on artificial intelligence.

"The government will actively support this mission," South Korean President Moon announced at a Samsung semiconductor plant.

"As you asked, Samsung will become the first in the non-memory sector as well as in the memory sector," Jay told President Moon.

Not everyone was bullish on the new initiative. Samsung, after all, was responding to an industry-wide crisis in low chip prices by investing in more and different chips—with the help of the Korean government, as it had in the past. And perhaps it was too late. The expansion would have made more financial sense when chip profits were bustling.

"Some experts think that Samsung's management isn't as innovative as it used to be," the *Hankyoreh* newspaper wrote about Semiconductor Vision 2030. "The son has parted ways with his father, who focused on blazing trails and shaking things up at every turn."

PROSECUTORS APPEALED JAY LEE'S bribery case to the Supreme Court.

The audience stood up and the cameras flickered as the thirteen justices took their seats on August 29. The big question? Whether or not the appellate court had given Jay Lee a properly rigorous trial, and used an adequate definition of the word "bribe" when it decided that Jay had given a lesser amount in bribes than the amount that resulted in the verdict of the first trial.

The Supreme Court justices ruled that "there was an error in the form of a misinterpretation of the legal doctrine on illicit solicitations." The lower court, the justices ruled, should have considered the possibility that three horses that were donated by Samsung—rather than one horse, as it was originally decided in Jay's very first trial two years earlier—were bribes.

The Supreme Court voided the appellate court's decision that Jay Lee was responding "passively" to government demands and sent his case back down for a retrial.

The verdict was a victory for the prosecutors. And Jay's problems were compounded. A retrial could open the way for even more charges against him—and more prison time.

"Samsung's Lee Faces a Retrial That Could Put Him Back in Jail," Bloomberg News reported.

Samsung asked the public for leniency and understanding.

"In this increasingly uncertain and difficult economic environment, we ask for support and encouragement," the company said in a statement, "so we can rise above the challenges and continue to contribute to the broader economy."

FIVE MONTHS AFTER DELAYING the release of the Galaxy Fold, Samsung decided the phone was ready. Released in America on September 28, reviews were skeptical and lackluster. The phone was heavy, expensive at $1980—and, by Samsung's indication, excessively fragile. Each device came with a warning label that read: "Do not press the screen with a hard or sharp object, such as a pen or fingernail. Do not place cards, coins, or keys on the screen. Do not expose the phone to liquids or small particles."

Samsung said the Fold was a luxury device for early adopters. It was too early for the Fold to go mainstream. "And there may yet be a time when the world is full of foldable displays," speculated *Wired* reviewer Lauren Goode, who gave the Fold five stars out of ten. That time, however, was not now.

Almost a month later, on October 25, Jay Lee returned to court for his retrial, where he got a lecture from the judge. "As the head of a company representing our country, I hope you will feel responsible and humbly accept the results of this trial," the judge said in the first hearing.

The judge added, "In 1993, then 51-year-old Lee Kun-hee declared he would abandon all outdated and flawed practices, then overcame crisis through innovation. . . . In 2019, what declaration should Lee Jaeyong [Jay Y. Lee] make, at the age of 51?"

One day later, Jay's board seat was up for renewal. He ceded it. But he continued to hold the title vice chairman and to manage the company.

After three years of court hearings and four trials, his bribery verdict was anticipated in early 2020. As this book went to bed, Jay Lee was still on trial.

Notes

I CONDUCTED MANY INTERVIEWS under the agreement that I would withhold the interviewees' name and identifying information. This was the only way to research a book on Samsung. Many former employees are still involved with the company, either as suppliers, contractors, or consultants, or have close friends and family at the company. I gave anonymous sources extra scrutiny, double-checking their claims with other interviewees and gathering written documentation wherever possible.

1: GALAXY DEATH STAR

3   "I heard some popping": Amber Powell, "Note 7 Owner Describes Phone Fire on Southwest Airlines Flight," *Wave 3 News*, August 2, 2018, https://www.wave3.com/clip/12785336/note-7-owner-describes-phone-fire-on-southwest-airlines-flight/.

4   Since late August: Jemima Kiss, "Samsung Galaxy Note 7 Recall Expanded to 1.9m Despite Only 96 Causing Damage," *The Guardian*, October 13, 2016, https://www.theguardian.com/technology/2016/oct/13/samsung-galaxy-note-7-recall-expanded.

4   Samsung had begun to recall: Paul Mozur and Su-Hyun Lee, "Samsung to

Recall 2.5 Million Galaxy Note 7s over Battery Fires," *The New York Times*, September 2, 2016, https://www.nytimes.com/2016/09/03/business /samsung-galaxy-note-battery.html.

4   When he punched the new phone's: Jordan Golson, "Replacement Samsung Galaxy Note 7 Phone Catches Fire on Southwest Plane," *The Verge*, October 5, 2016, https://www.theverge.com/2016/10/5/13175000 /samsung-galaxy-note-7-fire-replacement-plane-battery-southwest.

4   "I did everything": Powell, "Note 7 Owner Describes Phone Fire."

5   Green wondered when he'd hear: Powell, "Note 7 Owner Describes Phone Fire."

5   "Until we are able to retrieve": Golson, "Replacement Samsung Galaxy Note 7."

5   Citing "exigent circumstances": Jordan Golson, "Burned Galaxy Note 7 from Southwest Flight Seized by Federal Regulators for Testing," *The Verge*, October 11, 2016, https://www.theverge.com/2016/10/11/13241032 /samsung-galaxy-note-7-subpoena-seizure-cpsc.

5   "like pins and needles": Joe Augustine, "Farmington Teenager: Replacement for Recalled Samsung Phone Melted in Hand," KSTP, October 9, 2016, https://web.archive.org/web/20161009150630/http://kstp.com /news/samsung-replacement-phone-melted-zuis-farmington/4285759/.

5   Michael Klering and his wife: Monique Blair, "Nicholasville Man Injured by Replacement Samsung Phone," WKYT, October 8, 2016, https://www .wkyt.com/content/news/Nicholasville-mans-replacement-Samsung-Galaxy-Note-7-catches-fire-396431431.html.

6   "The most disturbing part": Jordan Golson, "Samsung Knew a Third Replacement Note 7 Caught Fire on Tuesday and Said Nothing," *The Verge*, October 9, 2016, https://www.theverge.com/2016/10/9/13215728 /samsung-galaxy-note-7-third-fire-smoke-inhalation.

6   "The evidence suggests that Samsung": Rhett Jones, "Man's Replacement Galaxy Note 7 Catches Fire, Samsung Accidentally Texts 'I Can Try and Slow Him Down,'" *Gizmodo*, October 9, 2016, https://gizmodo.com /mans-replacement-galaxy-note-7-catches-fire-samsung-ac-1787588341.

6   A woman in Taiwan was walking: Alan Friedman, "Replacement Samsung Galaxy Note 7 Explodes in Taiwan While User Was Walking Her Dog?" *Phone Arena*, October 8, 2016, https://www.phonearena.com/news /Replacement-Samsung-Galaxy-Note-7-explodes-in-Taiwan-while-user-was-walking-her-dog_id86347.

6   "It filled my bedroom": Jordan Golson, "A Fourth Replacement Galaxy Note 7 Caught Fire in Virginia This Morning," *The Verge*, October 9, 2016, https://www.theverge.com/2016/10/9/13218730/samsung-galaxy-note-7-fire-replacement-fourth-virginia.

6   Then, hours later, another: Jordan Golson, "Samsung Says It's 'Working

Diligently' as Fifth Replacement Note 7 Burns," *The Verge*, October 9, 2016, https://www.theverge.com/2016/10/9/13219878/samsung-galaxy-note-7-replacement-fire-fifth-statement.

6   AT&T announced on October 9: Jordan Golson, "AT&T Halting Samsung Galaxy Note 7 Sales Following Multiple Fires with Replacement Phones," *The Verge*, October 9, 2016, https://www.theverge.com/2016/10/9/13219054/att-samsung-galaxy-note-7-stop-sales.

7   "Samsung is confident": Jordan Golson, "Samsung Is 'Pausing' Shipments of the Galaxy Note 7 to Telstra, Says Internal Memo," *The Verge*, October 9, 2016, https://www.theverge.com/2016/10/9/13220618/samsung-galaxy-note-7-pause-shipments.

7   "In other words": Samuel Burke, "The Samsung Galaxy Note 7 Debacle: A Timeline," *CNNMoney*, October 11, 2016, https://money.cnn.com/video/technology/2016/10/11/samsung-galaxy-note-7-timeline.cnnmoney/index.html.

7   "As this is going to another company": Nilay Patel, "Samsung Is Sending Incomprehensible Emails to Note 7 Owners Looking for a Refund," *The Verge*, October 10, 2016, https://www.theverge.com/2016/10/10/13227058/samsung-galaxy-note-7-refund-support-email-incomprehensible.

8   "Does anyone here have": *The Late Show with Stephen Colbert*, "Donald Trump Asks the Terminally Ill for a Huge Favor," posted by YouTube user The Late Show with Stephen Colbert on October 7, 2016, https://www.youtube.com/watch?time_continue=396&v=f_Flwq_zUVY.

8   "Hey @sprint, what if": Sapna Maheshwari, "Samsung's Response to Galaxy Note 7 Crisis Draws Criticism," *The New York Times*, October 11, 2016, https://www.nytimes.com/2016/10/12/business/media/samsungs-passive-response-to-note-7s-overheating-problem-draws-criticism.html.

8   "I'm sitting in front of": Hacker News (web discussion forum), "Samsung Blocks Exploding Note 7 Parody Videos," October 21, 2016. The user "net-sharc" posted the message on October 21, 2016. More sarcastic comments can be found on social media posts on Samsung's official Facebook and Twitter pages, as well as Samsung product reviews on Best Buy's website, from September 2016 to December 2017.

8   Airlines around the world: Steve Dent, "Australian Airlines Ban Use of Samsung's Galaxy Note 7," *Engadget*, September 8, 2016, https://www.engadget.com/2016/09/08/australian-airlines-ban-use-of-samsungs-galaxy-note-7/.

8   Pranksters posted YouTube videos: Kyle Orland, "Samsung Doesn't Want You to See Video of This *GTA V* Exploding Phone Mod," *Ars Technica*, October 20, 2016, https://arstechnica.com/gaming/2016/10/samsung-doesnt-want-you-to-see-video-of-this-gta-v-exploding-phone-mod/.

9   "It appears Samsung took the easy path": Daniel Nazar, "Samsung Sets Its Reputation on Fire with Bogus DMCA Takedown Notices," Electronic

Frontier Foundation, October 26, 2016, https://www.eff.org/deep links/2016/10/samsung-sets-its-reputation-fire-bogus-dmca-takedown-notices.

9   "We really don't care": Student at Yonsei University, interview by the author, November 28, 2009.

10   "Everyone on this street corner": Former Samsung vice president, interview by the author, June 24, 2015.

11   "Samsung won't tell us anything": Former correspondent in Korea, discussion with the author, September 8, 2016.

11   "didn't want anyone touching anything": Samsung senior marketing consultant, interview by the author, February 9, 2017.

12   "These one-sided, sensational views": David Steel, email message to a reporter in Seoul, October 20, 2016.

12   "Can you please check": Samsung marketing manager, Facebook message to the author, October 11, 2016.

12   "everyone loves a good Apple story": Foreign correspondent, interview by the author, September 28, 2016.

13   "Stay calm and confident": Samsung marketing manager, Facebook message to the author, October 11, 2016.

14   taking payments for their protests: "The Mystery of South Korea's Elderly Protesters," *Korea Exposé*, May 2, 2016, https://www.koreaexpose.com /the-mystery-of-south-koreas-elderly-protesters/.

15   "At a recent meeting": Nam Gi-hyeon, "President Park, Orders to Scrutinize Affairs Regarding Galaxy Note 7," *Maeil Business Newspaper*, October 16, 2016, https://www.mk.co.kr/news/politics/view/2016/10/722105/. This source is in Korean. The author's researcher translated the headline and the quoted text into English.

15   "This is not just Samsung's trouble": Choe Sang-hun, "Galaxy Note 7 Recall Dismays South Korea, the 'Republic of Samsung,'" *The New York Times*, October 23, 2016. https://www.nytimes.com/2016/10/23/world/asia /galaxy-note-7-recall-south-korea-samsung.html.

15   "The interview is a go": Nam Ki-young, email message to the author, July 18, 2011.

15   inspired by the Korean craft: Kohn Pederson Fox, "Samsung Seocho," no date, https://www.kpf.com/projects/samsung-seocho.

16   Steve Jobs had initiated a slew: Florian Mueller, "List of 50+ Apple-Samsung Lawsuits in 10 Countries," *FOSS Patents*, April 28, 2012, http://www.foss patents.com/2012/04/list-of-50-apple-samsung-lawsuits-in-10.html.

17   "I feel like the movie star": Samsung Electronics, "Samsung Galaxy Unpacked 2016 Live Stream Official Replay," posted by YouTube user crazydeals on August 3, 2016, https://www.youtube.com/watch?v= iYTKrRzNmag.

17   "Instead of presenting a product": D.J. Koh, interview by the author, July 21, 2011.

18   Their research suggested that: Galaxy marketer, interview by the author, November 8, 2016.

18   "We strongly believe we have": D.J. Koh, interview by the author, July 21, 2011.

18   Steve Jobs had elected to use: Brian Merchant, *The One Device: The Secret History of iPhone* (New York: Hachette Book Group, 2017), p. 363.

18   In August 2012, a California court: Nick Wingfield, "Jury Awards $1 Billion to Apple in Samsung Patent Case," *The New York Times,* August 24, 2012, https://www.nytimes.com/2012/08/25/technology/jury-reaches-decision-in-apple-samsung-patent-trial.html.

18   Samsung, however, won legal victories: Charles Arthur, "Samsung Galaxy Tab 'Does Not Copy Apple's Designs,'" *The Guardian,* October 18, 2012, https://www.theguardian.com/technology/2012/oct/18/samsung-galaxy-tab-apple-ipad; Associated Press, "Samsung Wins Korean Battle in Apple Patent War," August 24, 2012, https://www.cbc.ca/news/business/samsung-wins-korean-battle-in-apple-patent-war-1.1153862; Mari Saito and Maki Shiraki, "Samsung Triumphs Over Apple in Japan Patent Case," Reuters, August 31, 2012, https://in.reuters.com/article/us-apple-samsung-japan/samsung-wins-over-apple-in-japan-patent-case-idINBRE87U05R20120831.

19   Steve Jobs was an admirer of Sony: Leander Kahney, "Steve Jobs' Sony Envy [Sculley Interview]," *Cult of Mac,* October 14, 2010, https://www.cultofmac.com/63316/steve-jobs-sony-envy-sculley-interview/.

19   Apple designers borrowed: Christina Bonnington, "*Apple v. Samsung:* 5 Surprising Reveals in Latest Court Documents," *Wired,* July 27, 2012, https://www.wired.com/2012/07/apple-reveals-for-monday-trial/.

19   "We are going to patent it all": Fred Vogelstein, *Dogfight: How Apple and Google Went to War and Started a Revolution* (New York: Sarah Crichton Books, 2013), p. 172.

19   calling their larger phones "Hummers": Chris Ziegler, "Apple's Steve Jobs: 'No One's Going to Buy' a Big Phone," *Engadget,* July 16, 2010, https://www.engadget.com/2010/07/16/jobs-no-ones-going-to-buy-a-big-phone/.

19   "I am talking to you on": Brian Wallace, interview by the author, January 6, 2016.

19   In February 2014, Samsung sued: Aaron Souppouris, "Samsung Sues Dyson Following 'Intolerable' Copycat Claims," *The Verge,* February 17, 2014, https://www.theverge.com/2014/2/17/5418616/samsung-sues-dyson-following-intolerable-copycat-claims.

20   In early 2016, D.J. Koh: Associated Press, "Dongjin Koh Is the New President of Samsung's Mobile Business," *Mashable,* December 1, 2015, https://mashable.com/2015/12/01/samsung-koh-dongjin/.

20 iPhone was slated to be a modest release: Yoolim Lee and Min Jeong Lee, "Rush to Take Advantage of a Dull iPhone Started Samsung's Battery Crisis," *Bloomberg,* September 19, 2016, https://www.bloomberg.com/news/articles/2016-09-18/samsung-crisis-began-in-rush-to-capitalize-on-uninspiring-iphone.

20 eye-catching glass and metal casing: Jessica Dolcourt, "Samsung Galaxy Note 7 Review," *CNET,* December 9, 2016, https://www.cnet.com/reviews/samsung-galaxy-note-7-review/.

20 So the Note was built with a: Yoolim Lee and Min Jeong Lee, "Rush to Take Advantage of a Dull iPhone."

20 Samsung Electronics settled on two suppliers: Yoolim Lee and Min Jeong Lee, "Rush to Take Advantage of a Dull iPhone."

20 Samsung Electronics owned a fifth of SDI: Lee Jae-eun, "20.15% of Samsung SDI Stocks, Owned by Samsung Electronics and Affiliated Persons," *Chosunbiz,* August 5, 2014, http://biz.chosun.com/site/data/html_dir/2014/08/05/2014080502438.html. This source is in Korean. The author's researcher translated the headline and text into English.

21 Hong Kong–based Amperex: Yoolim Lee and Min Jeong Lee, "Rush to Take Advantage of a Dull iPhone."

21 "We were sensitive to the iPhone's": Samsung mobile manager, interview by the author, September 27, 2016.

22 "Apple's taunts that Samsung": Yoolim Lee and Min Jeong Lee, "Rush to Take Advantage of a Dull iPhone."

22 "The pressure was tremendous": Samsung mobile manager, interview by the author, September 27, 2016.

22 repeatedly changing their minds: Yoolim Lee and Min Jeong Lee, "Rush to Take Advantage of a Dull iPhone."

22 "The pressure was huge": Samsung mobile manager, interview by the author, September 27, 2016.

22 They sent off models to carriers: Yoolim Lee and Min Jeong Lee, "Rush to Take Advantage of a Dull iPhone."

22 "We faced skeptics who doubted us": "Samsung Galaxy Unpacked 7 (Galaxy Note 7)—Live Event Steam (Recap)—August 2nd 2016!" posted by YouTube user Tech 101 on August 2, 2016, https://www.youtube.com/watch?v=fCE3xc9_dLQ.

23 "You know the kind": Samsung marketer, interview by the author, September 8, 2016.

23 Samsung had to push back launches: Robert Triggs, "Note 7 Delayed in Malaysia, Netherlands, Russia, Ukraine, and Maybe More," Android Authority, August 12, 2016, https://www.androidauthority.com/note-7-delayed-netherlands-ukraine-russia-malaysia-709516/.

23   "The Galaxy Note 7 is a beautiful": Dolcourt, "Samsung Galaxy Note 7 Review."

23   "took Samsung's best phone": Matt Swider, "Samsung Galaxy Note 7 Review," *TechRadar,* January 23, 2017, https://www.techradar.com/reviews/phones/mobile-phones/samsung-galaxy-note-7-1325876/review.

23   "The company wound up making": Chris Velazco, "Samsung Galaxy Note 7 Review," *Engadget,* August 16, 2016, https://www.engadget.com/2016/08/16/samsung-galaxy-note-7-review/.

23   "How can you believe that?": Samsung senior manager, email message to the author, October 10, 2016.

23   she didn't believe the reports at first: Brad Stone, Sam King, and Ian King, "Summer of Samsung: A Corruption Scandal, a Political Firestorm—and a Record Profit," *Bloomberg Businessweek,* July 27, 2017, https://www.bloomberg.com/news/features/2017-07-27/summer-of-samsung-a-corruption-scandal-a-political-firestorm-and-a-record-profit.

23   "One mobile division executive": Jonathan Cheng and John D. McKinnon, "The Fatal Mistake That Doomed Samsung's Galaxy Note," *The Wall Street Journal,* October 23, 2016, https://www.wsj.com/articles/the-fatal-mistake-that-doomed-samsungs-galaxy-note-1477248978.

24   The company formed a task force: Stone, King, and King, "Summer of Samsung."

24   The powerful enforcer: Cheng and McKinnon, "Fatal Mistake That Doomed Samsung's Galaxy Note."

24   "Now, speculation that": Moon Eun-hae, "[Galaxy Note 7 Recall] 'An Excruciating Sum' of 2 Trillion Won . . . Who Will Take Responsibility?" *EBN,* September 3, 2016, http://www.ebn.co.kr/news/view/848961. This source is in Korean. The author's researcher translated the title and the text into English.

24   called Choi a "terrorist": Former Samsung television vice president, phone interview by the author, February 3, 2017.

24   "The Tower is watching": Former Samsung mobile vice president, interview by the author, March 2, 2016. More information on the Tower can be found at Jessica E. Lessin, "Inside the Tower: Samsung's Mobile Power Brokers," *The Information,* February 10, 2014, https://www.theinformation.com/articles/Inside-the-Tower-Samsung-s-Mobile-Power-Brokers.

24   in a building called: Cheng and McKinnon, "Fatal Mistake That Doomed Samsung's Galaxy Note."

24   "It wasn't a definitive answer": Cheng and McKinnon, "Fatal Mistake That Doomed Samsung's Galaxy Note."

25   "By putting our top priority": Georgina Mitchell, "Samsung to Recall Galaxy Note 7 Worldwide After Battery Problems," *The Sydney Morning Herald,* September 2, 2016, https://www.smh.com.au/business/samsung-

to-recall-galaxy-note-7-worldwide-after-battery-problems-20160902-gr7v4w.html.

25   Samsung was set to recall : Samuel Gibbs and Alan Yuhas, "Samsung Suspends Sales of Galaxy Note 7 After Smartphones Catch Fire," *The Guardian*, September 2, 2016, https://www.theguardian.com/technology/2016/sep/02/samsung-recall-galaxy-note-7-reports-of-smartphones-catching-fire.

25   Samsung dismissed those reports as fake: Cheng and McKinnon, "Fatal Mistake That Doomed Samsung's Galaxy Note."

25   "It wasn't a recall at all": Matt Novak, "How One Outrageous Law Turned the Samsung Phone Recall into a Disaster," *Gizmodo*, October 11, 2016, https://gizmodo.com/how-one-outrageous-law-turned-the-samsung-phone-recall-1787659621.

25   "Without guidance from the CPSC": Novak, "How One Outrageous Law."

25   "In my mind": Christina Warren, "Samsung Finally Launches Official Recall of the Exploding Note 7," *Gizmodo*, September 15, 2016, https://gizmodo.com/feds-to-launch-official-recall-of-the-exploding-samsung-1786681573.

25   announcing its cooperation with the agency: "Samsung Confirms Engagement with Consumer Product Safety Commission in Response to Note7 Battery Issue," Samsung Newsroom, September 9, 2016, https://news.samsung.com/us/samsung-confirms-engagement-with-cpsc-consumer-product-safety-commission-in-response-to-note7-battery-issue/.

25   A Galaxy Note 7 that was being: Josh Cascio, "St. Pete Family Says Jeep Totaled by Exploding Note 7," *FOX 13 News*, September 7, 2016, http://www.fox13news.com/news/local-news/st-pete-familys-jeep-totaled-by-exploding-note-7.

26   "the power of communication": Park Soon-chan, "Samsung's Strength in Communication That Elicited Recall," *Chosun Biz*, September 5, 2016, http://biz.chosun.com/site/data/html_dir/2016/09/05/2016090500302.html. This source is in Korean. The author's researcher translated the title and the quoted text into English.

26   "When checking the information": "Second Report of Expert Meeting Regarding Galaxy Note 7," Korean Agency for Technology and Standards, September 19, 2016. Document is in the author's possession. This source is in Korean. The author's researcher translated the title and the quoted text into English.

26   "They're buying time": Park Chul Wan, interview by the author, November 28, 2016.

27   "In 2007 I was looking into": Park Chul Wan, interview by the author, November 28, 2016.

27   His body was found: Kim Ju-cheol, "<Incident> Mobile Phone Battery Thought to Have Exploded in Cheongwon, Fatal Accident Occurs,"

Chungju Shinmun, November 27, 2007, http://cjwn.com/sub_read.html ?uid=5950&section=&section2=. This source is in Korean. The author's researcher translated the title and the quoted text into English.

27   Another machinery operator admitted: Koo Jun-hoe, "Mobile Phone Battery Explosion Fatality? Turns Out to Be a Case of Accidental Manslaughter," *SBS,* November 30, 2007, https://news.sbs.co.kr/news/end Page.do?news_id=N1000343794. This source is in Korean. The author's researcher translated the title and the quoted text into English.

27   penning a popular column: Park Chul Wan, "Has the Gate of Lithium Ion Battery–Related Accidents Reopened?" *IT Chosun,* September 9, 2016. This source is in Korean. The author's researcher translated the title and the quoted text into English. Park Chul Wan published many opinion articles in the fall of 2016 stating his hypothesis on the Galaxy Note 7 fires. This is merely one example of such an article.

27   "A lot of Korean battery experts": Park Chul Wan, interview by the author, November 30, 2016. YouTube has many videos of lithium-ion-battery short circuits that Park Chul Wan described, resulting in a quick blast and a short, intense fire. One such lab experiment is available at https://www.youtube .com/watch?v=HCGtRgBUHX8.

28   "We're also observing Professor": Screenshot of KakaoTalk message from Lee Soo-hyung, in the author's possession.

28   "According to an acquaintance": Screenshot of KakaoTalk message from Park Chul Wan's colleague, in the author's possession.

28   "Now it is the time to act": Nina Criscuolo, "iPhone 7 Arrives Amid Samsung Galaxy Note 7 Recall," *WISH-TV,* September 16, 2016, https://www .wishtv.com/news/world/iphone-7-launch-unluckily-times-with-samsung-galaxy-note-7-recall_20180321053744804/1064305112.

29   "I knew Samsung hadn't solved it": Park Chul Wan, interview by the author, December 1, 2016.

29   "The executives told Mr. Lee": Cheng and McKinnon, "Fatal Mistake That Doomed Samsung's Galaxy Note."

29   "one of the toughest challenges": Cheng and McKinnon, "Fatal Mistake That Doomed Samsung's Galaxy Note."

30   "When the family spoke": Samsung executives, interviews by the author, January 2011–February 2018.

30   Four months later, on February 16: Kim Jeong-pil and Seo Young-ji, "Samsung Vice Chairman Lee Jae-yong Arrested on Charges of Bribing Pres. Park," *Hankyoreh,* February 17, 2017.

30   He'd suffered a heart attack: Sam Byford, "Samsung Chairman's Heart Attack Raises Questions About Son's Succession," *The Verge,* May 12, 2014, https://www.theverge.com/2014/5/12/5708884/lee-kun-hee-suffers-apparent-heart-attack.

2: SHADOW OF EMPIRE

33   "I had returned home late": Lee Byung-chul, *Autobiography of Hoam* (Paju, South Korea: Nanam, 2014), pp. 44–45. "Hoam" (pronounced Ho-Ahm) was B.C. Lee's pen name that he used in essays and articles. This source is in Korean. The author's researcher translated the title and the quoted text into English.

33   "Koreans have no solidarity": Lee, *Autobiography of Hoam*, p. 51.

33   he took time off: Ibid., p. 65.

33   he opened a vegetable: Ibid., pp. 65–66.

34   "In placing my hopes": Ibid., p. 66.

34   B.C. was probably eyeing the success: Anthony Michell, *Samsung Electronics and the Struggle for Leadership of the Electronics Industry* (Singapore: John Wiley & Sons [Asia], 2011), pp. 18–19.

34   The *zaibatsu* were some of: "Japan's *zaibatsu*: Yes, General," *The Economist*, December 23, 1999, https://www.economist.com/business/1999/12/23/yes-general.

35   Admired for their immense wealth: Hidemasa Morikawa, *Zaibatsu: The Rise and Fall of Family Enterprise Groups in Japan* (Tokyo: University of Tokyo Press, 1992).

35   "If this concentration of economic power": Eleanor M. Hadley, *Antitrust in Japan* (Princeton, NJ: Princeton University Press, 1970), p. 142.

35   "We met every week": Lee, *Autobiography of Hoam*, p. 79.

35   In 1947 he relocated: Ibid., p. 88.

35   epiphany while getting a haircut: Ibid., pp. 91–93.

36   South Korea followed with: Michell, *Samsung Electronics and the Struggle for Leadership*, pp. 47–49.

36   "Past noon and until evening": *Autobiography of Hoam*, p. 96.

36   Four days later, party officials: Ibid., p. 97.

36   They were brought before "People's Courts": "JTBC DocuShow Korean War Special: Memories of Seoul's Three Month-Long Occupation by the North Korean Army," *JoongAng Ilbo*, June 20, 2014, http://news.joins.com/article/15020744.

37   "I recognized that it was my car": Lee, *Autobiography of Hoam*, pp. 95–97.

37   "I loaded up company employees": Ibid., p. 103.

37   "[The Japanese] steadfastly valued loyalty": Ibid., pp. 167–68.

37   He personally sat in on: Mark Clifford, *Troubled Tiger: Businessmen, Bureaucrats, and Generals in South Korea* (New York: Routledge, 1998), p. 321.

37   100,000 in his tenure: Andrei Lankov, "Lee Byung-chull: Founder of Sam-

sung Group," *The Korea Times,* October 12, 2011, http://www.koreatimes
.co.kr/www/news/issues/2014/01/363_96557.html.

37   He hired professional physiognomists: Henry Cho, grandson of B.C. Lee,
     text message to the author, October 23, 2017.

37   "Be prudent in hiring someone": Lee, *Autobiography of Hoam,* p. 67.

38   A "Samsung Man" commanded respect: Sergey Konovalov, "Corporate
     Values in Korea: A Descriptive Study of 'Samsung Man' Phenomenon"
     (MBA thesis, Korea Development Institute, 2006).

38   "Samsung treats you the best": Kim Chun-hyo, *Samsung, Media Empire and
     Family: A Power Web* (New York: Routledge, 2016), p. 38.

38   "HR actually is the leading": Former Samsung human resources executive,
     interview by the author, December 23, 2015.

38   "They said *nunchi*": Scott Seungkyu Yoon, former Samsung employee, in-
     terview by the author, December 2, 2015.

39   More than 36,000 American servicemen: Steve Vogel, "Death Miscount
     Etched into History," *The Washington Post,* June 25, 2000, https://www
     .washingtonpost.com/archive/local/2000/06/25/death-miscount-etched-
     into-history/ab9d6830-b10d-429c-a3b0-cbdbaa3a23d1/?utm_term=.
     eca70b035d82.

39   Its GDP was about the size: "The Next Big Bet," *The Economist,* October 1,
     2011, https://www.economist.com/briefing/2011/10/01/the-next-big-bet.

39   "The Koreans . . . are the cruellest": Ian Fleming, *Goldfinger* (London: Jona-
     than Cape, 1959), p. 81.

39   B.C. built his first fortune: Sea-jin Chang, *Financial Crisis and the Transfor-
     mation of Korean Business Groups: The Rise and Fall of Chaebols* (Cambridge:
     Cambridge University Press, 2003), p. 47.

40   In the 1950s, people began: "Collusion Between Rising Chaebol and the
     Political Force," *Kyunghyang Shinmun,* August 10, 1958, https://newslibrary
     .naver.com/viewer/index.nhn?articleId=1958081000329202001&editNo=
     2&printCount=1&publishDate=1958-08-10&officeId=00032&pageNo=
     2&printNo=4061&publishType=00020. This source is in Korean. The au-
     thor's researcher translated the title and the quoted text into English.

40   As his wealth grew: Lee, *Autobiography of Hoam,* pp. 113–20.

40   By the end of the decade: Lankov, "Lee Byung-chull."

40   the man with the "golden touch": "South Korea's $500 Million Man,"
     Time, July 19, 1976, http://content.time.com/time/magazine/article
     /0,9171,914344,00.html.

40   "Mr. All-Wool": Lee, *Autobiography of Hoam,* p. 94.

40   Henry was a grandson: Paik Sul-hee and Na Byeong-hyun, "Who Is Lee
     In-hee, Advisor of Hansol?" *Business Post,* August 29, 2016. http://www
     .businesspost.co.kr/BP?command=naver&num=33051. This source is in

Korean. The author's researcher translated the headline and text into English.

40   "The family, they protect each other": Tom Casey, interview by the author, August 6, 2015.

41   his family's crest of honor: Henry Cho, interviews by the author, January 11, 2016, and March 25, 2016.

41   "At meals he would tear": Henry Cho, email message to the author, May 25, 2019.

41   "He was so cool": Henry Cho, interview by the author, March 25, 2016.

41   B.C. loved golf: Han Eun-koo. "Chairman Lee Byung-chul's Strong Rival Mentality," March 2008, *Korea Economic Daily Magazine,* http://magazine .hankyung.com/money/apps/news?popup=0&nid=02&c1=2007&nkey =2008022800034000532&mode=sub_view.

41   and Chinese calligraphy: Yi Hwang-hoe, "Songcheon Chung Ha-geon: Lee Byung-chull Was the Kind of Person Who Didn't Put Off His Calligraphy Even on the Morning of His Birthday," *Maeil Business Newspaper.* October 31, 2014, https://www.mk.co.kr/news/culture/view/2014/10/1376161/. This source is in Korean. The author's researcher translated the title and the quoted text into English.

42   "As far as politics was concerned": Henry Cho, interview by the author, March 25, 2016.

42   "B.C. arrives at his downtown Seoul office": "South Korea's $500 Million Man."

42   with his second wife, Kurata Michiko: So Jong-seob, "Samsung's Lee Byung-chul Family Opens the Era of Chaebol Intermarriages with LG founder Koo In-hoe," *Sisa Journal,* November 6, 2014, http://www.sisa journal.com/news/articleView.html?idxno=140341&replyAll=&reply _sc_order_by=I. Kurata Michiko's name is recalled by Henry Cho and in other historical Korean news articles on Samsung. This source is in Korean. The author's researcher translated the title and the quoted text into English.

42   "Critics . . . called us puppies": Daniel Lee, *Son of the Phoenix: One's* [sic] *Man's Story of Korea* (Seoul: Voice Publishing House, 2008), p. 142.

42   "They treated me like a god": Henry Cho, interview by the author, March 25, 2016.

3: DYNASTY ASCENDANT

43   "I think they [Samsung] are mimicking": Jisoo Lee, interview by the author, October 21, 2015.

43   B.C. had five daughters: Kim Chun-hyo, *Samsung, Media Empire and Family: A Power Web* (New York: Routledge, 2016), pp. 40–43.

44 Hong Jin-ki was a colorful lawyer: Myeong Jin-kyu, *Young Lee Kun-hee* (Bucheon, South Korea: Fandom Books, 2013), p. 257. This source is in Korean. The author's researcher translated the title and the quoted text into English.

44 In 1965, B.C. founded Samsung's newspaper: Kim Eun-hyang, "Who Is Hong Ra-hee? 'Mistress of the Samsung Household, Daughter of Former *JoongAng Ilbo* Chairman Hong Jin-ki,'" *Dong-A Ilbo,* March 6, 2017, http://news.donga.com/3/all/20170306/83185072/2. This source is in Korean. The author's researcher translated the title and the quoted text into English.

44 modeled on Japan's daily papers: Kim, *Samsung, Media Empire and Family,* p. 43.

44 "Mass communications are the best way": "B.C. Lee's World," *Time,* April 1967, http://content.time.com/time/magazine/article/0,9171,843698,00.html.

44 "He was so good at holding": Henry Cho, interview by the author, March 25, 2016.

44 "We shall rise up": *History of the Korean Military Revolution,* vol. 1 (Seoul: Compilation Committee for Korean Military Revolution History and the Supreme Council for National Reconstruction, 1963), p. 199.

44 "That day . . . around 7:00 A.M.": Lee, *Autobiography of Hoam,* p. 176.

45 "a fresh summer rain was pouring": Ibid., p. 180.

45 "Passing the secretary's office": Ibid., pp. 182–83.

46 "You can say anything": Ibid., p. 183.

46 $4,400,000 in unpaid taxes: "B.C. Lee's World."

46 an unconventional thinker: Henry Stokes, "He Ran South Korea, Down to Last Detail," *The New York Times,* October 27, 1979, https://www.nytimes.com/1979/10/27/archives/he-ran-south-korea-down-to-last-detail-he-ran-south-korea-down-to.html.

46 A good deal of Park's worldview: Carter J. Eckert, *Park Chung-hee and Modern Korea* (Cambridge, MA: Harvard University Press, 2016), pp. 180–88.

46 bowed before a picture of Hitler: Eckert, *Park Chung-hee and Modern Korea,* p. 102.

46 "Their old, dated ways": Kim Chung-yum, *From Despair to Hope: Economic Policymaking in Korea 1945–1979* (Seoul: Korea Development Institute, 2011), p. 580, https://www.kdi.re.kr/kdi_eng/publications/publication_view.jsp?pub_no=11820.

46 "What is urgently required": Park Chung-hee, "Speech at the Cornerstone-Laying Ceremony of Gimhae District Reclamation Work," June 1, 1965, excerpted from *Selected Speeches of President Park Chung-hee: January 1965–December 1965* (Seoul: Office of the President 1966), Presidential Ar-

chives, National Archives of Korea, file number B000080900000011, http://
dams.pa.go.kr:8888/dams/ezpdf/ezPdfReader.jsp?itemID=/DOCU
MENT/2009/11/26/DOC/SRC/01042009112641568000041568013535
.PDF. A non-PDF web text is available in the archive's catalog of speeches
at http://pa.go.kr/research/contents/speech/index.jsp. This source is in
Korean. The author's researcher translated the title and the quoted text
into English.

46   Korea retained parts: Stokes, "He Ran South Korea."

47   a pure and unadulterated bloodline: Henry H. Em, "*Minjok* as a Construct,"
      in *Colonial Modernity in Korea,* ed. Gi-wook Shin and Michael Edson Robin-
      son (Cambridge, MA: Harvard University Asia Center, 2000), pp. 339–61.

47   The regime seized private banks: Lee Byung-chun, *Economic Development
      Based on Dictatorship and the Park Chung-hee Era* (Paju, South Korea: Changbi,
      2003), p. 111. This source is in Korean. The author's researcher translated
      the title and the quoted text into English.

47   His regime set tough export quotas: "President Park to Continue
      'Export First' Policy Next Year," *Kyunghyang Shinmun,* December 24, 1975,
      https://newslibrary.naver.com/viewer/index.nhn?articleId=
      1975122400329201003. This source is in Korean. The author's researcher
      translated the title and the quoted text into English.

47   the torrent of foreign money: Lee, *Economic Development Based on Dictator-
      ship,* p. 302.

47   Japanese payments totaling $800 million: Chung Dae-ha, "Companies That
      Used Claimed Money from Japan Are Turning Away from Victims of Con-
      script," *Hankyoreh,* May 30, 2012. This source is in Korean. The author's
      researcher translated the title and the quoted text into English.

47   "This is a steel factory": Chung, "Companies That Used Claimed Money."

47   hurling an ashtray: Clifford, *Troubled Tiger,* p. 15.

47   Federation of Korean Industries: Eun Mee Kim and Gil-sung Park, "The
      Chaebol," in *The Park Chung-hee Era: The Transformation of South Korea,* ed.
      Byung-kook Kim and Ezra F. Vogel (Cambridge, MA: Harvard University
      Press, 2013), pp. 274–76.

48   "The government will offer": Lee, *Autobiography of Hoam,* p. 228.

48   "In the coastal city of Ulsan": "B.C. Lee's World."

48   "Eat this saccharin!": Ibid.

48   Lee Chang-hee, B.C.'s second son: "A Second Birth: Samsung Chaebol
      (6) The Young Blood Five," *Maeil Business Newspaper,* July 17, 1968,
      https://newslibrary.naver.com/viewer/index.nhn?articleId=
      1968071700099201007. This source is in Korean. The author's researcher
      translated the title and the quoted text into English.

48   was a "troublemaker": Henry Cho, interview by the author, March 25,
      2016. More stories about Lee Maeng-hee's "troublemaker" image can be

found in Lee Yong-woo, *Lee Maeng-hee, Prince Sado of Samsung* (Seoul: Pyungminsa, 2012). In one instance, Lee Maeng-hee shot a Browning hunting rifle in the air in front of a group of Samsung executives. They ran off.

48 Maeng-hee later admitted: Yoizi Ishigawa, *Lee Byung-chul and the Samsung Empire* (Seoul: Dolsaem, 1988), pp. 64–65. This source is in Korean. The author's researcher translated the title and the quoted text into English.

49 "But his father refused": Henry Cho, interview by the author, March 25, 2016. More information on the succession can be found in Korean in "The Difficult Task of Succession," *Maeil Business Newspaper,* July 29, 1969, https://newslibrary.naver.com/viewer/index.nhn?articleId=1969072900099201009.

49 So Chang-hee sent an anonymous tip: Kim Jin-cheol, "Anonymous Tip to the Blue House, LG Marital Conflicts . . . 40 Years of Rock-Bottom Samsung Drama," *Hankyoreh*, April 24, 2012, http://www.hani.co.kr/arti/economy/economy_general/529764.html. This source is in Korean. The author's researcher translated the title and the quoted text into English. Many of these details became clearer during testimony in three inheritance lawsuits, filed by Lee Maeng-hee and other siblings against Samsung chairman Lee Kun-hee in 2012.

49 "In the future": Lee Maeng-hee, *Stories I Have Kept* (Seoul: Chungsan, 1993), p. 284. This source is in Korean. The author's researcher translated the title and the quoted text into English.

49 "I cannot forget the sudden shock": Lee, *Stories I Have Kept,* p. 284.

49 Walking with a slight limp: Yoizi Ishigawa, *Lee Byung-chul and the Samsung Empire,* pp. 50–52.

49 "Actually, I was the one": Yoizi Ishigawa, *Lee Byung-chul and the Samsung Empire,* p. 67.

49 He was convinced: Ibid., pp. 101–102.

49 An unofficial South Korean biographer: Lee Yong-woo, *Lee Maeng-hee, Prince Sado of Samsung* (Seoul: Pyungminsa, 2012). This source is in Korean. The author's researcher translated the title and the quoted text into English.

4: MARCH OF THE SAMSUNG MEN

51 In 1974 Samsung CEO: Kang Jin-ku, *The Samsung Electronics Myth and Its Secret* (Seoul: Goryeowon, 1996), pp. 183–84. This source is in Korean. The author's researcher translated the title and the quoted text into English.

51 With the OPEC oil embargo under way: Lee Kun-hee, *Lee Kun-hee Essays: Looking at the World Critically* (Seoul: Dongailbosa, 1997), p. 14. This source is in Korean. The author's researcher translated the title and the quoted text into English.

51   "I had been thinking that": Kang Jin-ku, *The Samsung Electronics Myth and Its Secret*, p. 183.

52   Samsung managers warned against: Michell, *Samsung Electronics and the Struggle for Leadership*, pp. 19–20.

52   "The company made semiconductors": Lee, *Lee Kun-hee Essays*, p. 15.

52   "Everything appears well-organized": U.S. Embassy Seoul "OPIC Finance Projects" (telegram), July 24, 1975, Wikileaks canonical ID 1975SEOUL5571_b, https://wikileaks.org/plusd/cables/1975SEOUL05571_b.html.

52   agreed to finance part: Kang, *The Samsung Electronics Myth and Its Secret*, p. 195.

52   seized B.C.'s broadcasting station: "Closure of TBC," *Dong-A Ilbo*, November 29, 1980, https://newslibrary.naver.com/viewer/index.nhn?articleId=1980112900209202001&editNo=2&printCount=1&publishDate=1980-11-29&officeId=00020&pageNo=2&printNo=18202&publishType=00020. This source is in Korean. The author's researcher translated the title and the quoted text into English.

52   lobbied successfully through: Yoizi Ishigawa, *Lee Byung-chul and the Samsung Empire* (Seoul: Dolsaem, 1988), p. 1. This source is in Korean. The author's researcher translated the title and the quoted text into English.

53   "We are too late": Myeong Jin-kyu, *Young Lee Kun-hee* (Bucheon, South Korea: Fandom Books, 2013), p. 272. This source is in Korean. The author's researcher translated the title and the quoted text into English.

53   under increasing pressure from Hyundai: Michell, *Samsung Electronics and the Struggle for Leadership*, p. 53.

53   his son tried to persuade him: Myeong, *Young Lee Kun-hee*, p. 273.

53   "Our nation has a large population": Kwon Se-jin, "Samsung, the Company That Found Its Driving Force in Era-Appropriate Technological Development," *Premium Chosun*, April 1, 2016. The article repeats verbatim the original Tokyo declaration, published by *JoongAng Ilbo* on March 15, 1983. http://premium.chosun.com/site/data/html_dir/2016/03/29/2016032901912.html. More information can be found in Woo Eun-sik, "1983 Tokyo Declaration . . . Writing the Semiconductor Legend," *Newsis*, June 30, 2013, http://www.newsis.com/view/?id=NISX20130629_0012194301. These two sources are in Korean. The author's researcher translated the titles and the quoted text into English.

53   "The people I'm going to call": Nam-yoon Kim, interview by the author, September 30, 2015.

54   Samsung christened its first: "Samsung Breaking Ground of Semiconductor Factory in 100,000-pyeong Ground at Kiheung," *Kyunghang Sinumun*, September 12, 1983, https://newslibrary.naver.com/viewer/index.nhn?articleId=1983091200329205015&editNo=2&printCount=1&publishDate=1983-09-12&officeId=00032&pageNo=5&printNo=11679&publishType=

00020. "Pyeong" is a Korean unit of floorspace, equivalent to 35.5833 square feet. This source is in Korean. The author's researcher translated the title and the quoted text into English.

54 "A visitor to the Samsung plant": Clifford, *Troubled Tiger,* p. 315.

55 "The thought of going bankrupt": Arirang News, "Interview with Lee Yoon-woo from Samsung Electronics [Korea Today]," posted by YouTube user Arirang Issue on March 19, 2012, https://www.youtube.com/watch?v=qkN36Ne61UE.

55 The Samsung engineers were granted: Nam-yoon Kim, phone interview with the author's researcher Max Soeun Kim, March 24, 2017. The Micron operation has been written about in many Korean newspapers. More detail is available at Shin Dong-jin, "For 34 Years, Samsung's Giheung Plant Is Moved by the Chipmaker's Credo," *DongA Ilbo,* Jan 11, 2017, http://www.donga.com/news/article/all/20170110/82313144/1#csidx30dec045219d fd395e94d6895c0b800. This source is in Korean. The author's researcher translated the title and the text into English.

55 "Having had a lot of experience": Lee, *Lee Kun-hee Essays,* p. 16.

55 "The Japanese brought this precision": Nam-yoon Kim, interview by the author, September 30, 2015.

56 "He predicted that when he died": Henry Cho, interview by the author, March 25, 2016.

56 A Samsung advertisement: Magazine advertisement in the author's possession.

5: THE CONFUCIAN AND THE HIPPIE

57 "Samsung's research lab . . . reminded me": Ira Magaziner and Mark Patinkin, *The Silent War: Inside the Global Business Battles Shaping America's Future* (New York: Random House, 1989), pp. 22–24.

58 "as automated as any TV plant": Magaziner and Patinkin, *Silent War,* pp. 30–35.

58 there would be no American-owned: Barnaby J. Feder, "Last U.S. TV Maker Will Sell Control to Koreans," *The New York Times,* July 18, 1995, https://www.nytimes.com/1995/07/18/us/last-us-tv-maker-will-sell-control-to-koreans.html.

59 "Steve knew the future": Jay Elliot, interview by the author, January 9, 2014.

59 "a personal computer for children": Alan C. Kay, "A Personal Computer for Children of All Ages," Xerox Palo Alto Research Center, 1972, http://www.vpri.org/pdf/hc_pers_comp_for_children.pdf.

59 But that didn't deter Jobs: Alan Kay, "American Computer Pioneer Alan Kay's Concept, the Dynabook, Was Published in 1972. How Come Steve

Jobs and Apple iPad Get the Credit for Tablet Invention?" Quora, April 21, 2019, https://www.quora.com/American-computer-pioneer-Alan-Kay-s-concept-the-Dynabook-was-published-in-1972-How-come-Steve-Jobs-and-Apple-iPad-get-the-credit-for-tablet-invention/answer/Alan-Kay-11.

59   would need to be portable: Jay Elliot, interview by the author, January 9, 2014.

59   Jobs disembarked at the grimy: Jay Elliot, interview by the author, January 9, 2014.

60   Samsung began supplying Apple: Frank Rose, *West of Eden: The End of In-nocence at Apple Computer* (New York: Stuyvesant Street Press, 1989), p. 163.

60   "Steve was boasting": Jay Elliot, interview by the author, January 9, 2014.

60   "Jobs is the figure": "Jobs, Samsung Electronics Had 30-Year Love-Hate Re-lationship," *Dong-A Ilbo*, October 7, 2011, http://www.donga.com/English/List/Article/all/20111007/402284/1/Jobs-Samsung-Electronics-had-30-year-love-hate-relationship.

61   "iPhone killer": John M. Glionna and Jung-yoon Choi, "iPhone Is Invading South Korea, Home of Samsung's Galaxy S," *Los Angeles Times,* January 22, 2011, https://www.latimes.com/business/la-xpm-2011-jan-22-la-fi-iphone-korea-20110122-story.html.

61   "Samsung's Use of Apple Patents in Smartphones": Kurt Eichenwald, "The Great Smartphone War," *Vanity Fair,* June 2014, https://www.vanityfair.com/news/business/2014/06/apple-samsung-smartphone-patent-war.

6: THE FIFTH HORSEMAN

62   "Not only is it bigger": M. G. Siegler, "The Fifth Horseman: Samsung," *TechCrunch,* January 5, 2013, https://techcrunch.com/2013/01/05/the-fifth-horsemen-of-tech-samsung/.

63   With the company spending: "Samsung Plans to Outspend Iceland's GDP on Advertising and Marketing," *The Guardian,* November 28, 2013, https://www.theguardian.com/technology/2013/nov/28/samsung-plans-to-outspend-icelands-gdp-on-advertising-and-marketing.

63   Before long, Samsung would: Kenneth Rapoza, "Samsung Now Accounts for One in Three Smartphone Sales," *Forbes,* January 28, 2014, https://www.forbes.com/sites/kenrapoza/2014/01/28/samsung-now-accounts-for-one-in-three-smartphone-sales/#67bb23c938e7.

63   "The founder had a vision": Young-joon Gil, SAIT senior vice president, interview by the author, November 16, 2010.

63   "He wanted to make another miracle": Gordon Kim, Samsung human re-sources director, interview by the author, November 15, 2010.

64   "will not be possible": Executive vice president at Weber Shandwick, email message to the author, October 26, 2010.

64 When the chairman visited: Jungah Lee and Jason Clenfield, "Samsung Low-Profile Heir Poised to Succeed Father Seen as a God," *Bloomberg Technology,* August 26, 2014, https://www.bloomberg.com/news/articles/2014-08-26/samsung-low-profile-heir-poised-to-succeed-father-seen-as-god-.

64 "One employee was charged": Sumi Lim, former Samsung senior business development manager, email message to the author, December 25, 2015.

64 Samsung booked the entire fourth: Jang Jae-hyung, "I Worked Part Time for the Lee Kun-hee Reception Team," *OhMyNews,* January 16, 2005, http://www.ohmynews.com/NWS_Web/View/at_pg.aspx?CNTN_CD=A0000232206. This source is in Korean. The author's researcher translated the title and the quoted text into English.

65 belief in a shared, ancient bloodline: B. R. Myers, *The Cleanest Race: How North Koreans See Themselves—and Why It Matters* (New York: Melville House, 2010).

65 South Koreans, he argues: B. R. Myers, "South Korea's Collective Shrug," *The New York Times,* May 27, 2010, https://www.nytimes.com/2010/05/28/opinion/28myers.html.

65 "the second-most-nationalist country": "The Cleanest Race," C-SPAN, February 11, 2010, https://www.c-span.org/video/?292562-1/the-cleanest-race.

65 a punching fist and a formation of the word "victory": Numerous internal Samsung videos of the festivals from the 2000s and early 2010s, in the author's possession. The festivals have been written about in "Leaked Video Offers Window into Samsung Culture," *Hankyoreh,* June 21, 2007, http://english.hani.co.kr/arti/english_edition/e_national/217378.html. The *Hankyoreh* article describes two of these videos.

67 "South Koreans cannot seem": Choe Sang-hun, "Samsung Heirs Stage a Korean Soap Opera," *The New York Times,* April 24, 2012, https://www.nytimes.com/2012/04/25/business/global/samsung-heirs-stage-a-korean-soap-opera.html.

67 was "very apt": Bruce Cumings, email message to the author, April 21, 2016.

68 "Victorious fighting spirit!": Samsung internal video in the author's possession.

68 They'd been through boot camps: Kim Mi-young, "I Escaped Samsung, and Then I Was Happy," *Hankyoreh,* December 6, 2013, http://www.hani.co.kr/arti/economy/economy_general/614281.html. More information on Samsung's boot camp and hikes are available in an employee's blog post at Kim Il-kown, "My Samsung Chronicle: Physical Training Is the Heart of Newcomer Orientation," *Camera4u* (blog name) at blog website Tistory, December 8, 2013, https://camera4u.tistory.com/465. These sources are in Korean. The author's researcher translated the titles and the text into English. The author is in possession of more documents, a slide presenta-

tion, employee blog posts, and a case study on Samsung's training practices.

68   A trumpet blared: Samsung internal video in the author's possession.

69   "It was amazing, scary, and weird": Samsung manager, interview by the author, November 24, 2016.

70   "That's offensive": Samsung public relations vice president, in conversation with the author, April 20, 2016.

70   All have been pardoned: No single article or book mentions the full number of arrests, convictions, or pardons. The author keeps a tally of *chaebol* leaders convicted and pardoned since the 1990s.

70   From January 2015 to February 2016: Samsung Electronics, "Proxy Material, 2016 Annual General Meeting of Shareholders." Document is in the author's possession.

71   "According to crisis simulations": Sangin Park, "Can South Korea Survive Without Samsung Electronics?" *Korea Exposé*, October 28, 2016, https://www.koreaexpose.com/can-south-korea-survive-without-samsung-electronics/.

72   "Do you plan to criticize Samsung?" Samsung employee, interview by the author, March 14, 2015.

72   "I hope you aren't annoyed": Nam Ki-young, email message to the author, September 15, 2015.

72   In our subsequent meetings: Samsung public relations executive vice president, in conversation with the author, December 31, 2015.

7: THE SCION

73   "If God loves humans": Bae Myung-bok, "Has Lee Byung-chul Found the Answer?" *Korea JoongAng Daily,* December 20, 2011, http://koreajoongang daily.joins.com/news/article/article.aspx?aid=2945807.

73   He died of lung cancer: Seoul PLAN Department, *Lee Kun-hee New Management Philosophy: Give Up If It's Not World-Class* (Seoul: Podowon, 1993), p. 40. This source is in Korean. The author's researcher translated the title and the quoted text into English.

73   Twenty-five minutes after his death: Kang Jun-man, *Lee Kun-hee Era* (Seoul: Inmul and Sasang, 2005), p. 32. This source is in Korean. The author's researcher translated the title and the quoted text into English.

73   They unanimously voted in Lee: "Samsung Lee Kun-hee Inaugurated as Chairman," *JoongAng Ilbo,* December 1, 1987. This source is in Korean. The author's researcher translated the title and the quoted text into English.

74   purges of potential rivals: Henry Cho, interview by the author, March 25, 2016.

75   a video camera was set up: "Samsung Camera Surveillance Scandal," *Hankyoreh*, April 1, 1995, https://newslibrary.naver.com/viewer/index.nhn?articleId=1995040100289109003&editNo=5&printCount=1&publishDate=1995-04-01&officeId=00028&pageNo=9&printNo=2192&publishType=00010. This source is in Korean. The author's researcher translated the title and the quoted text into English.

75   claimed to police that it had footage: Cho Seulgi-na, "[Photos] Samsung Cars Trailing CJ Chairman Lee Jae-hyun," *Asia Business Daily*, February 23, 2012, https://www.asiae.co.kr/article/2012022309051673366. More information on the indictment is available at Song Jin-won, "Samsung Employees Who Tailed CJ Chairman Summarily Indicted," Yonhap News Agency, September 6, 2012, https://www.yna.co.kr/view/AKR20120906167700004.

75   "Second Foundation": "Conglomerate Aims for the Biggest Growth in the New Year," *Dong-A Ilbo*, December 25, 1987, https://newslibrary.naver.com/viewer/index.nhn?articleId=1987122500209204001&editNo=2&printCount=1&publishDate=1987-12-25&officeId=00020&pageNo=4&printNo=20379&publishType=00020. This source is in Korean. The author's researcher translated the title and the quoted text into English.

75   "I am going to found the second establishment": Seoul PLAN Department, *Lee Kun-hee New Management Philosophy*, p. 165.

75   he named his home Seungjiwon: Sung Ki-myeong, "What Is Seungjiwon—Lee Kun-hee's Office?" *No Cut News*, January 14, 2008, https://www.nocutnews.co.kr/news/400102. This source is in Korean. The author's researcher translated the title and the quoted text into English.

75   "emperor's son" and a "lucky heir": Seoul PLAN Department, *Lee Kun-hee New Management Philosophy*, p. 11.

75   "I was feeling bleak": Lee, *Lee Kun-hee Essays*, p. 56. This source is in Korean. The author's researcher translated the title and the quoted text into English.

75   It still left the company: "Early Entry into the Semiconductor Market," *Kyunghyang Shinmun*, February 10, 1988. https://newslibrary.naver.com/viewer/index.nhn?articleId=1988021000329206005&editNo=3&printCount=1&publishDate=1988-02-10&officeId=00032&pageNo=6&printNo=13037&publishType=00020. More information is available at "Victory of the 4 Megabit," *Dong-A Ilbo*, February 10, 1988, https://newslibrary.naver.com/viewer/index.nhn?articleId=1988021000209202002. These sources are in Korean. The author's researcher translated the titles and the quoted text into English.

76   Samsung executives took the first: Laxmi Nakarmi and Robert Neff, "Samsung's Radical Shakeup," *BusinessWeek*, February 28, 1994, https://www.bloomberg.com/news/articles/1994-02-27/samsungs-radical-shakeup.

76   "My nickname is 'the silent one'": Lee, *Lee Kun-hee Essays*, p. 18.

76   wanted to be a film director: Kang, *Lee Kun-hee Era*, p. 120.

76   Chairman Lee was something of a recluse: Kang, *Lee Kun-hee Era,* p. 120. Chairman Lee talks about his love of *Ben-Hur* in *Lee Kun-hee Essays.* More information about why he chose *Ben-Hur* as a favorite—because the film has many elements and complexities that he notices each time he watches it—is in "Lesson of *mokgye,*" *Korea Herald,* June 3, 2003 (no author). The article is in English and is not online. It is in the author's possession.

76   "He digs into one issue": Hwang Young-key, interview by the author, June 5, 2014.

76   "man the artist": Lee, *Lee Kun-hee Essays,* p. 63.

76   "Even in idle conversation": Ibid., p. 63.

76   "He did not strike me as": Ibid., p. 129.

77   sometimes not even in full sentences: Henry Cho, KaKaoTalk message to the author, October 30, 2017. More information on Chairman Lee's eccentric style of speech can be found in "Chairman Lee Kun-hee's 'Quotes' . . . Not Silver-Tongued Like Jobs but a Fastball Style That Gets to the Point," *Dong-A Ilbo,* November 21, 2012, http://www.donga.com/news/article/all/20121120/50990866/1. This source is in Korean. The author's researcher translated the headline and text into English.

77   the first pure-breeding: Lee, *Lee Kun-hee Essays,* pp. 189–90.

77   "After returning to Korea": Ibid., p. 189.

77   more than one high-profile crash: Henry Cho, interview by the author, March 25, 2016. Some of these crashes were reported in Korean newspapers in the 1980s; in the author's possession. Henry Cho said that Chairman Lee "wrecked my mother's car."

77   "Driving at 200 mph": Nakarmi and Neff, "Samsung's Radical Shakeup."

77   "A very interesting man": Henry Cho, interview by the author, March 25, 2016.

77   "Several years ago I walked": Lee, *Lee Kun-hee Essays,* p. 42.

78   The chairman denied tales: "Lee Kun-hee Samsung Group Chairman 'Aiming for the Top, a Time for Sweeping Change,'" *Dong-A Ilbo,* August 4, 1993, https://newslibrary.naver.com/viewer/index.nhn?articleId=1993080400209105001. This source is in Korean. The author's researcher translated the title and the quoted text into English. In this article, the chairman denies being addicted to drugs. Other articles, in the author's possession, have the chairman on record denying other rumors.

78   claimed he had gotten a vasectomy: Kang, *Lee Kun-hee Era.*

78   "He was on drugs": Henry Cho, interview by the author, March 25, 2016. Rumors about Chairman Lee's drug addiction have most recently been reported at Kim Hyun-il, "A Summary of Chaebol Family Fueds Based on Stockpiled X-Files," *Sisa Journal.* August 12, 2015, http://www.sisajournal.com/news/articleView.html?idxno=142177.

78 a drug dealer who set up: Kim Yong-chul, *Thinking of Samsung* (Seoul: Social Commentary, 2010), p. 254. This source is in Korean. The author's researcher translated the title and the quoted text into English.

78 while he proceeds to pay: "Allegations of Prostitution by Samsung's Lee Kun-hee . . . Is the Company Involved?" *Newstapa*, July 21, 2016. English subtitles available at https://news.kcij.org/42.

78 In April 2018, three people: Yeo Hyun-ho, "Group Who Extorted 900 Million Won with 'Lee Kun-hee Prostitution Allegations Video' Given Prison Sentences," *Hankyoreh*, April 12, 2018, http://www.hani.co.kr/arti/society/society_general/840278.html. This source is in Korean. The author's researcher translated the title and the quoted text into English.

79 "I always heard": Kim Kyung-rae, interview by the author, July 8, 2016.

79 "He didn't come really": Hwang Young-key, interview by the author, June 5, 2014. More information on Chairman Lee's quiet early years as Samsung's new leader can be found in Kim Sung-hong and Woo In-ho, "Do They Need a Higher Authority Than Chairman?" *The Korea Herald*, June 3, 2003. "In his early days as Samsung Group chairman," the authors write, "Lee left virtually all decision-makings [*sic*] to the presidents of each company, taking a seemingly aloof attitude concerning details of business operation. [*sic*] It appeared that all he was doing as the leader of a large business group was meeting guests at his home or at the corporate guesthouse Seungjiwon in Hannam-dong, Seoul." The article is in English and is not available online. It is in the author's possession.

79 "Do you know . . . how many": Hwang Young-key, interview by the author, June 5, 2014.

80 "Even in such a situation": Lee, *Lee Kun-hee Essays*, pp. 56–57.

80 According to a booklet Samsung published: Lee Won-bok, *Let's Change Ourselves First: A Comic Book About Samsung's New Management Story* (Seoul: Samsung Economic Research Institute, April 1994). This source is in Korean. The author's researcher translated the title and the quoted text into English.

81 "From the summer of 1992": Lee, *Lee Kun-hee Essays*, p. 57.

8: GLORIOUS CHAIRMAN!

82 "Come and see for yourself": Seoul PLAN Department, *Lee Kun-hee New Management Philosophy*, p. 12.

82 "But they didn't listen": Nakarmi and Neff, "Samsung's Radical Shakeup."

83 "Tonight you should tell me": Song-Hong Kim and In-Ho Woo, "Change Everything You Got, Except for Your Family," *The Korea Herald*, May 31, 2003. The article is in English and is not available online. It is in the author's possession.

83 "The chairman [asked] some": Park Tae-hee and Kim Jung-yoon, "Fukuda

Report Author Recalls Role in Reform of a Giant," *Korea JoongAng Daily*, December 1, 2012, http://koreajoongangdaily.joins.com/news/article/article.aspx?aid=2963241.

83   An expensive piece of testing equipment: Nakarmi and Neff, "Samsung's Radical Shakeup."

83   The Japanese delegation continued: Kim and Woo, "Change Everything You Got."

83   "Automobiles were the most important": Hwang Young-key, interview by the author, June 5, 2014.

84   "The outer shape is important": Park Tae-hee and Kim Jung-yoon, "Fukuda Report Author Recalls Role."

84   "Why is this happening?": Hwang Young-key, interview by the author, June 5, 2014.

84   What he saw appalled him: Nakarmi and Neff, "Samsung's Radical Shakeup."

84   Samsung would replay recordings: Sam Grobart, "How Samsung Became the World's No. 1 Smartphone Maker," *Bloomberg Businessweek*, March 29, 2013, https://www.bloomberg.com/news/articles/2013-03-28/how-samsung-became-the-worlds-no-dot-1-smartphone-maker.

84   Next the livid chairman ordered: Kim and Woo, "Change Everything You Got."

85   On the morning of June 7, 1993: *The New Management* (Samsung Electronics, 2013), internal documentary, 7 minutes 16 seconds. Video is in the author's possession.

85   The chairman wasn't sleeping: Nakarmi and Neff, "Samsung's Radical Shakeup."

85   "I have felt a cold sweat": Office of the Executive Staff of the Samsung Group, *Samsung's New Management: Change Begins with Me* (Seoul: Cheil Communications and Samsung Printing, May 1994), p. 19. The English edition of this book is in the author's possession.

86   "The Cold War has ended": Office of the Executive Staff of the Samsung Group, *Samsung's New Management*, p. 21.

86   "At Samsung, we must adhere": Ibid., pp. 64–65.

86   "Change everything except your wife": "Samsung: Waiting in the Wings," *The Economist*, October 1, 2014, https://www.economist.com/business/2014/10/01/waiting-in-the-wings.

86   "perpetual crisis": Michell, *Samsung Electronics and the Struggle for Leadership*, p. 15.

86   "absolutely unexpected by anybody": Hwang Young-key, interview by the author, June 5, 2014.

86   "expletives and vulgar words": Jung-taek Shim, *Collapse of Samsung* (Seoul:

Penguin Random House Korea, 2015), p. 38. This source is in Korean. The author's researcher translated the title and the quoted text into English.

86   "I'm sorry, sir, but quantity": Sung-Ho Kim and In-Ho Woo, "Do They Need Higher Authority Than Chairman?" *The Korea Herald,* June 3, 2003. The article is in English and is not available online. It is in the author's possession.

87   "After two weeks, we had to": Kim and Woo, "Change Everything You Got." The article is in English and is not available online. It is in the author's possession.

87   "I don't know how he did it": Henry Cho, interview by the author, March 25, 2016.

87   8,500 pages of transcripts: Grobart, "How Samsung Became the World's No. 1 Smartphone Maker."

9: CHURCH OF SAMSUNG

88   "These were not": Hwang Young-key, interview by the author, June 5, 2014.

88   *Change Begins with Me*: Office of the Executive Staff of the Samsung Group, *Samsung's New Management.*

88   "It was kind of like Chairman Mao's": Peter Skarzynski, interview by the author, December 28, 2016.

89   "This is my country": Lee, *Let's Change Ourselves First.*

90   In 1993 Samsung employees: Steve Glain, "Going for Growth: Plans of Samsung Chairman Strike Many as Too Risky," *The Asian Wall Street Journal,* March 3, 1995, http://online.wsj.com/public/resources/documents/Lee.Kun.Hee.biz.pdf.

90   "more like 7:00 A.M. to 10:00 P.M.": Nam-yoon Kim, interview by the author, September 30, 2015.

90   In March of that year: Frank Rose, "Seoul Machine," *Wired,* January 4, 2005, https://www.wired.com/2005/05/samsung/.

90   A few employees at the front: *The New Management* (Samsung Electronics, 2013), internal documentary, 7 minutes 16 seconds. Video is in the author's possession.

91   "covered the pile": Gordon Kim, interview by the author, November 15, 2010.

91   "If you continue to make": Grobart, "How Samsung Became the World's No.1 Smartphone Maker."

91   "as if their babies had died": Kim Seon-jeong, interview by the author, June 30, 2015.

91   a tranche of high-level executives: Kim Seon-jeong, interview by the author, June 30, 2015.

91   "That was the perfect tool": Jisoo Lee, interview by the author, October 21, 2015.

91   The plans got under way: Seo Bo-mi, "Samsung's 'Innovative' Inheritance Technique," *Hankyoreh*, January 5, 2013, http://www.hani.co.kr/arti/english_edition/e_business/568398.html.

91   "We knew that this transaction": Cho Seung-hyeon, interview by the author, March 10, 2015.

92   A South Korean court later ruled: Choe Sang-hun, "Court Upholds Conviction of 2 at Samsung," *The New York Times,* May 29, 2007, https://www.nytimes.com/2007/05/29/business/worldbusiness/29iht-samsung.1.5910969.html.

92   "They were all sort of falling": Kim Seon-jeong, interview by the author, June 30, 2015.

93   forfeiting their right: Seo, "Samsung's 'Innovative' Inheritance Technique."

93   "On June 7th, 1993, a meeting": Office of the Executive Staff of the Samsung Group, *Samsung's New Management,* p. 28.

10: GO WEST, YOUNG HEIR

94   "She is not like Korean men": Andrew Pollack, "Unlikely Credits for a Korean Movie Mogul," *The New York Times,* July 5, 1996, https://www.nytimes.com/1996/07/05/business/international-business-unlikely-credits-for-a-korean-movie-mogul.html.

95   "My lifetime obsession": Yoolim Lee, "Miky Lee Tries to Rise to Challenge at South Korea's CJ Group," *Bloomberg Markets* (syndicated in *The Washington Post*), February 21, 2014, https://www.washingtonpost.com/business/miky-lee-tries-to-rise-to-challenge-at-south-koreas-cj-group/2014/02/21/c91dcba6-98ba-11e3-80ac-63a8ba7f7942_story.html?utm_term=.61f34284096a.

95   Her family used another: Peter Arnell, interview by the author (via Skype), July 4, 2017. More information on Taki's career is in Tomio Taki and Adam Taki, *Zennovation: An East-West Approach to Business Success,* ed. Mortimer Feinberg (Hoboken, NJ: John Wiley & Sons, 2012).

95   "[Tomio] had been at Donna Karan": Peter Skarzynski, interview by the author, December 28, 2016.

95   in a tennis outfit: Peter Skarzynski, interview by the author, December 28, 2016.

95   "I did all the ad work": Peter Arnell, interview by the author (via Skype), July 4, 2017.

95   "New York's worst bosses": Emily Gould, "New York's Worst Bosses: Peter Arnell," *Gawker,* March 15, 2007, https://gawker.com/244608/new-yorks-worst-bosses-peter-arnell.

95   push-ups in front of clients: Daniel Lyons, "The Crazy Genius of Brand
      Guru Peter Arnell," *Newsweek,* March 27, 2009, https://www.newsweek
      .com/crazy-genius-brand-guru-peter-arnell-76137.

95   "But, in his favor": Gould, "New York's Worst Bosses: Peter Arnell."

96   "Not all of his ideas": Peter Skarzynski, interview by the author, December
      28, 2016.

96   "The universe expands exponentially": *BREATHTAKING Design Strategy*
      (New York: Arnell Group, 2008). Report is in the author's possession.

96   "I think that electronics today": M. H. Moore, "Arnell Fashions Look for
      Samsung," *Adweek,* September 18, 1995, http://jhchoistudio.com/arnell
      _group/press_details.html?id=5.

96   "It was before aesthetics": Peter Arnell, interview by the author, July 4,
      2017.

96   "Simply healthy": Moore, "Arnell Fashions Look for Samsung."

97   "It was a clear standout": Thomas Rhee, interview by the author, May 8,
      2017.

97   "People said this is strange": Thomas Rhee, interview by the author, May
      8, 2017.

97   One problem was that Samsung's: Gordon Bruce (Samsung design consul-
      tant, 1994–1998), "Samsung" (PowerPoint presentation), undated. Presen-
      tation is in the author's possession.

97   "Let's put it this way": Richard Linnett, "Samsung's $400 Million Ad
      Challenge," *Advertising Age,* August 28, 2000, adage.com/article/news
      /samsung-s-400-million-ad-challenge/57158/.

97   Sony had done the same: Paul Richter, "Sony to Pay $3.4 Billion for Colum-
      bia Pictures: Japanese Firm Willing to Offer High Price to Get Film, TV
      Software for Video Equipment It Makes," *Los Angeles Times,* September 28,
      1989, https://www.latimes.com/archives/la-xpm-1989-09-28-mn-361-story
      .html. A detailed and excellent account of Sony's acquisition of Columbia
      Pictures is in John Nathan, *Sony: A Private Life* (Boston: Houghton Mifflin,
      1999).

97   In 1995 and 1996, three partners: Evelyn Iritani, "New Name in Lights
      in S. Korea," *Los Angeles Times,* August 19, 1996, https://www.latimes.com
      /archives/la-xpm-1996-08-19-mn-35586-story.html.

97   Miky Lee heard from: Iritani, "New Name in Lights."

98   "The word 'semiconductor'": Richard Corliss, "Hey, Let's Put on a Show!"
      Time, March 27, 1995, http://content.time.com/time/subscriber/article
      /0,33009,982723,00.html.

98   "I kept on saying to them": Pollack, "Unlikely Credits for a Korean Movie
      Mogul."

98   "asking for too much freedom": Iritani, "New Name in Lights in S. Korea."

98   The DreamWorks founders instead went: Pollack, "Unlikely Credits for a Korean Movie Mogul."

98   Instead, Samsung invested a much: "Samsung to Buy a Stake in New Regency Productions," *The New York Times,* February 29, 1996, https://www.nytimes.com/1996/02/29/business/company-news-samsung-to-buy-a-stake-in-new-regency-productions.html.

98   The chairman also created a short-lived: Samsung, "Meeting the Challenge: Samsung Annual Report 1995," p. 82.

98   shares in AST Research fell: James Granelli, "Stock of AST Research Falls 27% After Firm Says It Will Post Loss," *Los Angeles Times,* September 2, 1994, https://www.latimes.com/archives/la-xpm-1994-09-02-fi-34040-story.html.

98   founded in 1980 by three immigrants: David Olmos, "Albert Wong, a Founder of AST, to Resign," *Los Angeles Times,* November 9, 1988, https://www.latimes.com/archives/la-xpm-1988-11-09-fi-64-story.html.

98   "The best technology they have": Dan Sheppard, former AST vice president for global product marketing, interview by author, October 23, 2015.

99   "a committed source of supply": Ross Kerber, "$450-Million Deal to Give Samsung 40% of O.C.'s AST," *Los Angeles Times,* February 28, 1995, https://www.latimes.com/archives/la-xpm-1995-02-28-mn-37076-story.html.

99   "My biggest concern": Dan Sheppard, interview by the author, October 23, 2015. More about Samsung's hands-off strategy can be found at "AST Appoints New President," *CNET,* August 27, 1996, https://www.cnet.com/news/ast-appoints-new-president/.

99   In April 1995, the first stage: "Samsung Sells AST Brand of Computers," *The New York Times,* January 12, 1999, https://www.nytimes.com/1999/01/12/business/international-business-samsung-sells-ast-brand-of-computers.html.

99   "Public companies are dictated": James S. Granelli and P. J. Huffstutter, "Bad News in the E-Mail for AST Workers," *Los Angeles Times,* December 4, 1997, https://www.latimes.com/archives/la-xpm-1997-dec-04-fi-60659-story.html.

99   Still, Samsung followed with a full: Greg Miller, "Samsung Bids $469 Million for Remaining Stake in AST," *Los Angeles Times,* January 31, 1997, https://www.latimes.com/archives/la-xpm-1997-01-31-fi-23904-story.html.

100   "That's when our worst fears": Dan Sheppard, interview by the author, October 23, 2015.

100   "Why don't you do it like this?": Ibid.

100   senior executives were summoned: John O'Dell, "Samsung Names New Chief Exec at AST," *Los Angeles Times,* April 29, 1997, https://www.latimes.com/archives/la-xpm-1997-04-29-fi-53679-story.html.

100   "Who's next?": Dan Sheppard, interview by the author, October 23, 2015.

100   after six years of losses: Granelli and Huffstutter, "Bad News in the E-Mail for AST Workers."

100   Even Korean Samsung employees: Dan Sheppard, interview by the author, October 23, 2015.

101   By the end, Samsung had sunk: "Samsung's Bid for AST," *CNET*, January 30, 1997, https://www.cnet.com/news/samsungs-bid-for-ast/.

101   sell the name and patents: "Packard Bell's Alagem Buys AST Research," *Los Angeles Times*, January 11, 1999, https://www.latimes.com/archives/la-xpm-1999-jan-11-fi-62428-story.html.

101   Moody's gave it Baa2: Steve Glain, "Going for Growth."

11: SEOUL SEARCHING

102   "Ever heard of Samsung?": Gordon Bruce, interview by the author, December 31, 2015.

102   renowned school of Bauhaus designers: Gordon Bruce, *Eliot Noyes: A Pioneer of Design and Architecture in the Age of American Modernism* (London: Phaidon Press, 2007). More information on Bruce is available at his personal website, http://www.gbrucedesign.com/.

103   "Seventeen executives and designers": Gordon Bruce, interview by the author, December 31, 2015.

103   renowned for its curriculum: "Meeting the Challenge: Samsung Annual Report 1995," Samsung, p. 79.

103   "Chairman Lee is hosting": Gordon Bruce, interview by the author, December 31, 2015.

103   "We wanted to emphasize": David Brown, interview by the author, September 12, 2016.

103   "Samsung's designers need to be different": Gordon Bruce, interview by the author, December 31, 2015.

104   "What persuaded me": David Brown, interview by the author, September 12, 2016.

104   "In Asia . . . form is everything": Gordon Bruce, interview by the author, December 31, 2015.

104   "Is that ours?": Ibid.

105   "Management tells us Sony is successful": Tom Hardy, interview by the author, May 23, 2017.

105   rather than develop a house style: Gordon Bruce, "Samsung" (PowerPoint presentation, undated). Presentation is in the author's possession.

105   "I saw the same effect": Gordon Bruce, interview by the author, December 31, 2015.

105   Old architectural treasures: John M. Glionna, "A Foreigner's Battle to Preserve South Korea's *Hanok* Houses," *Los Angeles Times,* October 17, 2010, https://www.latimes.com/archives/la-xpm-2010-oct-17-la-fg-south-korea-heritage-20101017-story.html.

105   seek rounder eyes and sculpted noses: Patricia Marx, "About Face," *The New Yorker,* March 16, 2015, https://www.newyorker.com/magazine/2015/03/23/about-face.

106   "The paparazzi rushed up": Gordon Bruce, interview by the author, December 31, 2015.

106   He called the ArtCenter designers "heroes": Peter Arnell, interview by the author, July 4, 2017.

106   "It's the chairman!": Gordon Bruce, interview by the author, December 31, 2015.

107   "The place was just shark-infested": Gordon Bruce, interview by the author, December 31, 2015. This account is also available in Gerardo R. Ungson and Yim-Yu Wong, *Global Strategic Management* (Abingdon, UK: Routledge, 2014), p. 354.

108   "What do you know about Korea?": Gordon Bruce, interview by the author, December 31, 2015.

108   He'd lost his brother: "James Noboru Miho" (video interview), Experiencing War: Stories from the Veterans History Project, American Folklife Center, Library of Congress, October 26, 2011, https://memory.loc.gov/diglib/vhp-stories/loc.natlib.afc2001001.66630/.

108   "I was appalled": Gordon Bruce, email message to the author, March 2, 2017.

108   On February 15, 1995: Gordon Bruce, email message to the author, February 15, 2017.

108   "Miky insisted that Miho": Gordon Bruce, interview by the author, December 31, 2015.

108   Bruce settled into his new office: Gordon Bruce, interview by the author, December 31, 2015.

109   "They interviewed every designer": Ted Shin, interview by the author, February 3, 2017.

109   On September 1, 1995: Gordon Bruce, email message to the author, February 15, 2017.

109   "Nature is the best designer": Mike Winder, "'Mother Nature Is the Best Designer': Highlights from Gordon Bruce's Spring 2014 Graduation Speech," *ArtCenter News,* April 24, 2014, http://blogs.artcenter.edu/dottedline/2014/04/24/gordon-bruce-graduation-speech-highlights/. The account in this book is based on another interview with Bruce on December 31, 2015.

110   When Bruce and Miho received: Gordon Bruce, interview by the author, December 31, 2015.

111   He approved the curriculum: Gordon Bruce, "Go East, Young Man: Design Education at Samsung," *Design Management Journal* 9, no. 2 (Spring 1998): 57, https://onlinelibrary.wiley.com/doi/abs/10.1111/j.1948-7169.1998 .tb00206.x.

111   "A mobile phone, after all": Gordon Bruce, interview by the author, December 31, 2015.

111   For weeks at a time: Gordon Bruce and James Miho, internal Samsung notes and itineraries, 1996. Notes are in the author's possession.

111   "Mercedes-Benz prospered": Ibid.

111   "I learned something": Ted Shin, interviewed by author, February 3, 2017.

112   One student told me: Samsung designer, interview by the author, December 31, 2015.

112   "Hey, why don't we design": Ted Shin, interview by the author, February 3, 2017.

112   "The left side was a workspace": Ted Shin, email message to the author, October 11, 2017.

12: DESIGN REVOLUTION

113   "The upcoming twenty-first century": Samsung Design, "1996: Declaring the 'Year of the Design Revolution,'" (undated) http://web.archive.org /web/20170107151248/http://www.design.samsung.com/global/index .html. This source is a timeline and history of design at Samsung that was formerly posted on Samsung Design's website, at http://design.samsung .com/.

113   the forces of "digital convergence": Michell, *Samsung Electronics and the Struggle for Leadership*, p. 35.

114   "design scrapbook": Hyunhu Jang, "Samsung's 'Design Revolution' Started in 1996 with Sony, Not Apple," *The Verge*, August 31, 2012, https://www .theverge.com/2012/8/31/3273695/samsungs-design-revolution-started- in-1996-with-sony-not-apple.

114   "Apple Computer realized that their goal": Hyunhu Jang, "Samsung's 'Design Revolution' Started in 1996 with Sony, Not Apple."

114   Bruce convinced Samsung to hire: Bill Breen, "The Seoul of Design," *Fast Company*, December 1, 2005, https://www.fastcompany.com/54877/seoul- design.

114   "I decided to use": Tom Hardy, interview by the author, May 23, 2017.

115   "Balance of Reason and Feeling": Breen, "Seoul of Design."

116   set up design centers: "Samsung Design Innovation Center," Global Design

Studio Series #1, undated, http://design.samsung.com/global/m/contents /sdic/.

116   "simplicity/feeling": Breen, "Seoul of Design."

116   developed Samsung's first "smart home": Former Samsung designer, interview by the author, January 5, 2017.

116   Lee Min-hyouk, who later designed: Youngjin Yoo and Kyungmook Kim, "How Samsung Became a Design Powerhouse," *Harvard Business Review,* September 2015, https://hbr.org/2015/09/how-samsung-became-a-design -powerhouse.

116   "I called it the I-Phone": Ted Shin, interview by the author, February 3, 2017. Ted Shin's blueprints and photographs of the I-Phone are in the author's possession.

117   "Miho and I were becoming disliked": Gordon Bruce, email message to the author, February 16, 2017.

117   Miky invested $300 million: Pollack, "Unlikely Credits for a Korean Movie Mogul."

117   But Samsung and CJ were completing: Kim Hee-jung, "Lee Jae-hyun's Samsung Complex," *Business Post,* May 12, 2014, http://www.business post.co.kr/BP?command=naver&num=1856. This source is in Korean. The author's researcher translated the title and the quoted text into English.

117   "was like slapping the chairman": Gordon Bruce, email message to the author, February 4, 2017.

118   close its struggling film company: Inkyu Kang, "The Political Economy of Idols," in *K-pop: The International Rise of the Korean Music Industry,* ed. Jungbong Choi and Roald Maliangkay (Abingdon, UK: Routledge, 2014).

118   At IDS the growing neglect: Gordon Bruce, email message to the author, February 15, 2017.

118   "I call it Spartan creativity": Rich Park, interview by the author, November 15, 2010.

118   being enmeshed in an investigation: Sonni Efron, "4 S. Korean Tycoons Get Prison in Bribery Case," *Los Angeles Times,* August 27, 1996, https://www .latimes.com/archives/la-xpm-1996-08-27-mn-38054-story.html.

119   his father's business seemed to get: "Prosecutors' Announcement on Outcomes of the Investigation into Roh Tae-woo's Illicit Wealth Accumulation," *Hankyoreh,* December 6, 1995, https://newslibrary.naver.com /viewer/index.nhn?articleId=1995120600289121001. This source is in Korean. The author's researcher translated the title and the quoted text into English.

119   "Back in the old days": Henry Cho, interview by the author, March 25, 2016.

119   "From Samsung Group Chairman Lee Kun-hee": "Result of Prosecutors'
       Investigation of Roh Tae-woo's Accumulation of Wealth by Illicit Means,"
       JoongAng Ilbo, December 6, 1995. The People's Solidarity for Participatory
       Democracy, a Korean activist group, has compiled a list of presidential
       bribery allegations involving Samsung: http://www.peoplepower21.org/?
       module=file&act=procFileDownload&file_srl=515049&sid=78a97d8ea7
       9ea5dc78ded9006ebf75be&module_srl=114226. This source is in Korean.
       The author's researcher translated the title and the quoted text into En-
       glish.

119   Chairman Lee showed up: Um Ki-young, Paik Ji-yeon, "Third Trial of Roh
       Tae-woo's Corruption Case, Defendants Pledged for Favorable Handling,"
       MBC News, http://imnews.imbc.com/20dbnews/history/1996/2009574
       _19466.html. January 29, 1996. This source is in Korean. The author's re-
       searcher translated the title and the quoted text into English.

119   "We customarily gave such money": Um Sang-ik, "Samsung Lee Kun-hee's
       Courtroom in 1996, Jay Y. Lee's in 2017," *Chosun Pub,* August 28, 2017,
       http://pub.chosun.com/client/news/viw.asp?cate=C03&mcate=M1003
       &nNewsNumb=20170825920&nidx=25921. This source is in Korean. The
       author's researcher translated the title and the quoted text into English.

120   two years in prison: Efron, "4 S. Korean Tycoons Get Prison."

120   "[The court] has taken into account: "Summary of Roh Tae-woo's Slush
       Fund Trial," *JoongAng Ilbo,* August 27, 1996, https://news.joins.com/article
       /3317418. This source is in Korean. The author's researcher translated the
       title and the quoted text into English.

120   "The verdicts are shocking": Efron, "4 S. Korean Tycoons Get Prison."

120   "out of consideration for": Kim Moon-kwon, "Pardoning of Businessmen
       Means Boosting Morales for Economic Revival," *Korea Economic Daily,*
       September 30, 1997, https://www.hankyung.com/news/article/199709300
       1781. This source is in Korean. The author's researcher translated the title
       and the quoted text into English.

120   "My uncle's very lucky": Henry Cho, interview by the author, March 25,
       2016.

121   "We knew something was wrong": Nam S. Lee, interview by the author,
       November 13, 2015.

121   "The Thai Gamble": Seth Mydans, "The Thai Gamble: Devaluing Cur-
       rency to Revive Economy," *The New York Times,* July 3, 1997, https://www
       .nytimes.com/1997/07/03/business/the-thai-gamble-devaluing-currency-
       to-revive-economy.html.

121   currencies were falling off a cliff: PBS, "Timeline of the Panic," *Frontline,*
       undated, https://www.pbs.org/wgbh/pages/frontline/shows/crash/etc
       /cron.html.

121   "I've studied the automobile industry": Lee, *Lee Kun-hee Essays,* p. 91.

121 Interest payments alone: Michell, *Samsung Electronics and the Struggle for Leadership,* pp. 29–30.

121 most of Samsung's 45,000 automobiles: Woonghee Lee and Nam S. Lee, "Understanding Samsung's Diversification Strategy: The Case of Samsung Motors Inc.," *Long Range Planning* 40, nos. 4–5 (August–October 2007): 490.

121 that could manufacture 80,000: Samsung, "Meeting the Challenge: Samsung Annual Report 1995," p. 79.

121 losing $6,000 per vehicle: Stephanie Strom, "Skepticism over Korean Reform; After Daewoo Intervention, Is There the Will for Austerity?" *The New York Times,* July 30, 1999, https://www.nytimes.com/1999/07/30/business/international-business-skepticism-over-korean-reform-after-daewoo-intervention.html.

121 fell precipitously to fewer than 7,000: James B. Treece, "Renault to Pump $300 Million into Ailing Samsung," *Automotive News,* June 19, 2000, https://www.autonews.com/article/20000619/ANA/6190749/renault-to-pump-300-million-into-ailing-samsung.

122 bought by Renault: Treece, "Renault to Pump $300 Million."

122 "There was a sense that": Louis Kraar, "The Man Who Shook Up Samsung," *Fortune,* January 2000, http://archive.fortune.com/magazines/fortune/fortune_archive/2000/01/24/272330/index.htm.

122 gold rings and trophies: "Koreans Give Up Their Gold to Help Their Country," *BBC News,* January 14, 1998, http://news.bbc.co.uk/2/hi/world/analysis/47496.stm.

122 the largest bailout in its history: Andrew Pollack, "Crisis in South Korea: The Bailout; Package of Loans Worth $55 Billion Is Set for Korea," *The New York Times,* December 4, 1997, https://www.nytimes.com/1997/12/04/business/crisis-south-korea-bailout-package-loans-worth-55-billion-set-for-korea.html.

122 Samsung Electronics laid off one-third: Kraar, "Man Who Shook Up Samsung."

122 Senior managers took a 10 percent: "Samsung Cuts Jobs and Pay as Koreans Brace for Pain," *The New York Times,* November 27, 1997, https://www.nytimes.com/1997/11/27/business/international-business-samsung-cuts-jobs-and-pay-as-koreans-brace-for-pain.html.

122 "Workers can no longer count on": Ihlwan Moon and Brian Bemmer, "Samsung: A Korean Giant Confronts the Crisis," *BusinessWeek,* March 23, 1998, p. 18.

122 Koreans abandoned the expectation: "The Notion of Lifetime Employment Has Crumbled," *Maeil Business Newspaper,* March 24, 1998, https://newslibrary.naver.com/viewer/index.nhn?articleId=1998032400099147001. This source is in Korean. The author's researcher translated the title and the quoted text into English.

122 "Think about it this way": Hwang Young-key, interview by the author, June 5, 2014.

13: MY BOSS THE SHIT KICKER

123 "What do you need to double": Peter Skarzynski, interview by the author, December 28, 2016.

123 where Samsung occupied four floors: Dan Shieder and Jennifer Rampey, "Samsung Looks at Richardson," *Dallas Business Journal,* October 7, 1996, https://www.bizjournals.com/dallas/stories/1996/10/07/story1.html.

123 the South Korean government had finished: In-soo Han, "Success of CDMA Telecommunications Technology in Korea: The Role of the Mobile Triangle," in *Innovation and Technology in Korea: Challenges of a Newly Advanced Economy,* ed. Jörg Malich and Werner Pascha (Heidelberg, Germany: Physica Verlag, January 2007), p. 287.

124 Sprint was going to be: Mike Mills, "Group Takes Control of Sprint Spectrum," *The Washington Post,* January 7, 1998, https://www.washington post.com/archive/business/1998/01/07/group-takes-control-of-sprint-spectrum/9ec5bdfc-7b4e-4159-88d6-fd45cb757a20/.

124 "I need a cheap Asian company": Peter Skarzynski, interview by the author, December 28, 2016.

124 $600 million deal to sell 1.7 million: "Samsung Electronics to Export PCS Units to the U.S. in $600 Million Deal," *Dong-A Ilbo,* September 21, 1996, https://newslibrary.naver.com/viewer/index.nhn?articleId=1996092100209108002. This source is in Korean. The author's researcher translated the title and the quoted text into English.

129 the launch went forward as planned: Reily Gregson, "Samsung, Sprint Debut CDMA Dual-Band Internet Phone," *RCRWireless News,* September 27, 1999, https://www.rcrwireless.com/19990927/archived-articles/samsung-sprint-debut-cdma-dual-band-internet-phone.

129 "The SCH-3500 became the flagship": Peter Skarzynski, interview by the author, December 28, 2016. A sales figure of "more than six million units" is confirmed at Samsung's history museum in its Suwon headquarters, where the author saw the phone and its description on display.

129 "You buy LG": Peter Skarzynski, interview by the author, December 28, 2016.

130 getting drunk with business partners: "Korea Cracks Down on Bribes in Brothels," *The Economist,* October 15, 2016, https://www.economist.com/asia/2016/10/15/korea-cracks-down-on-bribes-in-brothels.

130 "I know we can't stop all of this": Peter Skarzynski, interview by the author, December 28, 2016.

131 Samsung settled the lawsuit: Wojtek Dabrowski, "RIM and Samsung Settle

Suit over BlackJack Device," Reuters, February 8, 2007, https://www
.reuters.com/article/idUSN0840240420070208.

131   renamed the phone "Jack": Matt Buchanan, "Samsung Jack Completes
BlackJack Windows Mobile Trilogy for $99," *Gizmodo*, May 14, 2009,
https://gizmodo.com/samsung-jack-completes-blackjack-windows-
mobile-trilogy-5253918.

14: SONY WARS

132   "I will kill you": Ihlwan Moon, "The Samsung Way," *BusinessWeek*, October
15, 2003, https://www.bloomberg.com/news/articles/2003-06-15/the-
samsung-way.

132   "kimchi-eating American": Donald MacIntyre, "Eric Kim," *Time*,
November 24, 2002, http://content.time.com/time/magazine/article
/0,9171,393732,00.html.

133   "I'm Korean": Eric Kim, interview by the author, April 26, 2015.

133   "Eric was an engineer": Five former Samsung executives, interviews by the
author, March 2014–January 2017.

133   Before Samsung, he'd been: Intel Corporation, "Intel Executive Biography:
Eric B. Kim," https://web.archive.org/web/20061202050519/http://www
.intel.com/pressroom/kits/bios/kim.htm.

133   "Now, the CEO wanted me": Eric Kim, interview by the author, April 26,
2015.

133   Samsung had fifty-five advertising agencies: Seung-Joo Lee and Boon-
Young Lee, "Case Study of Samsung's Mobile Phone Business" (KDI
School of Public Policy & Management Paper No. 04-11, May 2004),
https://ssrn.com/abstract=556923.

133   "Eric wanted to start fresh": Thomas Rhee, interview by the author, May 8,
2017.

133   Kim and his new marketing office: Eric Kim, interview by the author, April
26, 2015.

134   "K.T. hated his guts": Peter Skarzynski, interview by the author, December
28, 2016.

134   "There's no substitute": John A. Quelch and Anna Harrington, "Samsung
Electronics Company: Global Marketing Operations" (Harvard Business
School Case 504-051, March 2004, revised January 2008), https://www.hbs
.edu/faculty/Pages/item.aspx?num=30954.

134   "Products don't sell themselves": Eric Kim, interview by the author, April
26, 2015.

134   "We had a brainstorming [session]": Thomas Rhee, interview by the author, May 8, 2017.

135   "soft porno": Samsung design consultant, phone interview by the author, May 23, 2017.

135   showed an almost-nude woman: Samsung Electronics, Samsung television commercial, Arnell Group, late 1990s. Commercial is in possession of the author.

135   Kim's DigitAll campaign was unleashed: Samsung Electronics, "Samsung Electronics Launches Digital Multimedia Platform Worldwide" (news release), November 10, 1999, https://www.samsung.com/us/news/news PreviewRead.do?news_seq=388.

135   "Peter kept Eric waiting": Thomas Rhee, interview by the author, May 8, 2017.

135   "We want to beat Sony": Heidi Brown, "Look Out, Sony," *Forbes*, June 11, 2001, https://www.forbes.com/forbes/2001/0611/096.html#55951f564aa4.

136   And so Kim was reprimanded: Heidi Brown, Stephane Fitch, and Brett Nelson, "Follow-Through," *Forbes,* October 4, 2004, https://www.forbes.com/forbes/2004/1004/054.html#79df6d5511ec.

136   "The article almost got me fired": Eric Kim, email message to the author, April 30, 2019.

136   "weird situation": Chi Young-cho (executive vice president for corporate strategy and corporate development), interview by the author, November 16, 2010.

136   "Let's not overspend on advertising": Quelch and Harrington, "Samsung Electronics Company."

136   "The product design and the product planning": Jay Solomon, "Samsung Vies for Starring Role in Upscale Electronics Market," *The Wall Street Journal,* June 13, 2002, https://www.wsj.com/articles/SB1023932566661953320.

136   "Once upon a time": *Sherwood 48 Associates v. Sony Corp. of America*, 213 F. Supp. 2d 376 (S.D.N.Y. 2002).

137   "It's Spider-Man!" *Spider-Man,* directed by Sam Raimi (Culver City, CA: Columbia Pictures, 2002).

137   "I . . . went ballistic": Eric Kim, email message to the author, April 28, 2017.

137   Samsung's legal team prepared: Solomon, "Samsung Vies for Starring Role."

137   "At the end of the day": Former Samsung lawyer, interview by the author, June 2, 2017.

137   the judge ruled in Sony's favor: Allison Fass, "Lawyers for Spider-Man Win a Fight over Times Square," *The New York Times,* August 13, 2002, https://www.nytimes.com/2002/08/13/business/media-business-advertising-lawyers-for-spider-man-win-fight-over-times-square.html.

137   "They tried to remove us": Solomon, "Samsung Vies for Starring Role."

138   receives a surprise call from Morpheus: *The Matrix,* directed by Lana Wachowski and Lilly Wachowski (Burbank, CA: Warner Bros., 1999).

138   "I knew immediately what was happening": Peter Weedfald, interview by
       the author, February 24, 2016.

138   "We'll have a Samsung product placement": Ibid.

138   "Brand is the refuge of the ignorant": Peter Weedfald, *Green Reign Leader-
       ship: Business Lessons to Ensure Leadership Dominance* (Paramus, NJ: Gen One
       Ventures, 2012), p. 62.

138   go back to his seat: Peter Weedfald, email to the author, November 19,
       2015.

139   "You could tell pretty quick": Peter Skarzynski, interview by the author,
       December 28, 2016.

139   "notorious for squashing people": Claire Atkinson, "Inside Samsung's $100
       Million 'Matrix' Deal," *Ad Age,* May 12, 2003, https://adage.com/article
       /news/inside-samsung-s-100-million-matrix-deal/37489.

139   K.T.'s team was annoyed: Peter Skarzynski, interview by the author, De-
       cember 28, 2016.

139   "It sounded to me like a gun cocking": Peter Skarzynski, interview by au-
       thor, October 10, 2017.

140   The product placement appeared eight times: *The Matrix Reloaded,* directed
       by Lana Wachowski and Lilly Wachowski (Burbank, CA: Warner Bros,
       2003). Samsung's *Matrix* phone appears at 2:11 (used by Trinity), 1:22:32
       (Neo), 1:22:56 (Trinity), 1:23:48 (Morpheus), 1:24:43 (Naomi), 1:30:06
       (Trinity), 1:48:48 (Link), and 1:49:34 (Link).

140   "That was a big success": Eric Kim, interview by the author, April 26, 2015.

140   Samsung came in at number: Interbrand, "Best Global Brands 2003 Rank-
       ings," https://www.interbrand.com/best-brands/best-global-brands/2003
       /ranking/. The global brand rankings were published in *BusinessWeek* until
       2009.

140   complacency and "Not Invented Here" syndrome: Sea-Jin Chang, *Sony vs.
       Samsung: The Inside Story of the Electronics Giants' Battle for Global Supremacy*
       (Hoboken, NJ: Wiley, 2008), pp. 25–65.

140   "for a report on what Samsung": Moon, "The Samsung Way."

140   forcing him to slash jobs: "The Best and Worst Managers of 2003—the
       Worst Managers: Nobuyuki Idei," *BusinessWeek,* January, 12, 2004, http://
       web.archive.org/web/20040202013445/http://www.businessweek.com
       /magazine/content/04_02/b3865725.htm.

141   a joint venture to manufacture LCDs: Moon Ihlwan, "Samsung and Sony's
       Win-Win LCD Venture," *Bloomberg,* November 29, 2006, https://www
       .bloomberg.com/news/articles/2006-11-28/samsung-and-sonys-win-win-
       lcd-venturebusinessweek-business-news-stock-market-and-financial-
       advice.

141   worst managers of the year: "Best and Worst Managers of 2003: Nobuyuki
       Idei."

141 one of the best: "The Top 25 Managers of the Year," *BusinessWeek*, January 11, 1999. The ranking is no longer available online. The physical magazine is in the author's possession.

15: BORDEAUX

142 "to put food on the table": Reuters, "Downfall of Ex–Samsung Strategy Chief Choi Gee Sung Leaves 'Salarymen' Disillusioned," *The Straits Times*, August 25, 2017, https://www.straitstimes.com/asia/east-asia/downfall-of-ex-samsung-strategy-chief-choi-gee-sung-leaves-salarymen-disillusioned.

142 "To overcome his weakness": Eun Y. Kim and Edward C. Valdez, *Samsung 3.0: Talent, Technology and Timing: 88 Things You Need to Know About Samsung* (Austin, TX: CEO International, 2013), p. 132.

142 "I was often impressed": James Sanduski, interview by the author, January 3, 2017.

143 G.S. was expected to mentor: "Lee Jae-yong and Choi Gee-sung: New Leadership Team at Samsung," *Hankyoreh*, December 16, 2009, http://english.hani.co.kr/arti/english_edition/e_business/393675.html.

143 "The whole room would stand": Peter Skarzynski, interview by the author, December 28, 2016.

143 direct, substantive statements: Max Kim, "The Woes of Jay Y: Can the Galaxy S6 Save Samsung's Crown Prince?" *The Guardian*, April 20, 2015, https://www.theguardian.com/technology/2015/apr/20/samsung-galaxy-s6-woes-of-samsung-crown-prince-lee-jae-yong.

143 "He's sort of like a god": Daren Tsui, interview by the author, January 10, 2017.

143 "thirty-year veteran Park Gwang-gi": Adam Lashinsky, "Can Samsung's New Leader Dethrone Apple?" *Forbes*, July 27, 2015, http://fortune.com/2015/07/27/samsung-jay-lee/.

143 "He's exactly the kind of guy": Peter Weedfald, interview by the author, February 24, 2016.

143 "I had the impression": T.J. Kang, interview by the author, January 29, 2016.

143 "He's nice . . . but he doesn't have": Samsung mobile employee, interview by the author, February 16, 2016.

144 He studied East Asian history: Miyoung Kim, "Jay Lee, Samsung's Unassuming Heir Apparent," Reuters, December 5, 2012, https://www.reuters.com/article/us-samsung-lee-newsmaker/jay-lee-samsungs-unassuming-heir-apparent-idUSBRE8B40A220121205.

144 There he wrote his thesis: Yoon Hee-seok, "The Warning Vice Chairman Lee Jae-yong Delivered About Manufacturing 20 Years Ago," *Electronic Time*, September 22, 2015, http://www.etnews.com/20150918000274.

This source is in Korean. The author's researcher translated the title and the quoted text into English.

144   he married (and later divorced): Park Hyeon-cheol and Kim Young-hee, "Lee Jae-yong: Lim Sae-ryung Mutual Divorce," *Hankyoreh*, February 18, 2009, http://www.hani.co.kr/arti/society/society_general/339527.html. This source is in Korean. The author's researcher translated the title and the quoted text into English.

144   without completing his doctorate: "Jay Y. Lee," *Forbes*, undated, https://www.forbes.com/profile/jay-y-lee/#2741dc6e24a2.

144   "He was not involved": Nam S. Lee, interview by the author, November 13, 2015.

144   "Jay Y. Lee was eager to participate": Kim Yong-chul, *Thinking of Samsung* (Seoul, South Korea: Social Commentary, 2010), pp. 201–4.

144   Jay became the chief shareholder: Huh In-jung, "e-Samsung Owner Explains Concept," *Chosun Ilbo* (English edition), July 11, 2000, http://english.chosun.com/site/data/html_dir/2000/07/11/2000071161386.html.

144   "But a Samsung style of management": Kim, *Thinking of Samsung*, p. 201.

145   Shareholders cried foul: Samsung SDI, "Results for First Quarter of 2001" (letter to shareholders), April 24, 2001.

145   e-Samsung was shut down: Lim Ju-young, Ahn Hee, Lee Han-seung, "Special Prosecutors' 'e-Samsung' Case: 28 Including Senior Vice President Lee Jae-yong Found Not Guilty," Yonhap News Agency, March 13, 2008, https://www.yna.co.kr/view/AKR20080313108700004. This source is in Korean. The author's researcher translated the title and the quoted text into English.

145   Samsung ordered nine Samsung affiliates: Kim Rahn, "Lee Jae-yong Cleared of e-Samsung Allegations," *The Korea Times*, March 13, 2008, https://www.koreatimes.co.kr/www/news/nation/2009/07/113_20666.html.

145   Merrill Lynch claimed that: Park Soon-bin, "Chaebol Ventures' Blind-eye Management," *Hankyoreh*, April 3, 2001, http://h21.hani.co.kr/arti/economy/economy_general/2112.html. This source is in Korean. The author's researcher translated the title and the quoted text into English.

145   "I dispatched two": Kim, *Thinking of Samsung*, p. 204.

145   "FTC investigators cannot show up": Ibid., p. 205.

145   fined Samsung Group affiliates six times: Park Min-ha, "Samsung, Repeated Obstruction of Investigation . . . 400 Million Won in Fines," *SBS*, March 19, 2012, https://news.sbs.co.kr/news/endPage.do?news_id=N1001122402. This source is in Korean. The author's researcher translated the title and the quoted text into English.

145   "Failing in cars, failing in ventures": Kim, *Thinking of Samsung*, p. 202.

146   "The only people who really knew him": Kim, "Woes of Jay Y."

146 "Hyun introduced us": Alan Plumb, interview by the author, August 20, 2014.

146 Rolls-Royce had just signed: "Rolls-Royce Trent 900 Clocks Up 6,000 Flying Hours on the A380," Defense Aerospace, July 17, 2006, http://www .defense-aerospace.com/article-view/release/71238/trent-900-engine-logs-6,000-hours-on-a380.html.

146 "Jay said he didn't understand": Alan Plumb, interview by the author, August 20, 2014.

146 In 2004 Jay asked a Japanese engineer: Lee and Clenfield, "Samsung Low-Profile Heir Poised to Succeed Father Seen as a God."

147 "Jay came into design meetings a lot": Former Samsung designer, phone interview by the author, January 30, 2017.

147 "Which device is going to be": Kim Gaeyoun, interview by the author, November 15, 2010.

147 "That became a fun rivalry": Bill Ogle, interview by the author, January 6, 2017.

147 "We started thinking about this": Peter Weedfald, interview by the author, February 24, 2016.

148 "It looks like a wineglass!": Karen Freeze and Kyung-won Chung, "Design Strategy at Samsung Electronics: Becoming a Top-Tier Company," *Harvard Business Review* (case no. DMI021-PDF-ENG), November 12, 2008, https:// hbr.org/product/design-strategy-at-samsung-electronics-becoming-a-top-tier-company/DMI021-PDF-ENG.

148 Samsung selected an elite team: Ihlwan Moon, "Camp Samsung," *Business-Week,* July 3, 2006, https://www.bloomberg.com/news/articles/2006-07-02/camp-samsung.

148 "The design appears to be": Freeze and Chung, "Design Strategy at Samsung Electronics."

148 Yunje Kang, the designer: Yunje Kang, email message to the author, November 24, 2010.

148 selling more than three million units: Kim In-cheol. "Samsung Bordeaux TV, on a Record-Breaking March," Yonhap News Agency, syndicated by *Korea Economic Daily,* March 13, 2007, https://www.hankyung.com/news /article/2007031348458. This source is in Korean. The author's researcher translated the title and the quoted text into English.

148 "The selling point is very clear": Kim Yoo-chul, "Slim Tops Key Trend for TV Makers," *The Korea Times,* August 25, 2008, https://www.koreatimes .co.kr/www/news/tech/2012/07/129_29939.html.

148 Samsung eclipsed Sony: Rasmus Larsen, "Samsung Dominates Global TV Market for 10th Straight Year," *FlatpanelsHD,* March 15, 2016, https:// www.flatpanelshd.com/news.php?subaction=showfull&id=1458017308.

148   no obvious role in product development: Song Jung-a, "Samsung Sets Suc-
      cession Plan," *Financial Times*, January 19, 2007, https://www.ft.com
      /content/20947efa-a7b1-11db-b448-0000779e2340.

149   Howard Stringer, a Brit–turned–American citizen: Richard Siklos and
      Martin Fackler, "Howard Stringer, Sony's Road Warrior," *The New York
      Times*, May 28, 2006, https://www.nytimes.com/2006/05/28/business
      /yourmoney/28sony.html.

149   The company experienced losses: Kevin Kelleher, "How Sony Got Up and
      Out of Its Death Bed," *Time*, May 21, 2015, http://time.com/3892591
      /sony-turnaround/.

149   In 2011 Sony exited: Hiroko Tabuchi, "Sony to Cease Its Flat-Screen Part-
      nership with Samsung," *The New York Times*, December 26, 2011, https://
      www.nytimes.com/2011/12/27/technology/sony-sells-stake-in-lcd-panel-
      joint-venture.html.

149   Stringer, unable to repair: "Sony Chairman Sir Howard Stringer to Retire,"
      BBC News, March 11, 2013, https://www.bbc.com/news/business-
      21738549.

149   "Running a big company": Jonathan Handel, "Ex–Sony CEO Howard
      Stringer on Sony's Failures and Time Inc.'s Big Challenges," *Hollywood
      Reporter*, October 25, 2014, https://www.hollywoodreporter.com/news
      /sony-ceo-howard-stringer-opines-743982.

16: UNHOLY ALLIANCE

150   "I met him with the solution": Hwang Chang-gyu, "Dear Young People, Be
      Crazy to Make Others Happy," *Chosun Ilbo*, February 10, 2010, http://
      news.chosun.com/site/data/html_dir/2010/02/10/2010021001829
      .html?Dep0=chosunmain&Dep1=news&Dep2=headline1&Dep3=
      h1_04. This source is in Korean. The author's researcher translated the title
      and the quoted text into English.

150   hailed from the southern port town: "Who Is Hwang Chang-gyu, the
      Figure Newly Poised to Lead KT?" Yonhap News Agency, December 16,
      2013, https://www.yna.co.kr/view/AKR20131216195600017. More infor-
      mation is available at "What Did Hwang Chang-gyu Do as a Researcher
      and Businessman?" *Scienceall*, April 1, 2016. These sources are in Korean.
      The author's researcher translated the title and the quoted text into En-
      glish.

151   His fascination with semiconductors: Sung Ki-young, "'Star Executive'
      Samsung Electronics President of Semiconductors Hwang Chang-gyu's
      Concerns for Country," *Shindonga*, July 29, 2004, http://shindonga.donga
      .com/Library/3/06/13/103652/1. This source is in Korean. The author's
      researcher translated the title and the quoted text into English.

151   "Just leave it to me": Cho Hyun-jae, Chun Ho-rim, and Lim Sang-gyoon,

Samsung Electronics, Digital Conquerer (Seoul: Maeil Business Newspaper, 2005), p. 558. This source is in Korean. The author's researcher translated the title and the quoted text into English.

151   "Hwang's law": Jung Kyung-joon, "Moore's Law and Hwang's Law," *Dong-A Ilbo*, May 9, 2011, http://www.donga.com/news/article/all/20110508 /37032822/1. This source is in Korean. The author's researcher translated the title and the quoted text into English.

151   "This is exactly what I wanted": Song Jung-a, "Monday Interview: Hwang Chang-gyu, KT Chief Executive," *Financial Times*, May 3, 2015, https:// www.ft.com/content/211ece46-e82d-11e4-894a-00144feab7de.

151   "It was the moment": Hwang, "Dear Young People, Be Crazy to Make Others Happy."

151   "Glass Meetings," "Green Meetings": Eichenwald, "The Great Smartphone War."

151   NYer, in reality, was Apple: Ibid.

152   "settle claims against it": "Plea Agreement," *U.S. v. Samsung Electronics and Samsung Semiconductor,* case no. 05-0643 PJH (U.S. District Court, Northern District of California, San Francisco Division), October 13, 2005.

152   In September of that year: "Plea Agreement," *U.S. v. Quinn,* case no. CR 06-0635 PJH (U.S. District Court, Northern District of California, San Francisco Division), October 5, 2006.

152   He was joined by: Department of Justice, "Sixth Samsung Executive Agrees to Plead Guilty to Participating in DRAM Price-Fixing Cartel" (news release), April 19, 2007, https://www.justice.gov/archive/atr/public/press _releases/2007/222770.htm.

152   In February 2006, David Tupman: Brian Merchant, *The One Device: The Secret History of iPhone* (New York: Hachette, 2017), pp. 358–63.

152   "as fast as they've": Ibid., p. 362.

152   the transistor count in each chip: Ibid., pp. 153–54.

153   it crashed due to an unknown: Ibid., p. 362.

153   "This is a day": "Steve Jobs Introducing the iPhone at Macworld 2007," posted by YouTube user superapple4ever on January 9, 2007, https:// www.youtube.com/watch?v=x7qPAY9JqE4.

153   the iPhone would not have come: Merchant, *The One Device,* p. 363.

17: THE EMPEROR HAS NO CLOTHES

154   The South Korean government banned: Myung Oh and James Larson, *Digital Development in Korea: Building an Information Society* (Abingdon, UK: Taylor & Francis, 2011), pp. 106–7.

154   it was a protectionist trade measure: Robert Kelly, "Uber and Classic Asian

Mercantilism," *The Diplomat,* July 25, 2014, https://thediplomat.com /2014/07/uber-and-classic-asian-mercantilism/.

154 "You don't understand!": Former Google software engineer, interview by the author, March 2, 2015.

154 "we were given a directive": Former staffer in Chairman Lee Kun-hee's office, interview by the author, February 14, 2016.

155 "I live in Korea": Kim Yong-chul, interview by the author, September 24, 2015.

156 bank accounts under the names: Choe Sang-hun, "Corruption Scandal Snowballs at South Korea's Samsung Group," *The New York Times,* November 6, 2007, https://www.nytimes.com/2007/11/06/business /worldbusiness/06iht-samsung.1.8210181.html. A list of Kim's allegations from each press conference can be found at http://www.hani.co.kr/arti /society/society_general/252735.html.

156 $215 million slush fund: Choe Sang-hun, "South Korean Leader Allows Inquiry into Samsung Bribes," *The New York Times,* November 28, 2007, https://www.nytimes.com/2007/11/28/business/worldbusiness /28samsung.html.

156 "When you see a pile of excrement": Choe Sang-hun, "Book on Samsung Divides Korea," *The New York Times,* April 25, 2010, https://www.nytimes .com/2010/04/26/technology/26samsung.html.

156 Samsung fired his law firm: Evan Ramstad, "Samsung Whistleblower Returns to the Public Eye," *The Wall Street Journal,* May 19, 2010, https:// blogs.wsj.com/digits/2010/05/19/samsung-whistleblower-returns-to-the- public-eye/.

156 Investigators rummaged through: Choe Sang-hun, "Samsung Chairman's Office Raided in Inquiry," *The New York Times,* January 15, 2008, https:// www.nytimes.com/2008/01/15/business/worldbusiness/15samsung .html.

156 An investigation by South Korean: "The Mystery of Korea's Elderly Protesters," *Korea Exposé,* May 2, 2016, https://www.koreaexpose.com/the- mystery-of-south-koreas-elderly-protesters/.

156 "no different from a gigolo": Joo Jin-woo, "The Catholic Priests' Association Is Sad and Hurting," *SisaIn.* April 29, 2008, https://www.sisain.co.kr /?mod=news&act=articleView&idxno=1838. The article is a round-up of what other media outlets were publishing on Kim Yong-chul. The original articles cited in this article appear to have been taken down and are no longer available online. The article quotes the conservative *Chosun Ilbo* as calling Kim Yong-chul a "traitor," and the *Dong-A Ilbo* as calling him "no different than a gigolo." This source is in Korean. The author's researcher translated the headline and text into English.

156 One angry protester: Photograph in the author's possession. More information on the Korean Parents Federation is at Lee Sang-hak, "Conservative

Group Reproaching Lawyer Kim Yong-chul and Catholic Priests Association for Justice," Yonhap News Agency, April 23, 2008. The article is in Korean. The author's researcher translated the headline and text into English.

157   indicted Chairman Lee: Choe Sang-hun, "Samsung's Chairman Is Indicted for Tax Evasion in Corruption Case," *The New York Times,* April 17, 2008, https://www.nytimes.com/2008/04/17/business/worldbusiness/17iht-samsung.4.12107507.html.

157   The chairman denied the allegations: Lee Hae-seong, "'Everything Was Due to My Inattention' . . . Chairman Lee Kun-hee Returns Home After 11 Hours of Questioning by Samsung Special Prosecutors," *Korea Economic Daily,* April 5, 2008, https://www.hankyung.com/society/article/2008040413961. This source is in Korean. The author's researcher translated the title and the quoted text into English.

157   Prosecutors didn't pursue the bribery charge: Choe, "Samsung's Chairman Is Indicted."

157   "I sincerely apologize": "Samsung Chairman Steps Down," *CNNMoney,* April 22, 2008, https://money.cnn.com/2008/04/22/news/international/samsung_resignation/index.htm. The full press conference is shown on video at Associated Press, "Samsung Chairman Says He Will Resign to Take Responsibility for Scandal," posted by YouTube user AP Archive on April 22, 2008, https://www.youtube.com/watch?v=C_gDCUP0y2g.

157   "It grieves me": "Samsung Chairman Steps Down," CNN, April 22, 2008, http://edition.cnn.com/2008/WORLD/asiapcf/04/22/samsung.chairman.resignation/index.html?eref=edition.

157   "It was just a show": Kim Yong-chul, interview by the author, September 24, 2015.

157   Nine others were implicated: "Samsung Chairman Indicted," *Dong-A Ilbo,* April 18, 2008, http://www.donga.com/en/article/all/20080418/258134/1/Samsung-Chairman-Indicted.

157   "I apologize for causing concern": "Former Samsung Chairman Lee Kun-hee's Trial Gets Underway," *Hankyoreh,* June 13, 2008, http://english.hani.co.kr/arti/english_edition/e_national/293086.html.

157   Two months later a Seoul: Choe Sang-hun, "Former Samsung Chief Is Convicted," *The New York Times,* July 17, 2008, https://www.nytimes.com/2008/07/17/business/worldbusiness/17samsung.html.

157   "I am sorry": Rhee So-eui, "Ex–Samsung Chief Given 3-Year Suspended Jail Term," Reuters, July 16, 2008, https://www.reuters.com/article/us-samsung-sentence/ex-samsung-chief-given-3-year-suspended-jail-term-idUSSEF00017220080716.

158   judge commuted his sentence: Choe, "Former Samsung Chief Is Convicted."

158   felt little sense of triumph: Kim Yong-chul, interview by the author, September 24, 2015.

158   "Samsung X-File": Roh Hoe-chan, interview by the author, September 23, 2015.

159   "Ah, and will there be any": Kim Seung-wook. "The National Intelligence Service 'X-file Incident' That Spanned Eight Years," Yonhap News Agency, February 14, 2013, https://www.yna.co.kr/view/AKR20130214171400004. The transcript was also recounted by Roh Hoe-chan in our interview.

159   "Our conclusions were groundbreaking": Roh Hoe-chan, interview by the author, September 23, 2015.

160   Texas's Anderson Cancer Center: Kim Sung-jin, "Samsung Chairman Forecast to Extend Sojourn in US," The Korea Times, September 21, 2005, http://www.koreatimes.com/article/20050921/270242.

160   "You're on the list": Roh Hoe-chan, interview by the author, September 23, 2015. A partial video of one of these hearings (without the segment involving the justice minister) is at https://news.sbs.co.kr/news/endPage.do?news_id=N1000010673.

160   The vice minister resigned: Roh Hoe-chan, interview by the author, September 23, 2015. More information is at Ko Woong-seok, "Kim Sang-hee, Vice Justice Minister, Tenders Resignation," Yonhap News Agency, syndicated by Maeil Business Newspaper, August 18, 2005, https://www.mk.co.kr/news/home/view/2005/08/305992/. This article is in Korean.

160   Chairman Lee's brother-in-law: Heo Won-soon, "Ambassador Hong Brought Down by X-file Scandal . . . Shortest Serving U.S. Ambassador at Five Months and Three Days," Korea Economic Daily, July 26, 2005, https://www.hankyung.com/politics/article/2005072601461. This source is in Korean. The author's researcher translated the title and the quoted text into English.

160   Prosecutors opened an investigation: "Court Finds Roh Hoe-chan Not Guilty in 'Samsung X-file' Disclosure Case," Hankyoreh, December 5, 2009, http://english.hani.co.kr/arti/english_edition/e_national/391645.html.

161   "Roh hoe-chan . . . we're going to have": Roh Hoe-chan, interview by the author, September 23, 2015.

161   violation of the country's anti-wiretapping law: "'X-file' Roh Hoe-chan Found Partially Guilty," Maeil Business Newspaper, May 13, 2011, https://www.mk.co.kr/news/society/view/2011/05/305239/. This source is in Korean. The author's researcher translated the title and the quoted text into English.

161   The broadcaster who first revealed: Lim Ju-young, Kim Tae-jong, "'X-file' Journalist Lee Sang-ho Found Guilty at Appeal Trial," Yonhap News Agency, syndicated by Korea Economic Daily, November 23, 2011, https://www.hankyung.com/society/article/2006112339408. This source is in Korean. The author's researcher translated the title and the quoted text into English.

161   "Any person who possessed": "Court Finds Roh Hoe-chan Not Guilty."

161 In 2013 Supreme Court judges ruled: Youkyung Lee, "South Korean Law-maker Loses Seat over Samsung Wiretaps," Associated Press, February 16, 2013, https://news.yahoo.com/skorean-lawmaker-loses-seat-over-samsung-wiretaps-090038078finance.html.

161 Lawyers and pundits were puzzled: Judicial Watchdog Center, "[28th Criticism of Ruling] The Ruling as Seen from Outside the Courtroom—Roh Hoe-chan's Guilty Verdict on the Spy Agency X-file Case," *People's Solidarity for Participatory Democracy*, February 18, 2013, http://www.peoplepower21.org/Judiciary/996452. This source is in Korean. The author's researcher translated the title and the quoted text into English.

162 He was stripped of his seat: Lee, "South Korean Lawmaker Loses Seat."

162 "I don't regret it": Roh Hoe-chan, interview by the author, September 23, 2015.

162 Following the verdict, Chairman Lee voluntarily: International Olympics Committee Ethics Commission, "Decision with a Recommended Provisional Measure," no. D/03/08 (case no. 4/2008), July 18, 2008.

162 Samsung agreed behind the scenes: Kim Min-kyoung, "Lee Administration Made Samsung Pay US Litigation Costs for DAS," *Hankyoreh*, July 11, 2018, http://english.hani.co.kr/arti/english_edition/e_national/852869.html.

162 granting special amnesty: Choe Sang-hun, "Korean Leader Pardons Samsung's Ex-Chairman," *The New York Times*, December 29, 2009, https://www.nytimes.com/2009/12/30/business/global/30samsung.html.

162 With his exoneration: Associated Press, "Lee Reinstated to IOC," *ESPN*, February 8, 2010, http://www.espn.com/olympics/news/story?id=4895452.

162 The bid was successful: Oh Young-jin, "Praising Lee Kun-hee," *The Korea Times*, July 12, 2011, http://www.koreatimes.co.kr/www/news/opinion/2011/07/137_90741.html.

162 "The latest pardon reconfirms": Choe, "Korean Leader Pardons Samsung's Ex-Chairman."

163 "Samsung . . . the world's largest conglomerate": Michael Breen, "What People Got for Christmas," *The Korea Times*, December 25, 2008. The article is no longer available online. A copy of it is in the author's possession.

163 "Samsung is really upset": Michael Breen, interview by the author, November 17, 2010.

163 The newspaper agreed to retract: "Correction," *The Korea Times*, January 29, 2010, http://m.koreatimes.co.kr/phone/news/view.jsp?req_news idx=59927.

163 "Did you research and check": Michael Breen, interview by the author, November 17, 2010. Breen has an account of the lawsuit in his book, *The New Koreans: The Story of a Nation* (New York: St. Martin's Press, 2017), pp. 231–33.

163   "Under Korean law . . . if somebody takes": Michael Breen, interview by the author, November 17, 2010. More information is available in Breen, *New Koreans,* pp. 231–33.

164   The ambassador called Samsung CEO: Breen, *New Koreans,* pp. 231–33.

164   "Michael Breen 'Mocked' Korea": Choi Jeong-yup, "Michael Breen 'Mocked' Korea, 'Did He Even Apologize?'" *EBN,* May 14, 2010, http://www.ebn.co.kr/news/view/437428. This source is in Korean. The author's researcher translated the title and the quoted text into English.

164   "The reason I'm being sued": John Glionna, "Samsung Doesn't Find Satirical Spoof Amusing," *Los Angeles Times,* May 10, 2010, https://www.latimes.com/archives/la-xpm-2010-may-10-la-fg-korea-samsung-20100510-story.html.

164   "As Samsung has withdrawn": Michael Breen, interview by the author, November 17, 2010.

164   published a bestselling memoir: Kim Yong-chul, *Thinking of Samsung* (Seoul: Social Commentary, 2010).

164   "One newspaper reported on its popularity": Choe, "Book on Samsung Divides Korea."

164   front-page apology: "We Will Be Strict in Our Reporting on Conglomerates," *Kyunghyang Shinmun,* February 23, 2010, http://m.khan.co.kr/view.html?art_id=201002231826025. This source is in Korean. The author's researcher translated the title and the quoted text into English.

165   "But I wouldn't say": Kim Yong-chul, interview by the author, September 24, 2015.

165   "Mr. Lee's wisdom and experience": "Lee Kun-hee Reinstated as Samsung Electronics Chairman," *Hankyoreh,* March 25, 2010, http://english.hani.co.kr/arti/english_edition/e_business/412228.html.

165   "Vision 2020": "The Next Big Bet," *The Economist,* October 1, 2011, https://www.economist.com/briefing/2011/10/01/the-next-big-bet.

165   Some questioned its feasibility: Jonathan Cheng, "Samsung's Outlook for Vision 2020 Gets Blurry," *The Wall Street Journal,* June 29, 2015, https://blogs.wsj.com/digits/2015/06/29/samsungs-outlook-for-vision-2020-gets-blurry/.

165   G.S. Choi was appointed CEO: Moon Ihlwan, "Apple Envy Drives Samsung Shakeup," *Bloomberg Businessweek,* December 15, 2009, https://www.bloomberg.com/news/articles/2009-12-15/apple-envy-drives-samsung-shakeup.

165   declared a "management crisis": "Lee Kun-hee Reinstated as Samsung Electronics Chairman."

166   The iPhone ban in South Korea: "iPhone Gets Official Government Approval in South Korea," *TechCrunch,* November 18, 2009, https://tech

crunch.com/2009/11/18/iphone-gets-official-government-approval-in-south-korea/.

18: GUARDIANS OF THE GALAXY

167   "[Our] quality isn't good": "Summary of Executive-Level Meeting Supervised by Head of Division (February 10)" (Plaintiff's Exhibit No. 40), *Apple vs. Samsung Electronics,* case no. 11-1846 (U.S. District Court, Northern District of California, San Jose Division, 2011).

167   "He was a very aggressive leader": Former Samsung vice president, interview by the author, September 18, 2015.

167   "He was semifamous": Kim Titus, interview by the author, February 16, 2016.

168   "Samsung had a revolutionary new screen": Steve Kovach, "How Samsung Won and Then Lost the Smartphone War," *Business Insider,* February 26, 2015, https://www.businessinsider.com/samsung-rise-and-fall-2015-2.

168   "Some customers burned the product": Miyoung Kim, "Samsung: 'Fast Executioner' Seeks Killer Design," Reuters, March 23, 2012, https://www.reuters.com/article/uk-samsung/samsung-fast-executioner-seeks-killer-design-idUSLNE82M01420120323.

168   "the difference is truly": "Summary of Executive-Level Meeting."

168   $885 million in revenue: Kim, "Samsung: 'Fast Executioner' Seeks Killer Design."

169   "single direction . . . all the way": Chang Dong-hoon, interview by the author, July 21, 2011.

169   little-watched product launch: Singapore Telecommunications Limited, "Singtel and Samsung Bring to Singapore the 'Smart Life' with Galaxy S" (news release), May 27, 2010, https://www.singtel.com/about-Us/news-releases/singtel-and-samsung-bring-singapore-smart-life-samsung-galaxy-s.

169   Terlato family's "Galaxy" red blend: Ed Ho, interview by the author, October 5, 2017.

169   It came in four variations: Kat Hannaford, "Samsung Galaxy S Known as Vibrant, Captivate and Fascinate with US Carriers," *Gizmodo,* June 28, 2010, https://gizmodo.com/samsung-galaxy-s-known-as-vibrant-captivate-and-fascin-5574325.

169   "thermonuclear war": Walter Isaacson, *Steve Jobs* (New York: Simon & Schuster, 2011), p. 512.

169   wary of endangering the relationship: Poornima Gupta, Miyoung Kim, and Dan Levine, "How the Apple-Samsung War Is Completely Different Than Any Other Tech Rivalry in History," Reuters, February 10, 2013, https://www.businessinsider.com/apple-and-samsung-2013-2.

169 Apple drafted a proposal to license: "Samsung-Apple Licensing Discussion" (Defendant's Exhibit No. 586.001), *Apple vs. Samsung Electronics*, case no. 11-1846 (U.S. District Court, Northern District of California, San Jose Division, 2011).

170 It demanded $2.5 billion in damages: Ashby Jones, "Apple's Pretrial Salvo," *The Wall Street Journal*, July 25, 2012, https://www.wsj.com/articles/SB10 0008723963904432954045775473522815 67654.

170 Samsung quickly countersued: Julia Kollewe, "Samsung Counter-sues Apple over Battle for £100bn Smartphone Industry," *The Guardian*, April 22, 2011, https://www.theguardian.com/technology/2011/apr/22/samsung-apple-lawsuits-smartphones.

170 "He likes to be mysterious": Former Galaxy marketer, interview by the author, October 9, 2017.

170 "I didn't like Todd's team": Samsung mobile employee, interview by the author, February 16, 2016.

171 "We need more creativity!": Galaxy marketer, interview by the author, November 8, 2016.

171 "The approach was very traditional": "Transcript of Proceedings," *Apple vs. Samsung Electronics*, case no. 5:12-cv-00530 (U.S. District Court, Northern District of California, San Jose Division, April 14, 2014), vol. 7, pp. 1596–1622.

171 "We were reading forecasts": "Transcript of Proceedings," p. 1612.

171 The "Dale Sohn regime": Galaxy marketer, interview by the author, November 8, 2016.

172 "When you worked with Dale": Paul Golden, interview by the author, February 17, 2016.

172 "He brought in a lot": Peter Skarzynski, interview by the author, December 28, 2016.

172 Saturdays from Memorial Day to Labor Day: Bill Ogle, interview by the author, January 6, 2017.

172 "He put a slogan up": Peter Skarzynski, interview by the author, December 28, 2016.

172 "If you wanted to talk": Galaxy marketer, interview by the author, November 8, 2016.

173 "I was very impressed": "Transcript of Proceedings," p. 1668.

173 "It was a dream job": "Transcript of Proceedings," pp. 1667–69.

173 "knits the whole organization together": Geraldine E. Willigan, "High-Performance Marketing: An Interview with Nike's Phil Knight," *Harvard Business Review*, July–August 1992, https://hbr.org/1992/07/high-performance-marketing-an-interview-with-nikes-phil-knight.

173 "We were trying to create": Phil Knight, *Shoe Dog: A Memoir by the Creator of Nike* (New York: Scribner, 2011), p. 250.

174   $130 million in revenue: Roy S. Johnson, "The Jordan Effect," *Fortune,*
      June 22, 1998, http://archive.fortune.com/magazines/fortune/fortune
      _archive/1998/06/22/244166/index.htm.

174   eight times the sales: Kurt Badenhausen, "How Michael Jordan Still Makes
      $100 Million a Year," *Forbes,* March 11, 2015, https://www.forbes.com
      /sites/kurtbadenhausen/2015/03/11/how-new-billionaire-michael-
      jordan-earned-100-million-in-2014/#12bdd8c2221a.

174   LeBron James to sign: American Advertising Federation, "Todd Pendleton,
      2013 Advertising Hall of Fame Induction," posted by YouTube user Ameri-
      can Advertising Federation on July 24, 2014, https://www.youtube.com
      /watch?v=DaQcWMK8VQA.

174   sign Kobe Bryant: Dow Jones Newswires, "Nike Signs Kobe Bryant to $40
      Million Contract," *The Wall Street Journal,* June 25, 2003, https://www.wsj
      .com/articles/SB105649507860753000.

175   "no one wanted to go": Brian Wallace, interview by the author, January 6,
      2016.

175   "old cubicle farm": Brian Wallace, email message to the author, December
      2, 2016.

175   "I was escorted into": Brian Wallace, interview by the author, January 6,
      2016.

175   photo of Jobs on his desk: Former Galaxy marketer, interview by the au-
      thor, October 23, 2017.

175   "Yes, I think I can": Brian Wallace, interview by the author, January 6, 2016.

175   They wanted to take down: Kovach, "How Samsung Won and Then Lost."

175   "No, no": Brian Wallace, interview by the author, January 6, 2016.

176   "That's such a step down": Ibid.

176   "Oh my god, there's this": Ibid.

176   The average turnover: Former Galaxy marketer, interview by the author,
      October 23, 2017.

176   "just product and voiceover": "Transcript of Proceedings," pp. 1671–82.

176   "telling a story around their network": "Transcript of Proceedings," pp.
      1671–82.

176   "Hello Kitty": Clyde Roberson, interview by the author, April 6, 2016.

176   "One commercial showed a woman": Bill Ogle, interview by the author,
      January 6, 2017.

177   "trendy, Eurotrash, rich, white people": Galaxy marketer, interview by the
      author, November 8, 2016.

177   "At Samsung, you had one day": Petey McKnight, interview by the author,
      January 24, 2017.

177   "he's an extremely good marketer": Galaxy marketer, interview by the au-
      thor, November 8, 2016.

178   six hundred pairs of Nike shoes: Michal Lev-Ram, "Samsung's Road to Global Domination," *Fortune,* January 22, 2013, http://fortune.com/2013/01/22/samsungs-road-to-global-domination/.

178   "he would point out things": Galaxy marketer, interview by the author, November 8, 2016.

178   "He was a total conspiracy theorist": Former Galaxy marketer, interview by the author, January 10, 2017.

178   a "ghost" who "didn't talk": Former Galaxy marketer, interview by the author, October 9, 2017.

178   "He said no more than he": Brian Wallace, interview by the author, September 28, 2016.

178   remove the letter *H*: Former Galaxy marketer, email message to the author, February 1, 2017.

178   "How much would it take": Brian Wallace, interview by the author, January 6, 2016.

178   "At Samsung, there's always": Galaxy marketer, interview by the author, November 8, 2016.

178   "a start-up with the billion-dollar": Former Galaxy marketer, interview by the author, January 10, 2017.

179   "You could be partying with Jay-Z": Former Galaxy marketer, interview by the author, January 10, 2017.

179   "We had to be somewhat insular": Galaxy marketer, interview by the author, November 8, 2016.

179   Dale provided air cover: Brian Wallace, interview by the author, January 6, 2016.

179   Dale shortened the time frame: Ibid.

179   In fact, the team: Kovach, "How Samsung Won and Then Lost."

19: CULT OF STEVE

180   "Samsung = relentless innovation": "Transcript of Proceedings," *Apple vs. Samsung Electronics,* case no. 5:12-cv-00530 (U.S. District Court, Northern District of California, San Jose Division, April 14, 2014), vol. 7, pp. 1672–73.

181   "Less stylish, less innovative": Former Samsung vice president for channel and launch marketing Petey McKnight, email message to the author, May 23, 2017.

181   "Android people consider": Galaxy marketer, interview by the author, November 8, 2016.

181   split up focus groups: Galaxy marketer, interview by the author, November 8, 2016.

181   "There was this growing base": Brian Wallace, interview by the author, January 6, 2016.

181   showed Samsung leading: Galaxy marketer, interview by the author, November 8, 2016.

182   black rectangle with rounded edges: Jody Akana et al., Portable Display Device, U.S. Design Patent No. US D670,286 S, filed November 23, 2011, issued November 6, 2012, https://assets.sbnation.com/assets/1701443/USD670286S1.pdf.

182   "Steven P. Jobs, 1955–2011": John Markoff, "Apple's Visionary Redefined Digital Age," *The New York Times*, October 5, 2011 (online version published a day earlier under a different title), https://www.nytimes.com/2011/10/06/business/steve-jobs-of-apple-dies-at-56.html.

182   "Outside the flagship Apple store": Matt Richtel, "Jobs's Death Draws Outpouring of Grief and Tributes," *The New York Times*, October 5, 2011, https://www.nytimes.com/2011/10/06/technology/jobss-death-prompts-grief-and-tributes.html.

182   "Here's to the crazy ones": Apple Inc., "Celebrating Steve," posted by YouTube user Marc Shulz on October 26, 2011, https://www.youtube.com/watch?v=Zj3x_3ZxA_8.

182   The only Asian executive: Simon Mundy, "Samsung Heir Apparent Lee Jae-yong Faces Tough Investor Test," *Financial Times*, September 30, 2013, https://www.ft.com/content/f36f3bc0-f800-11e2-87ec-00144feabdc0.

182   "Unfortunately, Steve Job's [*sic*] passing": "Subject: Use Google to Attack Apple?" (Plaintiff's Exhibit No. 215), *Apple vs. Samsung Electronics*, case no. 5:12-cv-00530 (U.S. District Court, Northern District of California, April 14, 2014).

183   "someone from headquarters": Former Galaxy marketer, interview by the author, February 20, 2016.

183   didn't want to anger Apple: Several former Samsung executives, interviews by the author, March 2015–August 2016.

183   American executives disagreed: Several former Samsung executives, interviews by the author, March 2015–August 2016.

183   "If it worked, we could": Galaxy marketer, interview by the author, November 8, 2016.

183   "Hey Michael, we are going": "Subject: Use Google to Attack Apple?"

183   take advantage of Jobs's death: Brian X. Chen, "Samsung Saw Death of Apple's Jobs as a Time to Attack," *The New York Times*, April 16, 2014, https://bits.blogs.nytimes.com/2014/04/16/samsung-saw-death-of-steve-jobs-as-a-time-to-attack/.

183   briefly held back: Former Galaxy marketer, email message to the author, October 29, 2017.

184 Trucks carrying fresh apples: Petey McKnight, interview by the author, January 24, 2017.

184 "We had one objective": Brian Wallace, interview by the author, January 6, 2016.

20: COKE PEPSI REDUX

185 "If this doesn't work": Brian Wallace, interview by the author, January 6, 2016.

185 The majority opted for Pepsi: Bernice Kanner, "Coke vs. Pepsi: The Battle of the Bubbles," *New York,* October 5, 1981, https://books.google.com /books?id=BuYCAAAAMBAJ&pg=PA21&source=gbs_toc_r&cad= 2#v=onepage&q&f=false.

186 Coke, in response, became convinced: Roger Enrico and Jesse Kornbluth, *The Other Guy Blinked: How Pepsi Won the Cola Wars* (New York: Bantam Books, 1986).

186 tone down the approach: Brian Wallace and two former Samsung marketing executives, interviews by the author, January 24, 2017–February 15, 2017.

186 "You never attack the people": Brian Wallace, interview by the author, January 6, 2016.

186 *The customers were Apple's victims:* Wallace and two other former marketing executives. interviews by author. January 6–February 25, 2016.

186 "I think this is done": Steve Kovach, "Samsung Is Going Right for Apple Fanboys' Jugular with Its Latest Commercial," *Business Insider,* November 22, 2011, https://www.businessinsider.com/samsung-galaxy-s-ii-commercial-2011-11.

186 "It was very confusing": Brian Wallace, interview by the author, September 28, 2016.

187 half-eaten ice cream cone: Brian Wallace, interview by the author, January 6, 2016.

187 "The Samsung [response]": Ibid.

187 They were aghast: Ibid.

187 marketing development funds: Petey McKnight, interview by the author, January 24, 2017.

187 He annoyed Samsung headquarters: Galaxy marketer, interview by the author, November 8, 2016.

188 "I just need some help": Bobby Hundreds, "Inside 72andSunny, Advertising's Bright Horizon," *The Hundreds,* August 29, 2014, https://thehundreds .com/blogs/bobby-hundreds/72andsunny.

188 a campaign called "Unhate": Riazatt Butt, "Benetton Tears Down Pope-

Kissing Ads After Vatican Legal Threat," *The Guardian,* November 17, 2011, https://www.theguardian.com/world/2011/nov/17/benetton-pope-kissing-ads.

188   "in the field of advertising": Butt, "Benetton Tears Down Pope-Kissing Ads."

188   "I expect us to be": Hundreds, "Inside 72andSunny, Advertising's Bright Horizon."

188   "high pressure": Ibid.

188   "We had to get concepts": Ibid.

188   He assigned Joanne Lovato: Petey McKnight and two other executives, interviews by the author, January 24, 2017–October 20, 2017.

188   thick technical tome: Petey McKnight, interview by the author, January 24, 2017.

189   Three times a week: Ibid.

189   ad-lib their parts: Former Galaxy marketer, interview by the author, October 23, 2017.

189   "We don't have a campaign here": Brian Wallace, interview by the author, September 28, 2016.

189   frantic all-nighter: Brian Wallace, interview by the author, September 28, 2016.

21: THE NEXT BIG THING

190   It started, as before: Samsung Electronics, "Samsung Galaxy S II (The Next Big Thing) Commercial," posted by YouTube user iAnimationProduction on November 22, 2011, https://www.youtube.com/watch?v=GWnunavN4bQ.

191   "God damn! . . . We've got a campaign!": Brian Wallace, interview by the author, September 28, 2016.

191   Five days later: Former Galaxy marketer, interview by the author, October 23, 2017.

191   They proceeded to leak the film: Effie Awards, "Samsung Mobile: The Next Big Thing Is Already Here," 2013, https://effie.org/case_database/case/NA_2013_6977. Full case study not available online; it is in the author's possession.

192   fastest-growing brands on Facebook: Association of National Advertisers, "Speaker Profile: Todd Pendleton," https://www.ana.net/ajax/speaker/id/29267.

192   "We are the fastest-growing brand": Samsung Electronics, "Samsung Galaxy S III Launch Press Conference," posted by YouTube user GamerStuff on June 21, 2012, https://www.youtube.com/watch?v=Mppharp0R3Y.

192 "Get ready to take out": Chenda Ngak, "Samsung Galaxy S II Ad Mocks Apple Fans," *CBS News,* November 23, 2011, https://www.cbsnews.com /news/samsung-galaxy-s-ii-ad-mocks-apple-fans/.

192 During the third quarter of 2011: Hyunjoo Jin, "Samsung Surges Past Apple in Smartphones, Upbeat on Q4," Reuters, October 28, 2011, https:// www.reuters.com/article/us-samsung/samsung-surges-past-apple-in-smartphones-upbeat-on-q4-idUSTRE79R0B620111028.

192 No longer was the smartphone war: Kovach, "How Samsung Won and Then Lost."

192 Concerned emails from South Korea: Brian Wallace, interview by the author, January 6, 2016.

192 Samsung, declared Ben Bajarin: Ben Bajarin, "The Flaw in Samsung's Anti-iPhone Commercial," *Tech.pinions,* November 23, 2011, https://techpin ions.com/the-flaw-in-samsungs-anti-iphone-commercial/4240.

192 In very short order: Brian Wallace, interview by the author, January 6, 2016.

192 "Todd and I were fired probably": Ibid.

192 More than a dozen former colleagues: Fourteen former colleagues and employees in the Samsung Telecommunications American marketing office or related offices, interviews by the author, January 2016–December 2017.

192 One remembered Pendleton telling: Former Galaxy marketer, interview by the author, October 23, 2017.

192 "Don't quit, don't quit": Brian Wallace, interview by the author, January 6, 2016.

192 "Todd went rogue": Galaxy marketer, interview by the author, November 8, 2016.

192 Frustrated that the U.S. team: Seven former Galaxy marketers, interviews by the author, January 2015–February 2017.

193 "Every time we broke a rule": Brian Wallace, interview by the author, January 6, 2016.

193 When Samsung's worldwide mobile: Brian Wallace, email message to the author, April 29, 2019.

193 "low-trust societies": Francis Fukuyama, *Trust: The Social Virtues and the Creation of Prosperity* (New York: Free Press, 1995), pp. 127–45.

193 "I'm thinking we are gonna": Brian Wallace, interview by the author, January 6, 2016.

193 "The US team was outperforming": Kovach, "How Samsung Won and Then Lost."

193 the more complicated their relationship: Kovach, "How Samsung Won and Then Lost." Fifteen members of the marketing team and related teams recalled the same events reported in the *Business Insider* article.

194   The Samsung auditorium was filled: Brian Wallace, interview by the author, January 6, 2016.

194   "The [Korean] executives told the employees": Kovach, "How Samsung Won and Then Lost."

194   It was a Samsung ritual: Galaxy marketer, interview by the author, November 8, 2016.

194   "The concept of limited good": Ibid.

194   "is the most bizarro place": Brian Wallace, interview by the author, January 6, 2016.

194   Fortunately, the heir apparent: Former Galaxy marketer, interview by the author, October 23, 2017.

22: GALAXY TRILOGY

195   Though they hadn't seen the phone: Fifteen former Galaxy marketers, interviews by the author, January 2016–February 2018.

195   "To be honest, it was": Dan Graziano, "Samsung Reveals the Strategy That Kept the Galaxy S III a Secret," *BGR,* June 14, 2012, https://bgr.com/2012/06/14/samsung-galaxy-s-iii-top-secret/.

196   But the U.S. marketing team's strategy: Fifteen former Galaxy marketers, interviews by the author, January 2016–February 2018.

196   "A lot of those features": Galaxy marketer, interview by the author, November 10, 2016.

196   "You know, Todd": Samsung Electronics, "Samsung Galaxy S III Launch Press Conference," posted by YouTube user GamerStuff on June 21, 2012, https://www.youtube.com/watch?v=Mppharp0R3Y.

196   selling nine million preordered units: Reuters, "Samsung Gets 9 Million Orders for New Galaxy Phone: Report," May 18, 2012, https://www.reuters.com/article/us-samsung/samsung-gets-9-million-preorders-for-new-galaxy-phone-report-idUSBRE84H00X20120518.

196   forty million units in the first: Luke Westaway, "Samsung Galaxy S3 Breezes Past 40 Million Sales Mark," *CNET,* January 15, 2013, https://www.cnet.com/news/samsung-galaxy-s3-breezes-past-40-million-sales-mark/.

196   *TechRadar* declared it: Gareth Beavis, "Number 1: Galaxy S3," *TechRadar,* July 27, 2012, https://web.archive.org/web/20120921031030/http://www.techradar.com/news/phone-and-communications/mobile-phones/20-best-mobile-phones-in-the-world-today-645440/page:21.

196   At the Mobile World Congress: Kevin Thomas, "Samsung Galaxy S3 Scoops Best Smartphone Award at MWC 2013," 3G, February 27, 2013, https://3g.co.uk/news/samsung-galaxy-s3-scoops-best-smartphone-award-at-mwc-2013.

196 "the Ferrari of the Android circuit": Natasha Lomas, "Samsung Galaxy S3 Review," *CNET*, July 24, 2012, https://www.cnet.com/reviews/samsung-galaxy-s3-review/.

196 "best phone of 2012": Luke Westaway, "Best Smart Phone of the Year, as Voted For by CNET UK Readers," *CNET*, December 20, 2012, https://www.cnet.com/news/best-smart-phone-of-the-year-as-voted-for-by-cnet-uk-readers/.

196 hitting $5.9 billion, a 79 percent increase: Richard Lawler, "Samsung's Q2 2012 Earnings Show $5.86 Billion Operating Profit, That's a Lot of Galaxy S IIIs," *Engadget*, July 26, 2012, https://www.engadget.com/2012/07/26/samsungs-q2-2012-earnings-record-profit/.

196 seven hundred highly technical questions: Nilay Patel, "Apple vs. Samsung: Inside a Jury's Nightmare," *The Verge*, August 23, 2012, https://www.theverge.com/2012/8/23/3260463/apple-samsung-jury-verdict-form-nightmare.

197 "We wanted to make sure": Dan Levine, "Jury Didn't Want to Let Samsung Off Easy in Apple Trial: Foreman," Reuters, August 25, 2012, https://www.reuters.com/article/us-apple-samsung-juror/jury-didnt-want-to-let-samsung-off-easy-in-apple-trial-foreman-idUSBRE87O09U20120825.

197 "Unless you're smoking crack": Deborah Netburn, "Judge Koh Snaps at Apple Lawyer, Asks if He's on Crack," *Los Angeles Times*, August 16, 2012, https://www.latimes.com/business/la-xpm-2012-aug-16-la-fi-tn-apple-lawyer-smoking-crack-20120816-story.html.

197 She later decided that: Christina Bonnington, "Judge Calls for New Trial in *Apple v. Samsung*, Slashes Apple's Award by 40 Percent," *Wired*, March 1, 2013, https://www.wired.com/2013/03/koh-slashes-apple-damages/.

197 A second trial between: Bonnington, "Judge Calls for New Trial."

197 Samsung took home some legal victories: Charles Arthur, "Samsung Galaxy Tab 'Does Not Copy Apple's Designs,'" *The Guardian*, October 18, 2012, https://www.theguardian.com/technology/2012/oct/18/samsung-galaxy-tab-apple-ipad; Associated Press, "Samsung Wins Korean Battle in Apple Patent War," August 24, 2012, https://www.cbc.ca/news/business/samsung-wins-korean-battle-in-apple-patent-war-1.1153862; Mari Saito and Maki Shiraki, "Samsung Triumphs over Apple in Japan Patent Case," Reuters, August 31, 2012, https://in.reuters.com/article/us-apple-samsung-japan/samsung-wins-over-apple-in-japan-patent-case-idINBRE87U05R20120831.

197 "Today we're taking it": Apple Inc., "Apple Special Event 2012—iPhone 5 Introduction," posted by YouTube user the unofficial AppleKeynotes channel on September 14, 2012, https://www.youtube.com/watch?v=82dwZYw2M00&t=911s.

197 "As the data flowed in": Michal Lev-Ram, "Samsung's Road to Global Domination," *Fortune*, January 22, 2013, http://fortune.com/2013/01/22/samsungs-road-to-global-domination/.

198   talking points pulled from social-media: Lev-Ram, "Samsung's Road to Global Domination."

198   "The headphone jack is going": Samsung Electronics, "Samsung Galaxy S3 Ad: The Next Big Thing Is Already Here," posted by YouTube user DYP WWI on September 19, 2012, https://www.youtube.com/watch?v=NgOA7qqDQbE.

198   a man named Ji: Former Galaxy marketer, interview by the author, January 10, 2017.

198   "Help them and be open": Brian Wallace, interview by the author, September 28, 2016.

198   "They were accused of bribing": Kovach, "How Samsung Won and Then Lost."

199   Pendleton managed to keep his head: Former Galaxy marketer, interview by the author, October 23, 2017.

199   The auditors found nothing out: Kovach, "How Samsung Won and Then Lost."

199   feeling of being under attack: Four Galaxy marketers, interviews by the author, February 2015–January 2017.

199   seventy million online views: Lev-Ram, "Samsung's Road to Global Domination."

199   "Samsung, the market leader in smartphones": Ian Sherr and Evan Ramstad, "Has Apple Lost Its Cool to Samsung?" *The Wall Street Journal,* January 28, 2013, https://www.wsj.com/articles/SB10001424127887323854904578264090074879024.

199   The article riveted the tech industry: Kovach, "How Samsung Won and Then Lost."

200   He had a new $15 million ad-libbed: Jason Evangelho, "With Hilarious 2-Minute Super Bowl Ad, Samsung Steals Cool Factor from Apple," *Forbes,* February 3, 2013, https://www.forbes.com/sites/jasonevangelho/2013/02/03/with-hilarious-2-minute-super-bowl-ad-samsung-officially-steals-cool-factor-from-apple/#130b5461326a.

200   "We actually can't say": "New Samsung Commercial Mocks Apple Lawsuits in SuperBowl Teaser Ad Feat. Odenkirk, Rudd & Rogen," posted by YouTube user Zef Cat on February 1, 2013, https://www.youtube.com/watch?v=vf2xRupwzoA.

200   "a barrage of not-so-subtle jabs": Evangelho, "With Hilarious 2-Minute Super Bowl Ad."

200   "We have a lot of work": Daisuke Wakabayashi, "Apple Considered Firing Longtime Ad Agency TBWA," *The Wall Street Journal,* April 4, 2014, https://www.wsj.com/articles/apple-considered-firing-longtime-ad-agency-1396647347.

200   "We feel it too and it hurts": Zac Hall, "Internal Emails Reveal Phil Schiller

Shocked by Response from Apple's Ad Agency over Marketing Direction," 9to5Mac, April 7, 2014, https://9to5mac.com/2014/04/07/internal-emails-reveal-phil-schiller-shocked-by-response-from-apples-ad-agency-over-marketing-direction/.

200 "To come back and suggest": Ibid.

201 "Consumers want what we don't have": "FY'14 Planning Offsite" (Defendant's Exhibit No. 413.001), *Apple vs. Samsung Electronics,* case no. 12-CV-00630-LHK (U.S. District Court, Northern District of California, San Jose Division, 2011), http://cdn3.vox-cdn.com/assets/4244447/DX-413 .pdf.

201 The spots were called "embarrassing": Sean Hollister, "Apple's New Mac Ads Are Embarrassing," *The Verge,* June 28, 2012, https://www.theverge .com/2012/7/28/3197951/apple-olympic-ads-genius-bar-embarrass.

201 Apple pulled them: Jim Edwards, "Apple 'Spends Millions of Dollars on Ads That Don't Even Run," *Business Insider,* June 11, 2013, https://www .businessinsider.com/apples-advertising-under-phil-schiller-2013-6.

201 In 2013 Apple raised: Spencer E. Ante, "Apple Closes U.S. Ad-Spending Gap with Samsung," *The Wall Street Journal,* April 8, 2014, https://blogs.wsj.com /digits/2014/04/08/apple-closes-u-s-ad-spending-gap-with-samsung/.

201 It brought its advertising work: Ann-Christine Diaz and Maureen Morrison, "For Apple, Marketing Is a Whole New Game," *Ad Age,* June 9, 2014, https://adage.com/article/agency-news/apple-marketing-a-game/293605.

23: THE ECOSYSTEM

202 "There is no obstacle": T.J. Kang, interview by the author, January 29, 2016.

202 "Readers Hub, Video Hub, Music Hub": Ibid.

202 "We had a Kindle": Ibid.

203 ThinkFree was too small: Eric Lai, "What's Stopping ThinkFree from Liberating Businesses from Microsoft Office?" Computerworld, June 15, 2007, https://www.computerworld.com/article/2541855/what-s-stopping-thinkfree-from-liberating-businesses-from-microsoft-office-.html.

203 "I felt that at Samsung": T.J. Kang, interview by the author, January 29, 2016.

203 T.J. had been recruited: Ho Soo Lee, interview by the author, September 5, 2016.

203 Samsung's Chairman Lee II: Yonhap News Agency, "Lee Kun-hee Espouses Importance of Software," *JoongAng Korea Daily,* August 18, 2011, http:// koreajoongangdaily.joins.com/news/article/article.aspx?aid=2940367.

204 "They came to the realization": T.J. Kang, interview by the author, January 29, 2016.

204   a five-hundred-person group housed in: T.J. Kang, email message to the author, August 24, 2017.

204   "to match what Apple was doing": T.J. Kang, interview by the author, January 29, 2016.

204   "I very much wanted T.J.": Ho Soo Lee, interview by the author, September 5, 2016.

204   "Why don't you come": T.J. Kang, interview by the author, January 29, 2016.

204   "Samsung had eight or nine": Ibid.

204   Basic applications like Skype: Joern Esdohr, "Petition: VoIP for Bada!" *Joern Esdohr,* September 13, 2010, http://www.joernesdohr.com/bada/petition-voip-for-bada/.

204   the GPS system was poor: Gareth Beavis, "Samsung Wave Review," *TechRadar,* August 8, 2010, https://www.techradar.com/reviews/phones/mobile-phones/samsung-wave-680092/review/9.

204   chicken-and-egg problem: T.J. Kang, interview by the author, January 29, 2016.

204   "You must have some idea": Ibid.

205   Shin thought that acquisitions: Eight former Samsung executives, interviews by the author, January 2015–February 2017.

205   "They didn't want to wait": T.J. Kang, interview by the author, January 29, 2016.

205   Samsung's cash pile would soon: Daisuke Wakabayashi and Min-Jeong Lee, "Samsung's 'Good' Problem: A Growing Cash Pile," *The Wall Street Journal,* May 8, 2013, https://www.wsj.com/articles/SB10001424127887323798104578454440307100754.

205   Chairman Lee II and his son: Yonhap News Agency, "Lee Kun-hee Espouses Importance of Software." T.J. Kang, Daren Tsui, and Ed Ho, interviewed by the author, added that Jay Lee came out in favor of software and M&A.

205   Samsung set aside $1.1 billion: Jonathan Cheng, Evelyn M. Rusli, and Min-Jeong Lee, "Samsung Plays Catch-up on Software," *The Wall Street Journal,* October 7, 2013, https://www.wsj.com/articles/a-weak-spot-for-samsung-1381190699.

205   began poaching software engineers: T.J. Kang, interview by the author, March 27, 2016.

205   Work would eventually begin: Sam Byford, "Take a Look Inside Samsung's New $300 Million Silicon Valley Campus," *The Verge,* February 27, 2013, https://www.theverge.com/2013/2/27/4034988/samsung-silicon-valley-campus-pictures.

205   "There was a rumor that Palm": T.J. Kang, interview by the author, January 29, 2016.

205   T.J. was prepared to negotiate: Ibid.

205   tryst with a female contractor: David Goldman, "Marc Hurd's Sex Scandal Letter Emerges," *CNNMoney,* December 30, 2011, https://money.cnn .com/2011/12/30/technology/hurd_letter/index.htm.

205   "There was a fight": T.J. Kang, interview by the author, January 29, 2016.

205   In Cannes at the music technology conference: Ibid.

206   Tsui and Ho became early movers: Peter Delevett, "Mercury News Interview: Daren Tsui, Co-founder, CEO of mSpot," *San Jose Mercury News,* June 24, 2011, https://www.mercurynews.com/2011/06/24/mercury-news-interview-daren-tsui-co-founder-ceo-of-mspot/.

206   "Until then, you stored": T.J. Kang, interview by the author, January 29, 2016.

206   Before the days of Spotify: Ibid.

206   "It's not just the idea": Ed Ho, interview by the author, October 5, 2017.

207   starting a platform called Pantheon: Delevett, "Mercury News Interview: Daren Tsui."

207   "door-to-door directions and Yellow Pages": Daren Tsui, interview by the author, January 10, 2017.

207   It was at Zip2: Ibid.

207   "This would be a great chance": Ibid.

207   "There's a team of people that sit": Ed Ho, interview by the author, January 3, 2017.

207   The only breaks were for lunch: Ibid.

208   As the pair sat through meeting after meeting: Ibid.

208   "The acquisition negotiations": Daren Tsui, interview by the author, January 10, 2017.

208   After nearly a year: Daren Tsui, email message to the author, April 21, 2019.

208   "We were overwhelmed": Daren Tsui, interview by the author, January 10, 2017.

208   "I kept pushing and pushing": T.J. Kang, interview by the author, January 29, 2016.

208   The deal went through in May 2012: Leena Rao, "Samsung Acquires Mobile Entertainment and Music Streaming Startup mSpot," *TechCrunch,* May 9, 2012, https://techcrunch.com/2012/05/09/samsung-acquires-mobile-entertainment-and-music-streaming-startup-mspot/.

208   "Our goal was to be able": Daren Tsui, interview by the author, January 10, 2017.

208   The Galaxy marketing team in Korea: T.J. Kang, interview by the author, January 29, 2016.

209   "After the launch": Ibid.

209 Samsung rushed out mSpot's: Daren Tsui, interview by the author, January 10, 2017.

209 to seven countries: Jon Fingas, "Samsung Expects Music Hub to Reach Competitors' Devices, More Countries," *Engadget*, January 27, 2013, https://www.engadget.com/2013/01/27/samsung-music-hub-to-reach-other-companies-devices-more-countries/.

209 Samsung's local sales offices complained: T.J. Kang, interview by the author, March 27, 2016.

209 "Samsung worked with Silicon Valley": Sumi Lim, interview by the author, December 17, 2015.

209 "You and what army": Vogelstein, *Dogfight*, p. 54.

210 Google paid an estimated $50 million: Donald Melanson, "Google Exec Calls Android Acquisition Its 'Best Deal Ever,'" *Engadget*, October 27, 2010, https://www.engadget.com/2010/10/27/google-exec-calls-android-acquistion-its-best-deal-ever/.

210 What a missed opportunity for Samsung: Six former Samsung executives and employees, interviews by the author, January 2015–February 2017.

210 In February 2013, seeing little progress: Sam Byford, "Samsung Finally Folding Bada OS into Tizen," *The Verge*, February 25, 2013, https://www.theverge.com/2013/2/25/4026848/bada-and-tizen-to-merge.

210 the two realized they needed: T.J. Kang, email message to the author, October 16, 2017.

210 hedge against Android: Hod Greeley, interview by the author, February 10, 2016.

210 In October 2013, Samsung put on: John Koetsier, "The Samsung Platform Emerges: Registration Opens for Samsung's First-Ever Developer Conference," *VentureBeat*, August 26, 2013, https://venturebeat.com/2013/08/26/the-samsung-platform-emerges-samsung-announces-first-ever-developer-conference/.

210 "We arrive the morning of": Hod Greeley, interview by the author, February 10, 2016.

210 "Being in hardware . . . they don't": Ibid.

211 Waze CEO Noam Bardin was candid: Waze executive, email message to the author, April 29, 2019. Hod Greeley recounted the same sequence of events in his interview with the author on February 10, 2016.

211 treated the prospect of a software partnership: Hod Greeley, interview by the author, February 10, 2016.

211 "It's a new feature": Ibid.

211 In 2013 Google acquired Waze: Julie Bort, "Waze Cofounder Tells Us How His Company's $1 Billion Sale to Google Really Went Down," *Business In-*

sider, August 13, 2015, https://www.businessinsider.com/how-google-bought-waze-the-inside-story-2015-8.

211 "sat there and looked at the devices": Hod Greeley, interview by the author, February 10, 2016.

211 In October 2014, Facebook acquired: David Rowan, "The Inside Story of Jan Koum and How Facebook Bought WhatsApp," *Wired UK,* May 1, 2018, https://www.wired.co.uk/article/whats-app-owner-founder-jan-koum-facebook.

211 By that time, Samsung had already: Richard Trenholm, "Samsung ChatOn Will Be Turned Off in February," *CNET,* December 19, 2014, https://www.cnet.com/news/samsung-chaton-will-be-turned-off-in-february/.

211 "A ten-person startup should be struggling": Hod Greeley, interview by the author, February 10, 2016.

212 Jobs was "ruthless": Universal Music Group executive, email message to the author, May 4, 2019. T.J. Kang confirmed being told these statements by Rob Wells in an email message to the author on July 23, 2019.

212 "We already knew back then": Universal Music Group executive, email message to the author, May 4, 2019.

212 "Right now you're beholden": Daren Tsui, interview by the author, January 10, 2017.

212 "We did everything properly": Ibid.

212 T.J. and Daren explained to Samsung's: T.J. Kang, interview by the author, January 29, 2016.

213 Samsung's hardware executives, led by: Ibid.

213 T.J. and Daren were disappointed: Ibid.

214 By the end, T.J. and Daren: Ibid.

214 "They [Samsung] blew it": Universal Music Group executive, email message to the author, May 4, 2019.

214 But what if Samsung didn't need: Daren Tsui, interview by the author, January 10, 2017.

24: WHITE GLOVE

215 appointed to head it: Joanne Lovato, LinkedIn profile, https://www.linkedin.com/in/joanne-lovato-24465b7/, retrieved January 4, 2016.

215 The Samsung executives were frustrated: Galaxy marketer, interview by the author, May 3, 2017.

215 "We would give products": Ibid.

216 When a member of the team noticed: Ibid.

216 Samsung executives in the White Glove: Shane Snow, "How Stars Like Jay Z and Martha Stewart End Up with Samsung Devices," *Fast Company,* Janu-

ary 13, 2014, https://www.fastcompany.com/3020215/the-real-reason-famous-people-like-jay-z-and-martha-stewart-use-samsung-phones.

216 "They're gonna give you a phone": Ibid.

216 "To some extent it was": Galaxy marketer, email message to the author, August 31, 2017.

216 Samsung hosted a celebrity dinner: Snow, "How Stars Like Jay-Z and Martha Stewart End Up with Samsung Devices."

216 "quietest influencers": Ibid.

217 "Why would you need that?": Samsung Electronics, "Samsung Unpacked 2013—Galaxy S4 Unpacked Event—FULL Presentation," posted by YouTube user ItsMobileTech on March 16, 2013, https://www.youtube.com/watch?v=iLmOEVijYOQ.

217 In shock at the cultural insensitivity: Former Galaxy marketer, interview by the author, February 20, 2016; Petey McKnight, interview by the author, January 24, 2017.

217 "In the middle of a red-hot": Molly Wood, "Samsung GS4 launch: Tone-Deaf and Shockingly Sexist," *CNET,* March 14, 2013, https://www.cnet.com/news/samsung-gs4-launch-tone-deaf-and-shockingly-sexist/.

218 young models to stand around: Samsung Electronics, "Samsung Electronics Announces Curved OLED in Korea" (news release), June 27, 2013, https://news.samsung.com/global/samsung-pioneers-an-era-of-flawless-picture-quality-with-curved-oled-tv.

218 a Korean ideal called *aegyo*: Aljosa Puzar and Yewon Hong, "Korean Cuties: Understanding Performed Winsomeness (*Aegyo*) in South Korea," *Asia Pacific Journal of Anthropology* 19 (2018): 333–49, https://www.tandfonline.com/doi/abs/10.1080/14442213.2018.1477826.

218 Pendleton's office was relieved: Galaxy marketer, interview by the author, November 8, 2016.

218 Samsung surpassed Apple: Brian X. Chen, "Samsung May Have Passed Apple in the U.S.—for Now," *The New York Times,* June 5, 2013, https://bits.blogs.nytimes.com/2013/06/05/samsung-may-have-surpassed-apple-in-the-u-s-for-now/.

218 Through his friend and sponsorship partner: Former Galaxy marketer, interview by the author, October 23, 2017.

218 With the help of Mitch Kanner: Snow, "How Stars Like Jay Z and Martha Stewart End Up with Samsung Devices."

218 "The soul of a hustler": Shawn Carter, Thomas Bell, Roland Chambers, and Kenneth Gamble, "What More Can I Say," track 3 on Jay-Z, *The Black Album,* Roc-A-Fella, 2003. The lyrics are available at https://genius.com/Jay-z-what-more-can-i-say-lyrics.

218 "I'm a business, man": Kanye West, Jon Brion, and Devo Springsteen, "Diamonds from Sierra Leone (Remix)," feat. Jay-Z, track 13 on Kanye West, *Late*

Registration, Roc-A-Fella, 2005. The lyrics are available at https://genius .com/Kanye-west-diamonds-from-sierra-leone-remix-lyrics.

218 Jay-Z was an entrepreneur: Alvin Hall, "Jay-Z: From Brooklyn to the Boardroom," *BBC News,* December 1, 2006, http://news.bbc.co.uk/2/hi /business/6160419.stm.

219 "First of all, we're in": Drake Baer, "The Heart of the Deal: Why Jay-Z Really Hooked Up with Samsung," *Fast Company,* August 5, 2013, https:// www.fastcompany.com/3015234/the-heart-of-the-deal-why-jay-z-really-hooked-up-with-samsung?jwsource=cl.

219 During a blazingly fast thirty days: Former Galaxy marketer, interview by the author, October 23, 2017.

219 Samsung agreed to pay: Andrew Hampp, "Jay-Z's New Blueprint: The Billboard Cover Story," *Billboard,* June 21, 2013, https://www.billboard.com /articles/news/1567999/jay-zs-new-blueprint-the-billboard-cover-story.

219 It was to turn Galaxys: Ibid.

219 "How do you navigate your way": Samsung Electronics, "Jay-Z 'Magna Carta Holy Grail' Samsung Commercial," posted by YouTube user PlugMatch on June 16, 2013, https://www.youtube.com/watch?v= lmZvjKPeAK0.

220 Viewers watching the NBA Finals: United Press International, "Jay-Z Releases Trailer for New Album 'Magna Carta Holy Grail,'" June 16, 2013, https://www.upi.com/Entertainment_News/2013/06/16/Jay-Z-releases-trailer-for-new-album-Magna-Carta-Holy-Grail/6091371437283/.

220 "We need to write the new rules": Samsung Electronics, "Jay-Z 'Magna Carta Holy Grail' Samsung Commercial."

220 "If 1 Million records gets SOLD": Jay-Z ("Mr. Carter," @S_C_), "If 1 Million records gets SOLD and billboard doesnt report it, did it happen? Ha. #newrules #magnacartaholygrail Platinum!!! VII IV XIII," Twitter, June 18, 2013, 2:34 A.M., https://twitter.com/s_c_/status/346682205168357377 ?lang=en.

220 "change music forever": "Jay-Z's $5 Million Samsung Deal Will Change Music Forever," *Business Insider,* July 4, 2013, https://www.businessinsider .com/jay-zs-5-million-samsung-deal-2013-7.

220 "Samsung made Jay-Z's new album profitable": Hampp, "Jay-Z's New Blueprint."

220 "1 million, 2 million, 3 million": Shawn Carter, Chauncey Hollis Jr., Darhyl Camper Jr., Michael Dean, "Somewhereinamerica," track 7 on Jay-Z, *Magna Carta Holy Grail,* Roc Nation, 2013. The lyrics are available at https://genius.com/Jay-z-somewhereinamerica-lyrics.

220 Pendleton and his team were out: Former Galaxy marketer, interview by the author, October 23, 2017.

221 At the stroke of midnight: Karissa Donkin, "Jay-Z's Magna Carta Holy

Grail Release a #Samsungfail," *The Star*, July 4, 2013, https://www.thestar
.com/entertainment/music/2013/07/04/jayzs_magna_carter_holy
_grail_release_a_samsungfail.html.

221 "Jay!" Pendleton called over: Former Galaxy marketer, interview by the author, October 23, 2017.

221 "#JayZ's sponsors at #Samsung": Hannah Bae (@hanbae), "#JayZ's sponsors at #Samsung prove themselves not only intrusive, but technically inept http://www.nytimes.com/2013/07/05/arts/music/jay-z-is-watching-and-he-knows-your-friends.html . . . #SamsungFail," Twitter, July 17, 2013, 10:31 A.M., https://twitter.com/hanbae/status/357311502623064064.

221 Pirated files went up online: Donkin, "Jay-Z's Magna Carta Holy Grail Release a #Samsungfail."

221 "Magna Carta Holy Fail?": Henry T. Casey, "Magna Carta Holy Fail? A Samsung/Jay-Z Release Postmortem," Contently, July 8, 2013, https://contently.com/2013/07/08/magna-carta-holy-fail-a-samsungjay-z-release-postmortem/.

221 Finally, around 2:00 A.M.: Donkin, "Jay-Z's Magna Carta Holy Grail Release a #Samsungfail."

221 Pendleton, as usual, managed: Former Galaxy marketer, interview by the author, October 23, 2017.

221 "There's a reason why Twitter": Casey, "Magna Carta Holy Fail?"

221 "disheartening" and "not cool": Breakfast Club Power 105.1 FM, "The Breakfast Club Classic—Jay Z Interview 2013," posted by YouTube user Breakfast Club Power 105.1 FM on July 11, 2013, https://www.youtube.com/watch?v=Y2TbLohPKI0.

222 the Recording Industry Association of America: David Greenwald, "RIAA Updates Rules Before Jay-Z's 'Magna Carta,'" *Billboard*, July 1, 2013, https://www.billboard.com/articles/news/1568753/riaa-updates-rules-before-jay-zs-magna-carta.

222 *Billboard* number one spot: Keith Caulfield, "Jay Z's 'Magna Carta' Debuts at No. 1 on Billboard 200 Chart," *Billboard*, July 16, 2013, https://www.billboard.com/articles/news/1924866/jay-zs-magna-carta-debuts-at-no-1-on-billboard-200-chart.

25: MILK

223 "I told you that wasn't going": T.J. Kang, interview by the author, March 27, 2016.

223 Samsung had established a task force: Ho Soo Lee, T.J. Kang, Daren Tsui, and Ed Ho, interview by the author, January 2016–August 2017.

223 "If you continue to crank out": T.J. Kang, interview by the author, March 27, 2016.

223 T.J., Daren, and Ed were determined: T.J. Kang, Daren Tsui, and Ed Ho, interviews by the author, March 2016–August 2017.

223 T.J. traveled back to San Jose: T.J. Kang, interview by the author, March 27, 2016.

224 "We had a terrible reputation": Han Kuk-hyun (director of Samsung UXCA), email message to the author's researcher Junyoub Lee, January 15, 2015.

224 The lab poached Bay Area talent: T.J. Kang, interview by the author, March 27, 2016.

224 "there was no mass production": Han Kuk-hyun, email to the author's researcher, January 15, 2015.

224 "We created this prototype": Neil Everette, email to the author, April 29, 2019. Everette published details of the dial on his personal website at https://www.neileverette.com/samsungmusic. The dial was also the subject of an untitled internal Samsung documentary that is in the author's possession.

224 "This is perfect for the radio": T.J. Kang, interview by the author, March 27, 2016.

224 "So look, this is our last chance": Daren Tsui, interview by the author, January 10, 2017.

224 "mom test": Ibid.

225 It helped that there was: T.J. Kang and Daren Tsui, interviews by the author, March 27, 2016, and January 10, 2017.

225 "How soon would you want": T.J. Kang, interview by the author, March 27, 2016.

225 An internal steering committee: Daren Tsui, email message to the author, October 13, 2017.

225 Daren didn't like the practice: Ibid.

225 The name "Milk" was controversial: Galaxy marketer, interview by the author, May 3, 2017.

225 But the committee went forward: Daren Tsui, email message to the author, October 13, 2017.

226 "Let's throw away the hardware-oriented": Simon Mundy et al., "Fresh Urgency for Samsung Reinvention Drive," *Financial Times*, January 27, 2014, https://www.ft.com/content/b6353d72-84eb-11e3-8968-00144feab7de.

226 T.J. delivered Milk Music: T.J. Kang, interview by the author, March 27, 2016.

226 Finally the software entrepreneur: Ibid.

26: THE SELFIE THAT SHOOK THE WORLD

227 A dedicated iPhone fan: Galaxy marketer, interview by the author, May 3, 2017.

227   Samsung was in its fifth year: Beth Snyder Bulik, "You Need to See This: Samsung's Oscars Campaign," *Ad Age,* March 1, 2014, https://adage.com /article/media/samsung-breaks-samsung-campaign-oscars/291938.

227   an estimated $20 million on ads: Suzanne Vranica, "Behind the Preplanned Oscar Selfie: Samsung's Ad Strategy," *The Wall Street Journal,* March 3, 2014, https://www.wsj.com/articles/behind-oscar-selfie-samsungs-ad-strategy-1393886249.

227   just off the main Oscars stage: Elizabeth Stamp, "The *Architectural Digest* Greenroom at the 2014 Oscars," *Architectural Digest,* January 31, 2014, https://www.architecturaldigest.com/gallery/oscar-greenroom-david-rockwell-slideshow.

227   the Twitter Mirror: Katherine Rosman, "Twitter Mirror: Celebrities' New Publicity Machine," *The Wall Street Journal,* December 10, 2013, https://www.wsj.com/articles/no-headline-available-1386699836.

228   "These things are like": Galaxy marketer, interview by the author, November 8, 2016.

228   "Todd provided the leadership": Galaxy marketer, interview by the author, May 3, 2017.

228   Mayo instructed her assistants: Former Galaxy marketer, interview by the author, January 10, 2017.

228   "Amber drove the brilliance": Galaxy marketer, interview by the author, May 3, 2017.

228   Mayo knew that no matter: Former Galaxy marketer, interview by the author, January 10, 2017.

228   "You have a loose idea": Former Galaxy marketer, interview by the author, January 10, 2017.

228   Samsung wasn't sure: Former Galaxy marketer, interview by the author, January 10, 2017.

229   "NOT IN TELEPROMPTER": "86th Annual Academy Awards, 2014" (script), act 6, March 2014. The beginning of the section reads "ELLEN IN AUDIENCE/RECORDING BREAKING SELFIE (ELLEN SITS NEXT TO MERYL STREEP) NOT IN TELEPROMPTER."

229   Amber's team worked with DeGeneres: Former Galaxy marketer, interview by the author, January 10, 2017; Galaxy marketer, interview by the author, May 3, 2017.

229   The spontaneity meant fewer guarantees: Former Galaxy marketer, interview by the author, January 10, 2017.

230   "I really hope that everyone jumps in": *The Ellen DeGeneres Show,* "Behind the Scenes at the Oscars," posted by YouTube user TheEllenShow on March 5, 2014, https://www.youtube.com/watch?v=7w4TMdCLNMM.

230   gray checkered bow tie: Unpublished photograph in the author's possession, March 2, 2014.

230 Some 43 million people: Emily Yahr, "How Many People Watched the 2014 Oscars? 43 Million, the Most in 10 Years," *The Washington Post*, March 3, 2014, https://www.washingtonpost.com/news/arts-and-entertainment /wp/2014/03/03/how-many-people-watched-the-2014-oscars-43-million-the-most-in-10-years/?utm_term=.0d56547662c9.

230 "We know that the most important": Academy of Motion Picture Arts and Sciences, "86th Academy Awards (2014 Oscars) FULL SHOW," posted by YouTube user Sebastian Vînătoru on April 7, 2014, https://www.youtube .com/watch?v=x_-lVPnytsA.

230 Cameras were conveniently not looking: *The Ellen DeGeneres Show*, "Jared Leto on Pizza and the Oscar Selfie," posted by YouTube user TheEllenShow on March 3, 2014, https://www.youtube.com/watch?v= 4zVtaGFU4lo.

230 "Happy" singer Pharrell Williams: Academy of Motion Picture Arts and Sciences, "86th Academy Awards."

231 "For those not at the event": Galaxy marketer, interview by the author, May 3, 2017.

231 "Meryl, here's my idea, okay?": Academy of Motion Picture Arts and Sciences, "86th Academy Awards."

233 "If only Bradley's arm was longer": Ellen DeGeneres (@TheEllenShow), "If only Bradley's arm was longer. Best photo ever. #oscars," Twitter, March 3, 2014, 12:06 P.M., https://twitter.com/theellenshow/status/44032 2224407314432?lang=en.

234 winners of ten Oscars who together: Harry Wallop, "Oscars 2014: The Most Famous 'Selfie' in the World (Sorry Liza)," *The Telegraph*, March 3, 2014, https://www.telegraph.co.uk/culture/film/oscars/10674655/Oscars-2014-The-most-famous-selfie-in-the-world-sorry-Liza.html.

234 tweeting from her usual iPhone: Chris Matyszczyk, "Ellen Does Samsung Oscars Selfie, Tweets from iPhone Backstage," *CNET*, March 2, 2014, https://www.cnet.com/news/ellen-does-samsung-oscars-selfie-tweets-from-iphone-backstage/.

234 Todd Pendleton was watching: Galaxy marketer, interview by the author, May 3, 2017.

234 "We just crashed Twitter": Academy of Motion Picture Arts and Sciences, "86th Academy Awards."

234 "Something is technically wrong": The Academy (@TheAcademy), "Sorry, our bad. #Oscars," Twitter, March 3, 2014, 12:15 P.M., https://twitter.com/ theacademy/status/440324450907475968.

234 "See, Meryl, what we did?": Academy of Motion Picture Arts and Sciences, "86th Academy Awards."

234 "Did you hear we broke Twitter?": *The Ellen DeGeneres Show*, "Behind the Scenes at the Oscars."

235 one million retweets: Caspar Llewellyn Smith, "Ellen DeGeneres' Oscar Selfie Beats Obama Retweet Record on Twitter," *The Guardian*, March 2, 2014, https://www.theguardian.com/film/2014/mar/03/ellen-degeneres-selfie-retweet-obama.

235 "It's not like you can": Russell Brand. "Truth Behind Oscar Selfie: Russell Brand The Trews (E07)," posted by YouTube user Russell Brand on March 7, 2014, https://www.youtube.com/watch?v=F78Bs9dwQZ0.

235 three million retweets: Eli Langer, "Ellen's Viral Selfie Leads to $3 Million Donation," *CNBC*, March 4, 2014, https://www.cnbc.com/2014/03/04/ellens-viral-selfie-leads-to-3-million-donation.html.

235 "It may seem odd": Wallop, "Oscars 2014: The Most Famous 'Selfie' in the World."

235 *Time* later included: "Oscar Selfie," in "100 Photos: The Most Influential Photographs of All Time," *Time*, October 18, 2016, http://100photos.time.com/photos/bradley-cooper-oscars-selfie.

235 "Hey, @TheEllenShow!": Lauren Jenkins (@jenks), "Hey, @TheEllen Show! We painted a picture of you at Twitter HQ. Come take a #selfie with us! @LizFiandaca @genaweave," Twitter, April 11, 2014, 6:18 A.M., https://twitter.com/jenks/status/454367882911031296.

235 "I thought it was a pretty": *The Ellen DeGeneres Show*, "Pres. Barack Obama on Ellen Breaking His Twitter Record," posted by YouTube user TheEllenShow on March 19, 2014, https://www.youtube.com/watch?v=8WmtUzftpS0.

235 "Four more years": David Jackson, "Obama's Victory Tweet Sets Record," *USA Today*, November 7, 2012, https://www.usatoday.com/story/theoval/2012/11/07/obama-election-tweet-election-2012/1688953/.

235 copyright owners: David Bugliari (Bradley Cooper's agent at the Creative Artists Agency), email message to the author, May 30, 2019.

236 "The earned media": MIPTV, "Keynote: Maurice Levy, Publicis Groupe—MIPTV 2014," posted by YouTube user mipmarkets on April 8, 2014, https://www.youtube.com/watch?time_continue=1605&v=1ipQk5DgM_s.

236 "It does seem high": "Ellen's Oscar Selfie: Worth $9 Billion?" *NBC News*, April 9, 2014, https://www.nbcnews.com/tech/social-media/ellens-oscar-selfie-worth-1-billion-n75821.

236 "king's ransom": "Ellen's Oscar Selfie: Worth $9 Billion?"

236 nine hundred Twitter mentions a minute: Vranica, "Behind the Preplanned Oscar Selfie."

236 "Samsung is telling better stories": Mark Bergen, "Forget the Selfie: Samsung Is Out-Innovating Apple in Marketing," *Ad Age*, April 11, 2014, https://adage.com/article/digital/fact-check-samsung-s-selfie-worth-a-billion/292625.

236   Samsung would ship 85 million: Eric Mack, "Samsung Is Undisputed King
       of Smartphones; Apple Still Distant Second," *Forbes*, April 30, 2014, https://
       www.forbes.com/sites/ericmack/2014/04/30/samsung-is-undisputed-
       king-of-smartphones-apple-still-distant-second/#30ae6ca31c61.

236   "If you look at that picture": Andrew McMains, "Samsung Tells the Story
       Behind the Selfie That Ate Hollywood (and Twitter)," *Adweek*, March 18,
       2014, https://www.adweek.com/brand-marketing/samsung-tells-story-
       behind-selfie-ate-hollywood-and-twitter-156385/.

27: RETURN TO TRADITION

237   They had fought, sacrificed, and succeeded: Steve Kovach, "How Samsung
       Won and Then Lost."

237   "I left Samsung destroyed": Brian Wallace, interview by the author, January
       6, 2016.

237   "We called that weight": Former Galaxy marketer, interview by the author,
       February 20, 2016.

237   The pressure from headquarters was rising: Ten former colleagues and em-
       ployees of Todd Pendleton, interviews and email correspondence with the
       author, January 2016–February 2018.

237   "One Samsung": Mark Bergen, "Samsung Restructures U.S. Marketing
       Team as Mobile Division Falters," *Ad Age*, November 4, 2014, https://adage
       .com/article/cmo-strategy/samsung-restructures-u-s-marketing-
       team/295385.

238   And Samsung brought back Jay-Z: Brad Wete, "Kanye West and Jay Z to
       Kick Off Samsung's SXSW Concert Series," *Billboard*, March 10, 2014,
       https://www.billboard.com/articles/events/sxsw/5930374/kanye-west-
       and-jay-z-to-kick-off-samsungs-sxsw-concert-series.

238   "Within a matter of hours": Daren Tsui, interview by the author, January
       10, 2017.

238   "Why are you guys doing this?": Ibid.

239   "They were tracking us": Ibid.

239   "Rosenberg suggested that": Daren Tsui, email to the author, August 1,
       2019.

239   in-house software called TouchWiz: Gordon Kelly, "How Google Used Mo-
       torola to Smack Down Samsung—Twice," *Forbes*, February 10, 2014,
       https://www.forbes.com/sites/gordonkelly/2014/02/10/how-google-
       used-motorola-to-smack-down-samsung-twice/#46f9912e21fa.

239   "a revamped interface that resembled": Brad Stone, "Google's Sundar Pi-
       chai Is the Most Powerful Man in Mobile," *Bloomberg Businessweek*, June 27,
       2014, https://www.bloomberg.com/news/articles/2014-06-24/googles-
       sundar-pichai-king-of-android-master-of-mobile-profile.

239   Google executives were furious: Ibid.

240   Pichai told Shin that Google: Ibid.

240   from nine to about twenty: Amir Efrati, "Google's Confidential Android Contracts Show Rising Requirements," *The Information,* September 26, 2014, https://www.theinformation.com/articles/Google-s-Confidential-Android-Contracts-Show-Rising-Requirements.

240   "set as the default": Ibid.

240   some Samsung executives thought: Four Samsung executives, interviews by the author, January 2016–October 2017.

240   "reminiscent of the": Ina Fried, "After Google Pressure, Samsung Will Dial Back Android Tweaks, Homegrown Apps," *Recode,* January 19, 2014, https://www.vox.com/2014/1/29/11622840/after-google-pressure-samsung-will-dial-back-android-tweaks-homegrown.

241   "An engagement ring": Brad Stone, Sam King, and Ian King, "Summer of Samsung: A Corruption Scandal, a Political Firestorm—and a Record Profit," *Bloomberg Businessweek,* June 27, 2017, https://www.bloomberg.com/news/features/2017-07-27/summer-of-samsung-a-corruption-scandal-a-political-firestorm-and-a-record-profit.

241   T.J. recalls a Google executive: T.J. Kang, interview by the author, March 27, 2016.

241   Google publicly announced the sale: Kelly, "How Google Used Motorola to Smack Down Samsung—Twice." Even though Google oversees the design and development of the Nexus phone, placing Google in competition with Samsung's hardware, the Nexus is manufactured by original equipment manufacturers (OEMs). Google's focus remains software, its search engine, and the Android operating system.

241   "G.S. Choi and Jay Y. Lee did not": T.J. Kang, interview by the author, March 27, 2016.

241   Samsung executives had a gnawing fear: Four former Samsung executives, interviews by the author, January 2016–February 2017.

242   "We have to try": T.J. Kang, interview by the author, March 27, 2016.

242   "The story of 1993": Cho Kwi-dong, "Former Samsung Adviser Fukuda, 'Time for Samsung to Forget About New Management and Reset,'" *Chosun Ilbo,* June 11, 2015, http://biz.chosun.com/site/data/html_dir/2015/06/11/2015061101667.html. This source is in Korean. The author's researcher translated the title and the quoted text into English.

242   "I cannot protect you guys anymore": T.J. Kang, interview by the author, March 27, 2016.

242   In May 2014, Chairman Lee suffered: In-Soo Nam, "Samsung Chairman Lee Kun-hee Regains Consciousness," *The Wall Street Journal,* May 25, 2014, https://blogs.wsj.com/digits/2014/05/25/samsung-chairman-lee-kun-hee-regains-consciousness/.

242   But then the chairman: Stone, King, and King, "Summer of Samsung."

28: VULTURE MAN

243 inheritance tax that could reach: Se Young Lee, "For Samsung Heirs, Little Choice but to Grin and Bear Likely $6 Billion Tax Bill," Reuters, June 5, 2014, https://www.reuters.com/article/us-samsung-group-succession-tax/for-samsung-heirs-little-choice-but-to-grin-and-bear-likely-6-billion-tax-bill-idUSKBN0EG2SC20140605.

243 "To put that in perspective": Ibid.

243 For almost twenty years: Shu-Ching Jean Chen, "Samsung's Lee Family Accused of Corrupt Dealings," Forbes, November 13, 2007, https://www.forbes.com/2007/11/13/samsung-corruption-investigation-face-markets-cx_jc_1113autofacescan01.html.

243 They stayed mere steps ahead: Seo Bo-mi, "Samsung's 'Innovative' Inheritance Technique," Hankyoreh, January 5, 2013, http://www.hani.co.kr/arti/english_edition/e_business/568398.html.

244 "Is it possible?": Samsung C&T shareholder, interview by the author, February 4, 2015.

244 Rumors were spreading: "Rumor of Samsung Group Chairman Lee Kun-hee's Death Raises Stock Prices," BusinessKorea, April 16, 2015, http://www.businesskorea.co.kr/news/articleView.html?idxno=10149.

244 wanted an end to the uncertainty: Eight Samsung shareholders, meetings and phone conversations with the author, May 2014–June 2016.

244 at a ratio of 1.5 for his mother: "Hong Ra-hee's Stock Valuation Soars . . . Emerging as the 'Nucleus' of Management Succession," Yonhap News Agency, March 21, 2017, https://www.yna.co.kr/view/AKR20170320161100008. This source is in Korean. The author's researcher translated the title and the quoted text into English.

244 My company sources told me: A dozen Samsung employees and recently departed employees, meetings and phone conversations with the author, May 2014–August 2015.

244 "Under the merger agreement": "The Samsung Group Is Merging Two Major Units as It Prepares to Transfer Power in the Founding Family," AFP, May 26, 2015, https://www.businessinsider.com/afp-samsung-to-merge-two-major-units-2015-5.

244 "That seems . . . odd?": Matthew Levine, "Samsung Group Is Doing a Cozy Merger," Bloomberg, July 2, 2015, https://www.bloomberg.com/opinion/articles/2015-07-01/samsung-group-is-doing-a-cozy-merger.

245 Jay Lee owned a 23 percent stake: Min-jeong Lee and Jonathan Cheng, "Samsung Heir Apparent Jay Y Consolidates Power with Merger," The Wall Street Journal, May 26, 2015, https://www.wsj.com/articles/samsung-heir-apparent-consolidates-power-with-merger-of-two-major-firms-1432603589.

245 "The attempt by the boards": Levine, "Samsung Group Is Doing a Cozy Merger."

245 "fight against charlatans": Max Abelson and Katia Porzecanski, "Paul Singer Will Make Argentina Pay," *Bloomberg,* August 7, 2014, https://www .bloomberg.com/news/articles/2014-08-07/argentinas-vulture-paul-singer-is-wall-street-freedom-fighter.

245 Elliott Management controlled $27 billion: Nate Raymond and Joseph Ax, "Manhattan U.S. Attorney's Top Deputy to Join Hedge Fund Elliott," Reuters, June 18, 2015, https://www.reuters.com/article/hedgefund-elliott-lawyer/manhattan-u-s-attorneys-top-deputy-to-join-hedge-fund-elliott-idUSL1N0Z423T20150618.

245 It was known for its: Sheelah Kolhatkar, "Paul Singer, Doomsday Investor," *The New Yorker,* August 20, 2018, https://www.newyorker.com/magazine /2018/08/27/paul-singer-doomsday-investor.

246 "Vote AGAINST the transaction": "Samsung C&T (KNX:000830): proposed merger with Cheil Industries (KNX:028260)," ISS Special Situations Research, July 3, 2015, https://www.issgovernance.com/file/publications /samsung_ct__merger_with_cheil_industries.pdf.

246 On June 3 Elliott bought: Se Young Lee and Joyce Lee, "U.S. Fund Elliott Challenges Samsung Group Restructuring Move," Reuters, June 3, 2015, https://www.reuters.com/article/us-cheil-industries-m-a-samsung-c-t /u-s-fund-elliott-challenges-samsung-group-restructuring-move-idUSKBN0OJ34C20150604.

246 Jay Lee was having breakfast: Adam Lashinsky, "Can Samsung's New Leader Dethrone Apple?" *Fortune,* July 27, 2015, http://fortune.com/2015 /07/27/samsung-jay-lee/.

247 On June 9 Elliott filed: "Chronology of Samsung C&T's Merger with Cheil Industries," Yonhap News Agency, July 17, 2015, https://en.yna.co.kr /view/AEN20150715010200320.

247 sold all of them for $608 million: "Samsung C&T Says to Sell 9 Mln Treasury Shares to KCC Corp," Reuters, June 10, 2015, https://www.reuters .com/article/samsung-ct-kcc-idUSL3N0YW3H020150610.

247 "The move is necessary": Kang Yoon-seung, "Elliott Seeks to Block Samsung C&T's Share Sale," Yonhap News Agency, June 11, 2015, https://en .yna.co.kr/view/AEN20150611002651320?input=rss. Elliott published its public statements and presentations for shareholders and media, in English and Korean, at http://www.fairdealforsct.com/.

247 The use of treasury shares: Kanga Kong, "Korea Inc. Ready to Kill Major Reforms No Matter Who Wins Vote," *Bloomberg,* April 27, 2016, https:// www.bloombergquint.com/markets/korea-inc-ready-to-kill-major-reforms-no-matter-who-wins-vote.

247 "deeply alarming": Elliott Management,"Elliott's Perspectives on Samsung C&T and the Proposed Takeover by Cheil Industries," June 2015, p. 6., http://www.fairdealforsct.com/present/.

247 The next, day, Elliott filed: Se Young Lee, "Hedge Fund, Samsung Group Locked in Fight over $8 Bln Merge," Reuters, June 11, 2015, https://www

.reuters.com/article/samsung-ct-kcc-elliott/update-1-hedge-fund-makes-fresh-legal-challenge-to-samsung-group-deal-idUSL3N0YX07120150611.

247 On July 1 a South Korean court: Choe Sang-hun, "Court Rejects Investor's Move to Disrupt Samsung Merger," *The New York Times*, July 1, 2015, https://www.nytimes.com/2015/07/02/business/dealbook/court-rejects-investors-move-to-disrupt-samsung-merger.html.

247 On July 3 Elliott appealed: Agence France-Presse, "US Fund Elliott Files Appeal Against Samsung Merger Ruling," July 3, 2015, https://www.aaj.tv/2015/07/us-fund-elliott-files-appeal-against-samsung-merger-ruling/.

247 more than a hundred newspapers: "Amid Crisis, Samsung C&T Makes Clear Its Intention to Merge . . . Advertising Extensively," Yonhap News Agency, July 13, 2015. Recordings of Samsung's TV spots are in the author's possession. https://www.yna.co.kr/view/AKR20150713028100003. This source is in Korean. The author's researcher translated the headline and text into English.

247 "Elliott is trying to defeat": Kim Byung-chul, "Samsung Really Wants to Succeed in the Merger," *Huffington Post Korea*, July 14, 2015, https://www.huffingtonpost.kr/2015/07/14/story_n_7790824.html. This source is in Korean. The author's researcher translated the headline and text into English.

248 sent out five thousand employees: Jo Gwi-dong, "Samsung C&T's Persuasion of Elliott-allied Foreign Shareholders Successful," *Chosun Biz*, July 17, 2015, http://biz.chosun.com/site/data/html_dir/2015/07/17/2015071701664.html. This source is in Korean. The author's researcher translated the headline and text into English. Information on the walnut cakes and watermelons are in Jen Wieczner. "Inside Elliott Management: How Paul Singer's Hedge Fund Always Wins," *Fortune*. December 7, 2017, https://fortune.com/2017/12/07/elliott-management-hedge-fund-paul-singer/.

248 "We beg of you": Jonathan Cheng and Min-Jeong Lee, "As Vote Nears, Samsung Pulls Out All the Stops," *The Wall Street Journal*, July 15, 2015, https://www.wsj.com/articles/as-vote-nears-samsung-pulls-out-all-the-stops-1436994473.

248 "That's shady": Representative of minority shareholder, interview by the author, July 16, 2015.

248 "Korea-related issues": Joo Jin-woo, "We Are Revealing the 'Samsung Jang Choong-gi' Text Messages in Full,'" *SisaIN*, August 9, 2017, https://www.sisain.co.kr/?mod=news&act=articleView&idxno=29814. This source is in Korean. The author's researcher translated the headline and text into English.

248 "lawyer actually handling this case": Ibid. More information is at Kim Eun-ji and Joo Jin-woo, "Messages of Favor That Samsung's Jang Choong-ki Received," *SisaIN*, August 16, 2017, https://www.sisain.co.kr/?mod=news&act=articleView&idxno=29863. This source is in Korean. The author's researcher translated the headline and text into English.

248 "'Jewish ISS [Institutional Shareholder Services]' blatantly": Haviv Rettig Gur, "Fight over One of the World's Largest Tech Companies Turns Anti-Semitic," *The Times of Israel*, July 9, 2015, https://www.timesofisrael.com/fight-over-one-of-the-worlds-largest-tech-companies-turns-anti-semitic/.

249 "According to a source": Ibid.

249 "Jews have too much power": Anti-Defamation League, "ADL Urges Korean Gov't to Condemn Anti-Semitism Following Pernicious Stereotypes in Media Outlets" (news release), July 10, 2015, https://www.adl.org/news/press-releases/adl-urges-korean-govt-to-condemn-anti-semitism-following-pernicious-stereotypes.

249 The fund had about $450 billion: Jung Jin-yeop, "2015 Accounts of National Pension Service," National Pension Service Management Committee, March 2016. This source is in Korean. The author's researcher translated the headline and text into English. The document is in the author's possession.

249 National Pension Service's investment committee: Seo Han-ki, "NPS: 'We Decided on Samsung C&T Merger,' . . . Seems Like They Will Agree," Yonhap News Agency, July 10, 2015, https://www.yna.co.kr/view/AKR20150710173053017. This source is in Korean. The author's researcher translated the headline and text into English.

249 "Pension Fund Could Be Samsung Kingmaker": Min-Jeong Lee and Jonathan Cheng, "Pension Fund Could Be Samsung Kingmaker," *The Wall Street Journal*, June 5, 2015, https://www.wsj.com/articles/pension-fund-could-be-samsung-kingmaker-1433502314.

249 To the consternation of media: Hwang Ye-rang, "The Secret of July 2015 Between Samsung-NPS-Blue House," *Hankyoreh 21*, November 2016, https://www.yna.co.kr/view/AKR20150713028100003. This source is in Korean. The author's researcher translated the headline and text into English.

249 "The NPS is shooting": Kim Woo-chan, interview by the author's researcher Max Soeun Kim, July 10, 2015.

250 "Our stake in Cheil Industries": Kim Myeong-ryol, "Looking into the Proceedings from NPS Meeting Regarding Samsung C&T Merger," *Money Today*, November 23, 2016, http://news.mt.co.kr/mtview.php?no=2016112316234472720. The NPS meeting minutes are in the author's possession. This source is in Korean. The author's researcher translated the headline and text into English.

250 "In order to counterbalance": Ibid.

250 Eight voted in favor: Yoo Hyun-min, "NPS Agrees on Merger Between Samsung C&T and Cheil Industries Based on Synergy," Yonhap News Agency, November 23, 2016, https://www.yna.co.kr/view/AKR20161123126400008. This source is in Korean. The author's researcher translated the headline and text into English.

250   "Construction? It's not looking good": Kim Seon-jeong (former top Samsung financial executive), interview by the author, June 14, 2015.

250   Samsung C&T put up a website: Ken Kurson, "Spat Between Samsung and NYC Hedge Fund Takes Nasty Detour into Jew-Baiting," *Observer*, July 13, 2015, https://observer.com/2015/07/spat-between-samsung-and-nyc-hedge-fund-takes-nasty-detour-into-jew-baiting/.

251   "The depiction of Jews": Ibid.

251   Samsung pulled its advertising: Ibid.

251   "I think it's a shame": Ryan Daly, "Samsung's Alleged Anti-Semitic Cartoons Pulled from Website," *Fortune*, July 16, 2015, http://fortune.com/2015/07/16/samsungs-anti-semitic-cartoons/.

251   the accusation was not yet proven: Kang Se-hoon, "Fervent Last-Minute Debate Regarding Samsung Merger," *Newsis*, July 14, 2015, http://www.newsis.com/view/?id=NISX20150714_0013791085. This source is in Korean. The author's researcher translated the headline and text into English.

251   "If this is for the 'national good'": Samsung C&T Minority Shareholders Group, "Public Statement to NPS," July 2015. Document is in the author's possession. This source is in Korean. The author's researcher translated the headline and text into English.

251   "How dare they attack Samsung": Stock analysts and minority shareholders, meetings with the author, July 2015.

252   "I've studied the pension service": Kang Dong-oh, interview by the author, July 15, 2015.

252   The 553 stakeholders present: Park Eun-jee, "Shareholders Vote to Approve Samsung Merger," *Korea JoongAng Daily*, July 18, 2015, http://korea joongangdaily.joins.com/news/article/article.aspx?aid=3006777.

252   One shareholder stormed the stage: Jonathan Cheng and Min-Jeong Lee, "Samsung Shareholders Back $8 Billion Merger, in Blow to U.S. Hedge Fund," *The Wall Street Journal*, July 17, 2015, https://www.wsj.com/articles/samsung-c-t-shareholders-approve-8-billion-merger-with-cheil-industries-1437105075.

253   "If the merger ratio was set": Park, "Shareholders Vote to Approve Samsung Merger."

253   69.53 percent of the shareholders: Chang Jae Yoo, "Q&A: NPS Embroiled in Korea's Political Scandal over Samsung Units' Merger," *Korea Economic Daily*, November 29, 2016, http://www.koreaninvestors.com/?p=1447.

253   "The approval is huge for us": Kim Yoo-chul, "Samsung Beats Elliott to Pass Merger Deal," *The Korea Times*, July 17, 2015, https://www.korea times.co.kr/www/news/tech/2015/07/133_182984.html.

253   "Elliott is disappointed": Cheng and Lee, "Samsung Shareholders Back $8 Billion Merger."

254 "Elliott has questions about the validity": Kim, "Samsung Beats Elliott to Pass Merger Deal."

254 "unable to speak since May": Anna Fifield, "In South Korea, Samsung Prepares to Crown Third-Generation 'Emperor,'" *The Washington Post,* October 9, 2014, https://www.washingtonpost.com/world/in-south-korea-samsung-prepares-to-crown-third-generation-emperor/2014/10/08/6ec0ba58-44bb-11e4-b437-1a7368204804_story.html.

254 voted through legal representatives: Kim, "Samsung Beats Elliott to Pass Merger Deal."

254 lost 40 percent of their value: "Samsung C&T Stocks Plunge as Lower Demand Hits Bottom Line," *The Korea Herald,* May 17, 2016, https://www.koreatimes.co.kr/www/biz/2019/04/488_204880.html.

254 suffered $500 million in losses: Park Han-na, "NPS Sees Big Losses After Samsung C&T Merger," *The Investor,* November 21, 2016, http://www.theinvestor.co.kr/view.php?ud=20161121000877.

254 "My observation on Samsung": Choi Kwang, interview by the author, May 7, 2017.

29: MY KINGDOM FOR A HORSE

256 "The president would like": "Lee Jae-yong's Feelings About the Park Meeting Revealed," *Maeil Business Newspaper,* August 6, 2017, https://www.mk.co.kr/news/society/view/2017/08/526943/. This source is in Korean. The author's researcher translated the headline and text into English.

256 "I am married to the nation": Choi Hoon, "Living in the Past," *Korea JoongAng Daily,* December 8, 2016, http://koreajoongangdaily.joins.com/news/article/article.aspx?aid=3027134.

256 "How is Chairman Lee Kun-hee's health?": Lee, "Park Geun-hye Administration."

257 "Samsung management succession situation": Lee Dong-hyun, "Blue House: Park Geun-hye Administration, Samsung Memo Written in August 2014," *Hankook Ilbo,* July 16, 2017, https://www.hankookilbo.com/News/Read/201707161519454085. This source is in Korean. The author's researcher translated the headline and text into English. Elliott Associates, the hedge fund that challenged the Samsung merger in a shareholder vote, published a comprehensive English account of evidence from Jay Lee's trial in a document intended for Elliott's later arbitration case against the South Korean government. The document is KL Partners, Three Crowns LLP, and Kobre & Kim, "Claimant's Amended Statement of Claim," *Elliott Associates, L.P. vs. Republic of Korea,* Arbitration Under the Arbitration Rules of the United Nations Commission on International Trade Law and the Free Trade Agreement between the Republic of Korea and the United States of America, April 4, 2019, https://pcacases.com/web/sendAttach/2594.

257   "I think I know what people": Kang Young-soo. "Lee Jae-yong: 'After Being Chastised by President Park, I See What People Mean About Her 'Laser Beam Gaze,'" *Chosun Ilbo,* April 7, 2017, http://news.chosun.com/site /data/html_dir/2017/04/07/2017040702451.html. This source is in Korean. The author's researcher translated the headline and text into English.

257   The Tower executives signed: Eun-Young Jeong, "His Kingdom for a Horse? Samsung Heir's Trial Hinges on an Equestrian Deal," *The Wall Street Journal,* July 4, 2017, https://www.wsj.com/articles/case-against-samsung-heir-rides-on-alleged-horse-trading-deal-1499088469. The Core Sports contract is in the possession of the author.

258   Soon-sil had befriended: Choe Sang-hun, "A Presidential Friendship Has Many South Koreans Crying Foul," *The New York Times,* October 27, 2016, https://www.nytimes.com/2016/10/28/world/asia/south-korea-choi-soon-sil.html.

258   "Rumors are rife that": Alexander Vershbow (U.S. ambassador to South Korea), cable sent on July 20, 2007, Public Library of U.S. Diplomacy, WikiLeaks, https://wikileaks.org/plusd/cables/07SEOUL2178_a.html.

258   Samsung bought an $830,000 racehorse: Kim Min-kyung, Chung Yoo-ra: "I Don't Think Samsung Was Unaware of Exchanging Horses," *Hankyoreh,* July 12, 2017. This source is in Korean. The author's researcher translated the headline and text into English.

258   Jay Lee sold off Samsung's corporate jets: "Samsung Private Jet's Last Flight to Carry George W. Bush," *Chosun Ilbo,* October 1, 2015, http://english .chosun.com/site/data/html_dir/2015/10/01/2015100101487.html.

258   raking in almost $6 billion: Min-Jeong Lee, "Samsung Sells Chemical Businesses to Rival Lotte," *The Wall Street Journal,* October 29, 2015, https:// www.wsj.com/articles/samsung-sells-chemical-businesses-to-rival-lotte-1446176583.

258   He sold the Samsung Life Insurance: Seo Ji-eun, "Two Years On, Samsung Heir Has Put His Brand on Group," *Korea JoongAng Daily,* May 10, 2016, http://koreajoongangdaily.joins.com/news/article/article.aspx?aid= 3018515.

258   under the initiative "Start-up Samsung": Kentaro Ogura, "Samsung Sheds Old Culture for New Growth," *Nikkei Asian Review,* June 30, 2016, https:// asia.nikkei.com/Business/Samsung-sheds-old-culture-for-new-growth.

259   The Samsung Summer Festival: Lee Suh-hee, "Samsung Cancels Summer Festival . . . Speeding Up the Process of Overthrowing Corporate Culture," *Hankook Ilbo,* March 22, 2016, https://www.hankookilbo.com/News /Read/201603220479056587. This source is in Korean. The author's researcher translated the headline and text into English.

259   The company cut down: Kentaro Ogura, "Samsung Sheds Old Culture for New Growth."

259 "People felt liberated": Samsung marketer, email message to the author, December 2, 2016.

259 The older Samsung Men: Fifteen Samsung vice presidents and former vice presidents, interviews by the author, March 2016–September 2017.

259 "It looks like any other": Nam S. Lee, interview by the author, November 13, 2015.

259 "For the older generation": Ho Soo Lee, interview by the author, September 5, 2016.

259 Pendleton's spirits remained high: Former Galaxy marketer, interview by the author, October 23, 2017.

260 "A patented design may be": Samsung Electronics, "Petition for a Writ of Certiorari to the United States Court of Appeals for the Federal Circuit," *Samsung Electronics v. Apple,* December 14, 2015, https://www.scribd.com /doc/293255771/Samsung-vs-Apple-Samsung-s-Appeal-to-the-Supreme-Court.

260 "Isn't it simply unworthy": Florian Mueller, *"Apple v. Samsung*: Petition for Supreme Court to Take First Look at Design Patent Case in 122 Years," *Foss Patents,* December 14, 2015, http://www.fosspatents.com/2015/12/apple-v-samsung-petition-for-supreme.html.

260 "gonna have me skateboard down": Casey Neistat, "I'm a Male Model," posted by YouTube user CaseyNeistat on March 2, 2016, https://www .youtube.com/watch?v=88N-811loAo.

260 decline and closure of Milk Music: Dan Rys, "Samsung's Milk Music Officially Shuts Down U.S. Service," *Billboard,* September 22, 2016, https:// www.billboard.com/articles/business/7518662/samsung-milk-music-officially-shuts-down-us.

261 "I'm just disappointed": Daren Tsui, interview by the author, January 10, 2017.

261 "It may be the worst code": Peter Bright, "Samsung's Tizen Is Riddled with Security Flaws, Amateurishly Written," *Ars Technica,* April 4, 2017, https:// arstechnica.com/gadgets/2017/04/samsungs-tizen-is-riddled-with-security-flaws-amateurishly-written/.

261 "The JTBC reporting team": "Choi Soon-sil Computer Obtained . . . President's Speeches Have Been Downloaded," *JTBC,* October 24, 2016. http:// news.jtbc.joins.com/article/article.aspx?news_id=NB11340632. This source is in Korean. The author's researcher translated the headline and text into English.

261 "daughter of Korea's Rasputin": Lee Ha-kyeong, "A Bloody Similarity," *Korea JoongAng Daily,* January 14, 2017, http://koreajoongangdaily.joins .com/news/article/article.aspx?aid=3028421.

262 President Park's approval rating collapsed: Ju-min Park and Se Young Lee, "S. Korea President's Approval Rating Falls to 5 Percent: Gallup," Reuters,

November 4, 2016, https://www.reuters.com/article/us-southkorea-politics-poll-idUSKBN12Z04Y.

262 Swarms of protesters cried foul: Choe Sang-Hun, "South Koreans 'Ashamed' over Leader's Secretive Adviser," *The New York Times*, November 5, 2016, https://www.nytimes.com/2016/11/06/world/asia/south-koreans-ashamed-over-les-secretive-adviser.html.

262 "It's unbelievable": Korean lawyer, interview by the author, November 14, 2016.

262 prosecutors detained Choi: Choe Sang-Hun, "Choi Soon-sil, at Center of Political Scandal in South Korea, Is Jailed," *The New York Times*, October 31, 2016, https://www.nytimes.com/2016/11/01/world/asia/south-korea-park-geun-hye-choi-soon-sil.html.

262 "It is hard to forgive myself": Ju-min Park and Tony Munroe, "South Korea's Park Says 'Hard to Forgive Myself' for Political Crisis," Reuters, November 3, 2016, https://www.reuters.com/article/us-southkorea-politics-idUSKBN12Z05N.

262 On November 8 prosecutors raided: "Prosecutors Raid Samsung's Headquarters, Grill Exec over Scandal," Yonhap News Agency, November 8, 2016, https://en.yna.co.kr/view/AEN20161108001354315.

262 Two weeks later, prosecutors descended: Song Jung-a, "South Korean Prosecutors Make New Raids over Political Scandal," *Financial Times*, December 21, 2016, https://www.ft.com/content/b2986b0a-c744-11e6-8f29-9445cac8966f.

263 "We got phone calls from people": National Pension Service staffer, WhatsApp message to the author, April 28, 2017.

263 The head of the pension service's: "Ex–Health Minister Sentenced to 2.5-Year Jail Term in Corruption Scandal," Yonhap News Agency, June 8, 2017, https://en.yna.co.kr/view/AEN20170608007251315.

263 "The fact that a Health Ministry official": Michael Katz, "Former Korean Minister Found Guilty of Pressuring Pension," Chief Investment Officer, June 12, 2017, https://www.ai-cio.com/news/former-korean-minister-found-guilty-pressuring-pension/.

263 "Jay Lee?": Analyst at Wall Street investment house, meeting with the author, December 22, 2016.

263 "Are you indeed an accomplice?": SBS News, "[SBS LIVE] Park Geun-hye—Choi Soon-sil Gate Live Broadcast," SBS, December 5, 2016, https://www.youtube.com/watch?time_continue=1&v=gkhXFaPbNSk. The conversation beginning with "Are you indeed an accomplice?" plays at 1:20:57. The quote "Jay Y. Lee seems like he has memory problems" plays at 5:05:07. Jay Lee says "I don't know" and "I don't remember" many times throughout the hearing. A full transcript of these proceedings is in National Assembly of the Republic of Korea, "Notes from Special Investigative Committee for Private Persons' Interference in Park Geun-hye Administration," Decem-

ber 6, 2016. Document is in the author's possession. This source is in Korean. The author's researcher translated the headline and text into English.

264   According to prosecutors, Jay bribed: Republic of Korea, "Prosecutors' Report on Private Persons' Interference in Park Geun-hye Administration," p. 18. Document is in the author's possession. This source is in Korean. The author's researcher translated the headline and text into English.

264   The Samsung vice chairman passed through: Bridge Economy, "Jay Lee Attends the Special Prosecutors' Office and the Court for Warrant Hearing, Unresponsive to Reporters' Questions," posted by YouTube user BridgeEconomy on January 17, 2017, https://www.youtube.com/watch?v=W3a5Qri3PIg. This source is in Korean. The author's researcher translated the headline and text into English.

264   Lee was interrogated for twenty-two hours: Jonathan Cheng and Timothy W. Martin, "Samsung Heir Meets Prosecutors: 22 Hours of Questioning and a $5 Lunch," *The Wall Street Journal*, January 15, 2017, https://www.wsj.com/articles/samsung-heir-meets-prosecutors-22-hours-of-questioning-and-a-5-lunch-1484305005.

264   the judge rejected the prosecutor's: Sherisse Pham, "Judge Rejects Arrest Warrant for Samsung Heir," *CNNMoney*, January 18, 2017, https://money.cnn.com/2017/01/17/investing/samsung-jay-lee-arrest-warrant-court/index.html.

265   "You've been summoned": JTBC News, "Jay Lee Attends His Second Warrant Hearing," posted by YouTube user JTBC News on February 16, 2017, https://www.youtube.com/watch?v=Ww0F4UJq5MQ. This source is in Korean. The author's researcher translated the headline and text into English.

265   two black Hyundai cars pulled up: OhMyNewsTV, "Lee Jae-yong 'Fateful Day,' Scene of Transportation to Seoul Detention Center," posted by YouTube user OhMyNewsTV on February 16, 2017, https://www.youtube.com/watch?v=mV_ktOwH3Vs. This source is in Korean. The author's researcher translated the headline and text into English.

265   "We acknowledge the cause": Hyunjoo Jin and Joyce Lee, "Samsung Chief Lee Arrested as South Korean Corruption Probe Deepens," Reuters, February 16, 2017, https://www.reuters.com/article/us-southkorea-politics-samsung-group/samsung-chief-lee-arrested-as-south-korean-corruption-probe-deepens-idUSKBN15V2RD.

265   Jay Lee was taken into custody: Ibid.

266   in a cell with a mattress: Eun-Young Jeong and Timothy W. Martin, "Samsung Heir's Prison Life: Seven Hours of TV on an LG Screen, $1.25 Meals," *The Wall Street Journal*, February 28, 2017, https://www.wsj.com/articles/samsung-heirs-prison-life-7-hours-of-tv-1-25-meals-1488282647.

266   "There are concerns about": Ju-min Park, "Mattress on Cell Floor, Toilet in the Corner for Samsung Scion," Reuters, February 17, 2017, https://www

.reuters.com/article/us-southkorea-politics-samsung-group-pri/mattress-on-cell-floor-toilet-in-the-corner-for-samsung-scion-idUSKBN15W0J4.

266 "This is a new experience": Brad Stone, Sam King, and Ian King, "Summer of Samsung: A Corruption Scandal, a Political Firestorm—and a Record Profit," *Bloomberg Businessweek,* July 26, 2017, https://www.bloomberg .com/news/features/2017-07-27/summer-of-samsung-a-corruption-scandal-a-political-firestorm-and-a-record-profit.

266 The Chairman's cousin Miky Lee: "History Repeats Itself," *Korea Herald,* April 8, 2018, http://www.koreaherald.com/view.php?ud=20180408000033.

266 citing health concerns: Ahn Sung-mi, "CJ Vice Chairwoman Signals Return with Rare Public Appearance," *The Investor,* December 6, 2016, http:// www.theinvestor.co.kr/view.php?ud=20161206000638.

266 She, too, had been targeted: Christine Kim, "The World According to Park: A Blacklist and Placenta Shots in South Korea," Reuters, January 19, 2017, https://www.reuters.com/article/us-southkorea-politics-blacklist-idUSKBN15314D.

266 Before dawn on March 30: Han Yeong-hye, "Seoul Detention Center, Where Former President Park Geun-hye Is Detained," *JoongAng Ilbo,* March 31, 2017, https://news.joins.com/article/21425038. This source is in Korean. The author's researcher translated the headline and text into English.

267 "We look around, and our Samsung": Former Samsung vice president, interview by the author, September 1, 2017.

267 Some 200,000 Galaxy Note 7s: Geoffrey A. Fowler and Joanna Stern, "Why Samsung's Battery Fix Gets a Grade C, For Now," *The Wall Street Journal,* January 22, 2017, https://www.wsj.com/articles/why-samsungs-battery-fix-gets-a-grade-c-for-now-1485133200.

268 CEO D.J. Koh took the stage: Samsung Electronics, "Samsung Finally Reveals Why the Galaxy Note 7 Kept Exploding," posted by YouTube user Tech Events on January 23, 2017, https://www.youtube.com/watch?v= Iu18CykEH9o.

269 "The rather poor way they handled": Paul Mozur, "Galaxy Note 7 Fires Caused by Battery and Design Flaws, Samsung Says," *The New York Times,* January 22, 2017, https://www.nytimes.com/2017/01/22/business /samsung-galaxy-note-7-battery-fires-report.html.

269 gave Samsung's battery fix: Fowler and Stern, "Why Samsung's Battery Fix Gets a Grade C."

269 "Samsung claimed that the problem": Park Chul Wan, email message to the author's researcher Max Soeun Kim, April 21, 2017.

269 "the most beautiful phone ever": Jessica Dolcourt, "Samsung Galaxy S8 Review," *CNET,* May 26, 2017, https://www.cnet.com/reviews/samsung-galaxy-s8-review/.

269 "brilliant phone": Gareth Beavis, "Samsung Galaxy S8 Review," *TechRadar,*

January 17, 2019, https://www.techradar.com/reviews/samsung-galaxy-s8-review.

269 sales of the Galaxy S8 weren't: Former Galaxy marketer, interview by the author, October 9, 2017. More reasons for lackluster sales are addressed in Kim Yong-won, "Samsung Electronics Galaxy S8 Sales Below Expectations, Dark Clouds Hanging over Long-Term Popularity," *BusinessPost*, April 26, 2017, http://www.businesspost.co.kr/BP?command=naver&num=49245. This source is in Korean. The author's researcher translated the headline and text into English.

270 stock market to all-time highs: Peter Wells, "Samsung Pushes South Korean Stocks Toward Record High," *Financial Times*, May 2, 2017, https://www.ft.com/content/62df3254-5ea6-3ea6-b2bb-fb90399b42a0.

270 most profitable tech company: Jacky Wong, "The World's Most Profitable Tech Company? Not Apple," *The Wall Street Journal*, July 6, 2017, https://www.wsj.com/articles/the-worlds-most-profitable-tech-company-not-apple-1499398067. This is not by the same reporter who interviewed the author of this book, but the article shows the soaring profits of the time.

270 "It's a kind of tragedy": "Dr. Oh-Hyun Kwon, Vice Chairman & CEO, Samsung Electronics Co., Ltd.," posted by YouTube user The Economic Club of Washington, D.C., on October 19, 2017, https://www.youtube.com/watch?v=uCkoRV4fOUs.

271 "What did Jay Y. Lee do wrong?": Stone, King, and King, "Summer of Samsung."

271 "In retrospect I had a lot": Jeyup S. Kwaak and Paul Mozur, "Mastermind or Naïf? Samsung Heir's Fate Hinges on the Question," *The New York Times*, August 23, 2017, https://www.nytimes.com/2017/08/23/business/samsung-jay-lee-trial-lee-jae-yong.html.

272 "Mastermind or Naïf?": Ibid.

272 "Back when Chairman Lee": "Jay Lee: 'Park Geun-hye Complained About Hong Seok-hyun in Private,'" Yonhap News Agency, August 2, 2017, https://www.yna.co.kr/view/AKR20170802175300004. This source is in Korean. The author's researcher translated the headline and text into English.

272 "Jay Y. Lee is not the group's": Joyce Lee and Hyunjoo Jin, "Downfall of Ex–Samsung Strategy Chief Leaves 'Salarymen' Disillusioned," Reuters, August 25, 2017, https://www.reuters.com/article/us-samsung-lee-choi/downfall-of-ex-samsung-strategy-chief-leaves-salarymen-disillusioned-idUSKCN1B50UX.

272 "None of us will ever forget": Samsung Electronics, "Samsung Galaxy Note 8 Unpacked—Full Replay," posted by YouTube user samsungnewzealand on September 10, 2017, https://www.youtube.com/watch?v=W2YErBpcB-w.

273   "A year ago, I wrote": Dan Seifert, "Samsung Galaxy S8 Review: One for the Fans," *The Verge,* September 5, 2017, https://www.theverge.com/2017/9/5/16253308/samsung-galaxy-note-8-review.

273   "direct favors": Seoul Central District Court, 27th Criminal Division, "Ruling on 2017 (Goh-hap) 194," August 25, 2017, pp. 50–105. The Korean term *Goh-hap* has no literal translation in English. It signifies criminal cases decided by a panel of three judges. Document is in the author's possession. This source is in Korean. The author's researcher translated the headline and text into English.

273   "The defendant, as a de facto head": *Newbc,* "Stenographic Notes from Lee Jae-yong's First Trial Obtained." August 25, 2017, http://www.newbc.kr/bbs/board.php?bo_table=comm1&wr_id=4151.

273   "The defendant also provided false witness": Seoul Central District Court 27th Criminal Division. "Ruling on 2017 (Goh-hap) 194," August 25, 2017, p. 19.

274   "I cannot accept as a lawyer": Kim Min-jung, "After Announcing Five-Year-Sentence, Jay Lee Seems Calm, Audience Gasps," *Hankook Ilbo,* August 25, 2017, https://www.hankookilbo.com/News/Read/201708251849861053. This source is in Korean. The author's researcher translated the headline and text into English.

274   "The iPhone X is basically": Matthew Miller, "The iPhone X Is Basically Samsung's Note 8 Plus Animojis," ZDNet, September 12, 2017, https://www.zdnet.com/article/iphone-x-no-thanks-samsungs-note-8-is-all-i-ever-wanted-in-a-smartphone-and-more/.

274   "iPhone X Features": Aatif Sulleyman, "A Leap Forward for Apple but Samsung Is Still Ahead," *The Independent,* September 15, 2017, https://www.independent.co.uk/life-style/gadgets-and-tech/features/iphone-x-vs-samsung-s8-features-note-8-android-years-ahead-ios-11-a7947206.html.

274   "That pisses me off": Former Galaxy marketer, interview by the author, October 9, 2017.

275   "These are two of the largest": Timothy W. Martin and Tripp Mickle, "Why Apple Rival Samsung Also Wins If iPhone X Is a Hit," *The Wall Street Journal,* October 2, 2017, https://www.wsj.com/articles/why-apple-rival-samsung-also-wins-if-iphone-x-is-a-hit-1506936602.

275   "Recently . . . Geoffrey Cain had a conspiracy": "Facebook Shows Another Foreign Spy Hidden Behind CNRP to Topple the Government," *Fresh News,* August 24, 2017, http://freshnewsasia.com/index.php/en/63667-facebook-2.html. The article is in Khmer, the language of Cambodia. A Cambodian translator completed this translation for the author.

275   just arrested an Australian filmmaker: Erin Handley and Niem Chheng, "Filmmaker James Ricketson Charged," *The Phnom Penh Post,* June 9, 2017, https://www.phnompenhpost.com/national/filmmaker-james-ricketson-charged.

275   the opposition leader, was also arrested: "Latest Updated: CNRP Leader Kem Sokha Arrested for 'Treason,'" *The Cambodia Daily,* September 3, 2017, https://www.cambodiadaily.com/news/nrp-leader-kem-sokha-arrested-treason-134249/.

EPILOGUE

277   "[The] previous 10 years": Anthony Cuthberson, "Galaxy Fold: Inside Samsung's Struggle to Deliver a Foldable Phone—and Why the Future of Smartphones Hinges on It," *The Independent,* July 1, 2019, https://www.independent.co.uk/life-style/gadgets-and-tech/features/samsung-galaxy-fold-foldable-phone-release-date-when-explained-a8980056.html.

277   For almost a decade: Aatif Sulleyman, "The Galaxy Fold Timeline: Samsung's Incredible Journey from 2011 to Foldgate," *Trusted Reviews,* April 23, 2019, https://www.trustedreviews.com/news/samsung-fold-timeline-3652069.

277   Then patents: Steve Dent, "Samsung Files Patent for a Bizarre Foldable Smartphone," *Engadget,* November 10, 2016, https://www.engadget.com/2016/11/10/samsung-foldable-smartphone/.

277   "It is difficult to talk": Shin Ji-hye, "Samsung Mobile Chief Not Confident About Foldable Phones This Year," *The Korea Herald,* January 9, 2018, http://www.koreaherald.com/view.php?ud=20180109000915.

278   that pretty much only offered marginal updates: Cuthberson, "Galaxy Fold: Inside Samsung's Struggle to Deliver a Foldable Phone."

278   President Moon appointed: Sohee Kim, Sam Kim, and Bruce Einhorn, "The Critic Tapped to Be Korea's Top Cop," Bloomberg News, May 18, 2017, https://www.bloomberg.com/news/articles/2017-05-18/samsung-threw-out-critic-tapped-to-be-korea-s-top-business-cop.

278   The KFTC was Korea's trustbuster: Korea Fair Trade Commission, *Annual Report 2017,* http://www.ftc.go.kr/solution/skin/doc.html?fn=bd60e97df64ecfb0bf2578c58ef3a70fc41f48786446f5f55aed59088afd80fd&rs=/fileup-load/data/result/BBSMSTR_000000002404/.

278   *chaebol* sniper: Eun-Young Jeong, "South Korea Names 'Chaebol Sniper' to Watchdog Role," *The Wall Street Journal,* May 17, 2017, https://www.wsj.com/articles/south-korea-names-chaebol-sniper-to-watchdog-role-1495020959.

278   In 2004, he heckled: Kim, Kim, and Einhorn, "The Critic Tapped to Be Korea's Top Cop."

278   "They were born": Kim Jaewon and Sotaro Suzuki, "Korea's 'Chaebol Sniper' Says Family Bosses Should Yield to Pros," *Nikkei Asian Review,* January 17, 2019, https://asia.nikkei.com/Editor-s-Picks/Interview/Korea-s-chaebol-sniper-says-family-bosses-should-yield-to-pros.

278   "While there's no change": "South Korea's New President Vows to Take on the Country's Huge Family-Run Conglomerates," Reuters, May 22, 2017, https://fortune.com/2017/05/22/south-korea-antitrust-samsung-hyundai-chaebol-regulations-oversight/.

279   In February 2018, Jay had already: Joyce Lee and Haejin Choi, "Samsung Scion Lee Walks Free After Jail Term Suspended, Faces Leadership Challenges," Reuters, February 4, 2018, https://www.reuters.com/article/us-samsung-lee/samsung-scion-lee-walks-free-after-jail-term-suspended-faces-leadership-challenges-idUSKBN1FO0R9.

279   the court reduced his five-year prison: Sam Kim, "Samsung's Jay Y. Lee Set Free in Unexpected Court Reversal," Bloomberg News, February 5, 2018, https://www.bloomberg.com/news/articles/2018-02-05/samsung-heir-jay-y-lee-goes-free-after-court-suspends-jail-term.

279   "The past year has been": Lee and Choi, "Samsung Scion Lee Walks Free After Jail Term Suspended, Faces Leadership Challenges."

279   Four months later, in June 2018: Jack Nicas, "Apple and Samsung End Smartphone Patent Wars," *The New York Times,* June 27, 2018, https://www.nytimes.com/2018/06/27/technology/apple-samsung-smartphone-patent.html.

280   "As Elliott's claims are groundless": Edward White and Kang Buseong, "Elliott's $718m Claim Against South Korea Poses Risk for Moon," *Financial Times,* May 3, 2019, https://www.ft.com/content/1a972668-6ca9-11e9-80c7-60ee53e6681d.

280   "It is baffling that the Republic": Representative of Elliott Management in New York City, email to the author, May 13, 2019.

280   The government needed: Song Gyung-hwa, "[News analysis] Lee Jae-yong's Disturbing Display of Confidence Ahead of Supreme Court Ruling," *Hankyoreh,* May 5, 2019, http://www.hani.co.kr/arti/PRINT/892707.html.

280   In September 2018, President Moon: Sherisse Pham, "Why Samsung's Billionaire Chief Is Headed to North Korea," *CNNMoney,* September 17, 2018, https://money.cnn.com/2018/09/17/technology/south-korea-samsung/index.html.

281   "Along with becoming a regular": Song Gyung-hwa, "[News analysis] Lee Jae-yong's Disturbing Display of Confidence Ahead of Supreme Court Ruling," *Hankyoreh.*

281   Meanwhile, financial investigators probed: Song Jung-a, "Regulator Says Samsung BioLogics Breached Accounting Rules," *Financial Times,* May 2, 2018, https://www.ft.com/content/4e37a1a6-4db4-11e8-8a8e-22951a2d8493.

281   Samsung BioLogics: Song Jung-a, "Two Samsung Employees Arrested over Alleged Cover-Up," *Financial Times,* April 29, 2019, https://www.ft.com/content/20347230-6a95-11e9-80c7-60ee53e6681d.

281 "Some analysts said the accounting": Song, "Regulator Says Samsung BioLogics Breached Accounting Rules."

281 The inflated value of Samsung BioLogics: Song Su-hyun, "[Newsmaker] How Much Trouble Is Samsung BioLogics In?" *The Korea Herald,* November 22, 2018, http://www.koreaherald.com/view.php?ud=2018112200 0745.

281 The investigation sparked: Joyce Lee, "Accounting Concerns Wipe $6 Billion Off Samsung BioLogics Market Value," Reuters, May 1, 2018, https://www.reuters.com/article/us-samsung-biologic-accounting-samsung-c/accounting-concerns-wipe-6-billion-off-samsung-biologics-market-value-idUSKBN1I301U.

281 the KFTC suspended trading: Kim Jaewon, "Samsung BioLogics Trading Suspended After Fraud Ruling," *Nikkei Asian Review,* November 14, 2018, https://asia.nikkei.com/Business/Companies/Samsung-BioLogics-trading-suspended-after-fraud-ruling.

282 Samsung denied the accusations: Arlene Weintraub, "Samsung BioLogics Strikes Back, Filing Lawsuit over Criminal Penalties Imposed by Korean Regulators," *FiercePharma,* November 28, 2018, https://www.fiercepharma.com/pharma/samsung-biologics-strikes-back-filing-lawsuit-over-criminal-penalties-imposed-by-korean.

282 Samsung BioLogics avoided being delisted: Choonsik Yoo and Yuna Park, "Samsung BioLogics Shares to Remain Listed, Resume Trade on Dec. 11," Reuters, December 10, 2018, https://www.reuters.com/article/samsung-biologics-accounting/samsung-biologics-shares-to-remain-listed-resume-trade-on-dec-11-idUSS6N1X502C.

282 eight executives were arrested: Edward White and Song Jung-a, "Eighth Samsung Employee Arrested over Biotech Fraud Case," *Financial Times,* June 5, 2019, https://www.ft.com/content/6d6ea99c-8720-11e9-a028-86cea8523dc2.

282 one of the suspects: "Court Rejects Arrest Warrant for Samsung BioLogics CEO," Yonhap, July 20, 2019, http://www.koreaherald.com/view.php?ud=20190720000005.

282 Investigators raided two plants: "Evidence Destroyed from Samsung BioLogics CEO's Computers: Prosecutors," Yonhap, June 3, 2019, https://en.yna.co.kr/view/AEN20190603006900325.

282 "They [Samsung employees] deleted all": Ibid.

282 "We deeply regret": Lim Jeong-yeo, "Samsung BioLogics Apologizes for Destructing Evidence of Alleged Fraud," *The Korea Herald,* June 14, 2019, http://www.koreaherald.com/view.php?ud=20190614000628.

282 "prosecutors raided": Chang Chung-hoon and Song Kyoung-son, "Supreme Court Ruling Deals Samsung Wild Card," *Korea JoongAng Daily,* September 2, 2019, http://koreajoongangdaily.joins.com/news/article/article.aspx?aid=3067449&cloc=joongangdaily%7Chome%7Cnewslist1.

282   Samsung senior vice president Justin Denison: "Watch Samsung Unveil Its Foldable Phone—The Galaxy Fold," posted by YouTube user Tech Insider, February 20, 2019, https://www.youtube.com/watch?v=sHR8efUn3SY.

283   "The screen on my Galaxy Fold": Mark Gurman (@murkgurman), "The screen on my Galaxy Fold review unit is completely broken and unusable just two days in. Hard to know if this is widespread or not," Twitter, April 17, 2019, 8:58 P.M., https://twitter.com/markgurman/status/11185744672 55418880?lang=en.

283   "My Galaxy fold screen broke": Dieter Bohn, "My Samsung Galaxy Fold Screen Broke After Just a Day," *The Verge,* April 17, 2019, https://www .theverge.com/2019/4/17/18411510/samsung-galaxy-fold-broken-screen-debris-dust-hinge-flexible-bulge.

283   "Now [the Galaxy Fold] is turning into": Chris Fox, "Samsung Galaxy Fold: Broken Screens Delay Launch," BBC, April 22, 2019, https://www.bbc .com/news/technology-48013395.

283   Samsung attributed some of the malfunctions: Chaim Gartenberg and Dieter Bohn, "Samsung Responds to Galaxy Fold Screen Damage: 'We Will Thoroughly Inspect These Units,'" *The Verge,* April 17, 2019, https://www .theverge.com/2019/4/17/18412572/samsung-galaxy-fold-screen-damage-statement-inspect-screen-protector.

283   But CNBC's Steve Kovach: Hudson Hongo, "Look at All These Fucked Up Galaxy Folds," *Gizmodo,* April 17, 2019, https://gizmodo.com/look-at-all-these-fucked-up-galaxy-folds-1834118147.

283   On April 22, four days before: Chris Welch, "Samsung Delays Galaxy Fold Indefinitely: 'We Will Take Measures to Strengthen the Display,'" *The Verge,* April 22, 2019, https://www.theverge.com/2019/4/22/18511170 /samsung-galaxy-fold-delay-indefinitely-statement-screen-display-broken-issues.

283   "It was embarrassing": Cuthberson, "Galaxy Fold: Inside Samsung's Struggle to Deliver a Foldable Phone—and Why the Future of Smartphones Hinges on It."

283   She attributed part of the failure: Ibid.

283   One week later, Samsung's profits plummeted: Ju-min Park, "Samsung Electronics First-Quarter Profit Falls 60% On-Year as Weak Chip Prices Bite," Reuters, April 29, 2019, https://www.businessinsider.com/samsung-electronics-first-quarter-profit-falls-60-on-year-as-weak-chip-prices-bite-2019-4.

283   Samsung announced a new expansion: "'Creative Ecosystems' Is Key to Non-Memory Chip Business," *Dong-A Ilbo,* April 25, 2019, http://www .donga.com/en/article/all/20190425/1711196/1/Creative-ecosystems-is-key-to-non-memory-chip-business.

284   "The government will actively support": Park Tae-hee, "Who Triggered

the Chip Crisis?" *Korea JoongAng Daily,* July 16, 2019, http://koreajoongang daily.joins.com/news/article/article.aspx?aid=3065513.

284 "Some experts think that Samsung's": Song Gyung-hwa, "[News analysis] Lee Jae-yong's Disturbing Display of Confidence Ahead of Supreme Court Ruling," *Hankyoreh.*

284 The audience stood up: "[Full Video] 'Influence-Peddling Scandal' Supreme Court Sentencing," YTN, August 29, 2019, https://www.ytn .co.kr/_ln/0301_201908291453019091_001. The source is in Korean. The author's researcher translated the source into English. An alternative video is available at https://www.youtube.com/watch?v=3h39mMR99IM&feat ure=youtu.be.

284 The Supreme Court justices ruled: "[Full Video] 'Influence-Peddling Scandal' Supreme Court Sentencing," YTN.

284 The lower court, the justices ruled: Sohee Kim, "Samsung's Lee Faces a Retrial That Could Put Him Back in Jail," Bloomberg News, August 29, 2019, https://www.bloomberg.com/news/articles/2019-08-29/south-korea -orders-retrial-of-samsung-s-lee-in-bribery-scandal.

284 The Supreme Court voided: Ibid.

285 "In this increasingly uncertain": Ibid.

285 "And there might yet be": Lauren Goode, "Review: Samsung Galaxy Fold," *Wired,* October 24, 2019, https://www.wired.com/review/samsung-galaxy- fold/amp.

285 "In 1993, then 51-year-old": Sohee Kim, "Billionaire Samsung Heir Endures Lecture from Judge in Bribery Trial," Bloomberg News, October 24, 2019, https://www.bloomberg.com/amp/news/articles/2019-10-25/billionaire- samsung-heir-endures-tongue-lashing-in-graft-trial.

285 Jay's board seat was up: Sohee Kim, "Samsung Billionaire to Cede Board Seat Before Bribery Probe," Bloomberg News, October 26, 2019, https://www .bloomberg.com/amp/news/articles/2019-10-08/samsung-billionaire-heir- to-cede-board-seat-before-legal-probe. The most up-to-date board roster is available at Samsung Electronics, "Board of Directors," https://www .samsung.com/global/ir/governance-csr/board-of-directors/.

Index

ABOUT THE AUTHOR

Geoffrey Cain is a foreign correspondent and author who's covered Asia and technology for *The Economist, The Wall Street Journal, Time, The New Republic,* and other publications. A resident of South Korea for five years and a Fulbright scholar, he studied at the School of Oriental and African Studies in London and the George Washington University. He is a term member of the Council on Foreign Relations.

ABOUT THE TYPE

This book was set in Dante, a typeface designed by Giovanni Mardersteig (1892–1977). Conceived as a private type for the Officina Bodoni in Verona, Italy, Dante was originally cut only for hand composition by Charles Malin, the famous Parisian punch cutter, between 1946 and 1952. Its first use was in an edition of Boccaccio's *Trattatello in laude di Dante* that appeared in 1954. The Monotype Corporation's version of Dante followed in 1957. Though modeled on the Aldine type used for Pietro Cardinal Bembo's treatise *De Aetna* in 1495, Dante is a thoroughly modern interpretation of that venerable face.